Application of Nanotechnology for Resource Recovery from Wastewater

Editors

Jitendra Kumar Pandey
Research and Development Department
University of Petroleum and Energy Studies
Dehradun, Uttarakhand
India

Syed M. Tauseef
Department of Health Safety Environment & Civil Engineering
University of Petroleum and Energy Studies
Dehradun, Uttarakhand
India

Suvendu Manna
School of Engineering
University of Petroleum and Energy Studies
Dehradun, Uttarakhand
India

Ravi Kumar Patel
Research and Development Department
University of Petroleum and Energy Studies
Dehradun, Uttarakhand
India

Vishal Kumar Singh
Research and Development Department
University of Petroleum and Energy Studies
Dehradun, Uttarakhand
India

Ankit Dasgotra
Research and Development Department
University of Petroleum and Energy Studies
Dehradun, Uttarakhand
India

CRC Press
Taylor & Francis Group
Boca Raton London New York

CRC Press is an imprint of the
Taylor & Francis Group, an **informa** business
A SCIENCE PUBLISHERS BOOK

Cover image is created by editors.

First edition published 2024
by CRC Press
2385 NW Executive Center Drive, Suite 320, Boca Raton FL 33431

and by CRC Press
4 Park Square, Milton Park, Abingdon, Oxon, OX14 4RN

CRC Press is an imprint of Taylor & Francis Group, LLC

Library of Congress Cataloging-in-Publication Data (applied for)

ISBN: 978-1-032-00913-1 (hbk)
ISBN: 978-1-032-00914-8 (pbk)
ISBN: 978-1-003-17635-0 (ebk)

DOI: 10.1201/9781003176350

Typeset in Times New Roman
by Radiant Productions

Foreword

It gives me immense pleasure to write this foreword letter for the book entitled "Application of Nanotechnology for Resource Recovery from Wastewater" edited by Professor Dr. Jitendra Kumar Pandey, Dr. Syed M. Tauseef, Dr. Suvendu Manna, Dr. Ravi Kumar Patel, Mr. Vishal Kumar Singh, and Mr. Ankit Dasgotra, published by CRC press.

Since last few decades, industries are looking for sustainable and eco-friendly options for resource recovery from waste. Recycle and reuse of waste is of tremendous importance for managing huge amount of industrial waste for circular economy. Most of the time, industrial wastes contain recoverable resources that would be useful in other applications. For example, greywater has enough nutrient to support the growth of microalgal biomass that are useful for biofuel production. Similarly, solid waste generated in metal extraction industries often contain high concentration of other metals that could be extracted using various processes. Since its inception, nanotechnology has been contributing to every aspect of human life. Researchers have used several nanotechnological processes for extracting not only resources compounds from wastes, but they have also been utilized to extract energy, nutrients, and catalysts from liquid waste as well. Extensive research has made it possible to utilize all the byproducts in waste management.

This book presents a critical overview on the current nanotechnologies that are being utilized for extraction of wealth from various industrial and domestic wastes. This book presents research, reviews and case studies on the extraction of metal, organic compounds, energy and nutrients from waste through nanotechnological interventions. I would like to congratulate the entire editorial team members for conceptualizing such an emerging topic to discuss in their book and wish them all the success. I would also like to congratulate CRC press as well for foresight in planning this book and wish for its commercial successes. I am sure that this book will find in literature of universities and research institute and researchers will enrich their knowledge. I wish for the success of the book.

Professor Devesh Kumar Avasthi
Dean, R&D
UPES, Dehradun, India

Preface

Rapid urbanization and industrial generating wastewater with new pollutants those are difficult to separate using the conventional wastewater treatment procedures. Thus, researchers are trying to develop new alternative methods in the form of technologies or materials to separate those pollutants. Current global trends on wastewater management indicate that most of the wastewater can be reused in some way or some of the ingredients that can be recovered from the wastewater can be reused for the similar product development. To support the never-ending demands, various industries have already started extracting valuable materials from wastewater and using it for producing products. Such activities minimize the need of new raw materials and they are also useful to achieve the United Nations sustainable development goals. Use of nanomaterials and nano-materials based advanced technologies for various applications is a current trend globally. Nano-technologies are already in use for wastewater management and pollutants separation.

This book provides our readers a comprehensive overview of the development on such research. Researchers, academicians, and industrial personnel share their research, reviews, and case studies in this book as book chapters. Extraction of heavy metals, organic components, and energy from wastewater using direct and indirect nano-technological intervention have been critically discussed in this book. The drawbacks, technical difficulties, and future research possibilities are also mentioned in most of the chapters.

We hope this book would be guidance to all those budding researchers who want to start research on this innovative topic. We would like to thank all of the contributing authors for associating with this project. We express deep sense of gratitude to University of Petroleum and Energy Studies, Dehradun for allowing us to publish this book. We also would like thank Vice-Chancellor, Pro-Vice-Chancellor, and Registrar of our university for their continuous support and motivation. We highly appreciate the efforts of all of our reviewers for their valuable time to voluntarily review the manuscripts and provide their comments and suggestions timely. We also would like appreciate the efforts of CRC press for accepting our proposal and publishing the book in a timely manner.

Dr. Jitendra Kumar Pandey
Dr. Syed M. Tauseef
Dr. Suvendu Manna
Dr. Ravi Kumar Patel
Mr. Vishal Kumar Singh
Mr. Ankit Dasgotra

Contents

1

Nanotechnology

An Important Tool for Wastewater Treatment

Binoyargha Dam, Nikita Chakraborty and
*Bhisma Kumar Patel**

1.1 Introduction

The sustainability of Mother Earth is determined by the availability of natural resources like oil, natural gas, fuel, petroleum, air and clean water. Almost every human action starting from common household tasks to complex agricultural and industrial processes require clean water. Available sources of water which can be directly accessed are very limited. Over the last few years, continuous growth of the global population, rapid urbanization, and incessant industrial and agricultural activities increased the demand for supply of clean water. To satisfy these growing demands, new technologies and methods are needed to be researched and developed.

Water, one of the natural resource of Earth, is also called as the universal solvent. Availability of water in a pure state is pivotal for all living creatures. In absence of water, concept of life is unconceivable. If adversely contaminated by bacteria, microorganisms, industrial wastes, organic pollutants, inorganic substances like ammonia, metal particles become unsuitable for use and are termed as waste water. Wastewater is generated from various agricultural, residential, commercial and industrial sources. List of various water pollutants, their origins, and effects are discussed in Table 1.

Dependence on only surface water sources to meet increasing demands of fresh water supply seems to be very much insufficient. Long term extraction of ground water can also have negative impacts like land subsidence. Therefore, recycling and reusing of water has emerged as an urgent social and environmental issue.

Department of Chemistry, Indian Institute of Technology Guwahati, North Guwahati, 781039, Assam, India.

* Corresponding author: patel@iitg.ac.in

Table 1: Various water contaminants, their resources and their outcomes.

Name of contaminants	Source	Effects	References
Agricultural impurities	Chemicals.	It pollutes fresh water sources directly.	Tang et al. 2016
Industrial pollutants	Water from industries.	Causes air and water pollution.	Liu et al. 2018
Inorganic pollutants	Trace elements, heavy metals, mineral acids, metal compounds.	Aquatic life, public health.	Sizmur et al. 2017
Macroscopic pollutants	Marine remains.	Plastic pollution.	Longwane et al. 2019
Nutrient pollutants	Fertilizers, plant remains.	Effect on process of eutrophication.	Ma et al. 2018
Organic pollutants	Herbicides, insecticides, detergents.	Cacogenic, aquatic life.	Wang et al. 2019a
Pathogens	Bacteria and viruses.	Water borne diseases.	Umar et al. 2019
Radioactive pollutants	Various isotopes.	Teeth, bones, skin.	Bayoumi and Saleh 2018
Sewage and infected water	Household waste water.	Water borne diseases.	Rajasulochana and Preethy 2016
Suspended solids and sediments	Mining operations, land cultivation.	Affecting aquatic life, damaging spawning of fish.	Richter and Ayers 2018

Given the significance of potable water in both developing and developed countries, there is need for the development of pioneering materials and newer strategies, whereby challenges related to the provision of safe potable water can be addressed. Several traditional materials and technologies are not capable enough to treat wastewater consisting of surfactants, pharmaceuticals, and industrial additives. Results of these conventional treatment procedures are also limited owing to high cost and poor treatment efficiency. To counter these problems, new approaches are being continuously examined. Nanotechnological approaches are being considered as one of the emerging fields in this context because of potential advantages like high efficiency in removing contaminants, large surface area, high reactivity, porous nature, lower expense and reusability (Baruah et al. 2016).

Nanotechnology is one of the most revolutionary technologies in the modern world. The term nanotechnology describes a wide range of technologies performed at nano scale with extensive applications in various industries. Nanotechnology is a multi-disciplinary field that looks at how we can manoeuver at atomic and molecular levels. Nanostructured technology is an area that covers creation of new materials from nano sized building blocks. Nanoparticles when used as nanofiltration membrane, nanosized zero-valent ions or adsorbents cause pollutant removal or pollutant separation from water whereas nanoparticles, when used as the catalyst for photochemical oxidation, affect the destruction of pollutants (Tyagi et al. 2012).

1.2 Potential nanotechnological processes for waste water treatment

1.2.1 Nanofiltration

Filtration through a membrane is very useful in removing several kinds of contaminants and ensures a better water purification. Recently, processes involving membrane waste water treatment have gained lots of importance because of their high efficiency towards elimination of solid waste materials, ions and pathogens.

Nanofiltration with reverse osmosis (RO) is a high-pressure membrane treatment process. But, it needs much lesser drive pressure than RO techniques, and hence much lower energy consumption is required. In nanomembranes, for generation of pressure and circulation of waste water, centrifugal pumps are used. There are large numbers of modules in the plant and within each module, there are different membrane configurations. Water produced by nanofiltration meets strict requirements in terms of water reuse. This process is very much efficient for the removal of bacteria, viruses, inorganic and organic substances. Removal rates of a few contaminants are shown in Table 2.

Various kinds of membranes offered by nanotechnology are:

- Aquaporin + polymer
- Carbon nanotube + polymer
- Zeolites + polymer-based membrane
- Ceramic membranes for nanofiltration.

Table 2: Contaminant removal effectiveness by nanofiltration (Tchobanoglous et al. 2014).

Contaminant	Removal Efficiency	Unit
Arsenic	< 40	%
Atrazine	85–90	%
Bacteria	3–6	Log
Calcium Chloride	10–50	%
Colour	90–96	%
Fluorides	10–50	%
Hardness	80–85	%
Magnesium Sulphate	80–95	%
NaCl	10–50	%
Nitrates	80–85	%
Proteins	3–5	Log
Protozoa	> 6	Log
Sodium Sulphate	80–95	%
Viruses	3–5	Log

Various types of membranes used in nanofiltration technique can be categorised into the following categories (Pendergast and Hoek 2011):

- Biologically inspired membranes.
- Nano ceramic membranes.
- Inorganic-organic membranes.

The same research group has also developed procedures to assess certain nanomaterial properties like porosity, potency and applications. According to their reports, it has been found that biologically inspired membranes have the greatest potential for improvement, but are far from commercial use. This trend may change with time as technology involving biologically-inspired membranes gets developed (Lens et al. 2013).

1.2.2 *Nanophotocatalyst*

Photoreaction accelerated in presence of a catalyst is termed photocatalysis. Photocatalysts get activated by ultra violet light and it oxidizes organic contaminants into carbondioxide and water. Photocatalysis proceeds via adsorption of photon, whose energy is superior than the energy of the band gap between the valence shell and the conduction shell of the semiconductor. When photon falls on the catalytic surface, negatively charged electrons get excited from the valence shell to the empty conduction shell and leave a positively charged hole behind it. This electron-hole pair generates extremely reactive radicals, which bind with the pollutants and decomposes them (Bora and Dutta 2014). The detailed mechanism is shown in Fig. 1.

Bandgap energies of various nanophotocatalysts are given in Table 3 (Bhatkhande et al. 2002).

1.2.2.1 Titanium dioxide nanomaterials as photocatalyst

Because of superior photoactivity, large band gap and non-lethal nature of titanium dioxide (TiO_2), it is considered as the most favourable material for heterogeneous photocatalytic processes. Xie and co-workers studied photocatalytic property of TiO_2 at three different temperatures, i.e., at 120, 160 and 200°C and found that the highest activity is observed at 160°C (Xie et al. 2010). They also doped titanium dioxide with silicon and observed that the photocatalytic activity of the catalyst does not improve on doping. However, reports suggest that doping of titanium dioxide with transition metals like iron, bismuth, silver, vanadium or with rare metal ions enhances its photocatalytic activity. Pang and Abdullah (2013) reported that TiO_2 nanotubes doped with iron show very high efficiency for purifying textile waste water as compared to titanium dioxide nanotubes and titanium dioxide powder.

1.2.2.2 Zinc oxide nanomaterials as photocatalyst

Zinc oxide (ZnO) is another highly favourable nanomaterial that has gained great deal of attention because of its photocatalytic activity in degradation of organic pollutants. Various morphologies of zinc oxide exhibit numerous degrees of photocatalytic activities (Ma et al. 2011, Zhai et al. 2012). They have also reported photocatalytic activities of zinc oxide nano-rods and zinc oxide nanodisks in decomposing organic

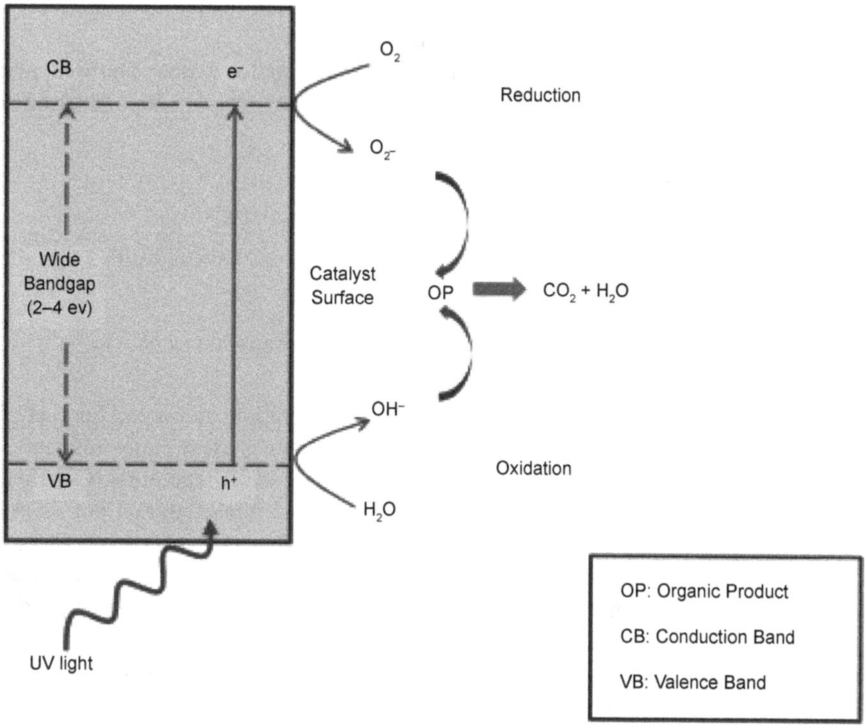

Fig. 1: Photocatalysis on the surface of the semiconductor.

Table 3: Band gap energies of various nanophotocatalyst.

Nanophotocatalyst	Bandgap energy (eV)
Zinc Oxide	3.2
Titanium dioxide (rutile)	3.0
Zinc Sulphide	3.7
Tin (IV) Oxide	3.5
Iron (III) Oxide	2.2
Silicon	1.1
Cadmium Sulphide	2.4
Tungsten trioxide	2.7
Strontium titanate	3.4
Tungsten diselenide	1.2
α- Iron (III) Oxide	3.1

dyes. It is reported that doping of metals like silver, copper and iodine increases the photocatalytic activities of ZnO nanomaterials. Mohan et al. (2012) in their work demonstrated decomposition of resazurin dye by using photocatalytic activity of pure and copper doped zinc oxide nanorods. Qiu et al. (2008) also reported that zinc oxide nanomaterials on doping with Co^{2+} suppressed the photodegradation of rhodamine B dye under ultra violet irradiation.

Lastly, even though the application of photocatalysts for waste water treatment possesses many advantages, many technical challenges need to be addressed to broaden the practical application of the process, like:

- Catalyst optimization in presence of light energy.
- Separation of nano catalyst after application and its reusability.
- Less economic viability due to the high operating cost required for providing ultra violet radiation.

1.2.2.3 *Advantages, disadvantages, and future prospectives of nanophotocatalysts*

Nanophotocatalysts play a major role in the detoxification of water (Tahir et al. 2019). Nanophotocatalysts with nano size of 1–100 nm possess many advantages like low cost, chemical stability, easy accessibility and eco-friendliness (Ciambelli et al. 2019). Among all, photocatalysts TiO_2 have good photostability, but several nanophotocatalysts possess relatively low stability because of photocorrosion (Weng et al. 2019). As light shines on the surface of photocatalyst, they get oxidised or reduced depending on materials and their oxidation state gets altered by generating electrons and holes. This leads to the decomposition of photocatalysts and as a result, efficiency of photocatalyst degrades. A major benefit of nano-size is related to the quantum size effect, as it enhances energy band gap and reduces particle size (Rajabi et al. 2016). Although nanophotocatalysts possess many advantages, at the same time they suffer from certain drawbacks such as difficulty in the recycling of catalyst from the reaction mixture and high toxicity. These issues limit the scope and application of nanophotocatalysts (Mahmoodi and Arami 2009). So, now scientists are focussing on various nanocomposites of different materials which will have less toxicity if used in the water treatment. To overcome the problem of recyclability, magnetically recyclable nanophotocatalysts can be used. At last, a photocatalytic treatment with brilliant effectiveness, better solar treatment and fewer site area requirements can be a boost in this field of research.

Nanomaterials have recently gained tremendous interest in dilapidation as well as in the mineralization of toxic organic pollutants owing to their significant physiochemical nature. In the process of photocatalysis, mainly there are two processes: degradation of organic pollutants and mineralization (Umar et al. 2013, 2012). During degradation, organic pollutants get decomposed to various products, while complete degradation of organic pollutants into carbon dioxide, water and other inorganic ions takes place during mineralization.

After degradation of pollutants in the presence of light by nanomaterials, GC-MS or HPLC-MS are being carried out to analyze newly formed degraded products. It is important to analyse these products at this stage because it is pivotal to conclude whether these new products are more or less toxic than the parent compound. Mineralization pathway simply refers to complete photodegradation of the compound into water and carbon dioxide, but sometimes other minerals like

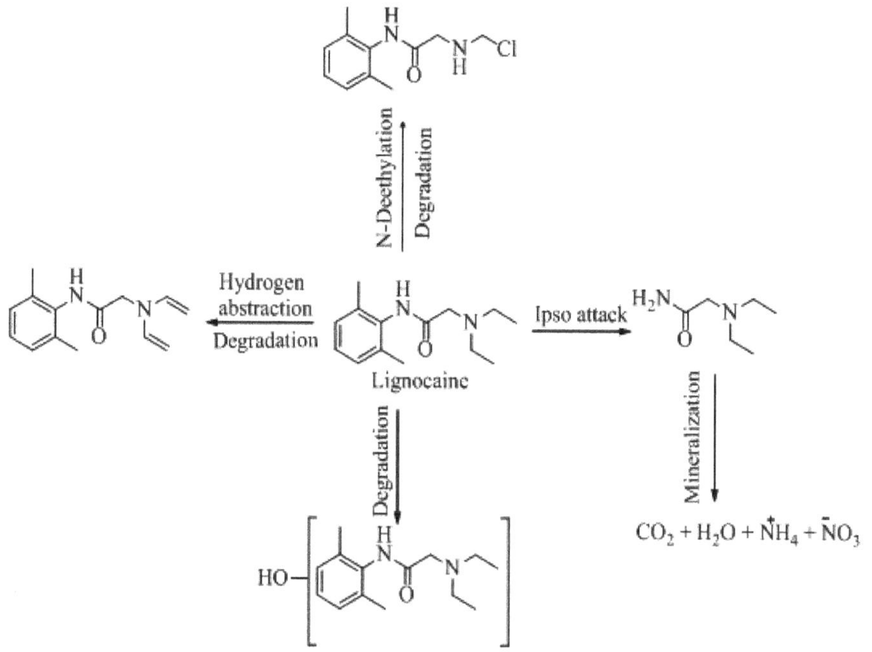

Fig. 2: Degradation and mineralization of products of Lignocaine.

ammonia, sulphide, sulphite, fluoride, chloride, phosphate and nitrate are also being released (Rasolevandi et al. 2019). The rate of mineralization is comparatively lesser than the rate of degradation, and this may be due to the formation of stable intermediates during this process. Therefore, long irradiation is required to remove the total organic carbon (TOC). TOC is the total quantity of bounded carbon in any organic compound, and it is analysed by a TOC analyser. Various degradation, and mineralization products of Lignocaine are shown in Fig. 2 (Rayaroth et al. 2018).

1.2.3 Nanomembranes

Nanomembranes are membranes that are formed by various nanofibres and are employed in the elimination of unwanted materials present in the aqueous phase. Several works have been reported on membrane nanotechnology to produce multifunctional membranes. This is achieved by the application of various nanomaterial substances in several polymer-based membranes. Membranes contain porous support with composite layers. For significant practice, the composite layer is graphene oxide/carbon nanotubes (CNT) dispersed in a polymeric matrix. CNTs possess anti-microbial nature which lessen fouling and biofilm formation (Hogen-Esch et al. 2019). Doping of antimicrobial materials like silver metal on polymeric membranes was also found useful as they reduce the production of biofilm on the surface of the membrane (Saleh et al. 2019, Kochkodan and Hilal 2015). There are few major challenges associated with nanomembranes such as membrane fouling and membrane clogging. These problems can be overcome by the addition of super hydrophilic nanoparticles in thin-film nanocomposite membrane. Metal oxide nanomaterials like Al_2O_3 and TiO_2 nanoparticles, antimicrobial nanomaterials like silver nano and carbon nanotubes are useful materials to overcome membrane fouling and clogging issues because of the high porosity and high hydrophilicity. Recently, some nanophotocatalysts are also being incorporated on the surface of nanocomposite membranes to make them fit for decomposition of organic contaminants. Titanium dioxide impregnated films and nanomembranes are applied for the decomposition of various organic pollutants and the deactivation of several microorganisms (Gopalakrishnan et al. 2018). Several research groups have also reported works on immobilization of metallic nanoparticles on membranes like chitosan, polysulfones cellulose acetate and its application on dechlorination and degradation of various toxic substances.

1.2.3.1 Advantages and disadvantages of nanomembranes

Main purpose of using membranes for purification is to separate toxic materials from water resources. There are certain advantages in applying nanomembranes as compared to traditional methods. In traditional methods, magnesium and calcium require another ion such as Na^+ ions to compensate as an exchanger. But in the case of nanomembrane, there is no need for an exchanger (Bhat et al. 2018, Muntha et al. 2017).

There are certain limitations involved in the application of nanomembrane which reduces its efficiency over conventional approaches. Some of these limitations are:

- Decomposition of nanomembrane after being used for some time. This is one of the chief issues because of which this approach becomes more pricey and incompetent.
- Nanomembrane does not remain stable for a longer period, so to get excellent results changing nanomembrane is required. But this gives rise to several other issues like impurity due to changes and high cost (Bellon et al. 2019). Some nanomembrane types, their pros, cons and uses are summarized in Table 4.

1.2.3.2 Some future perceptions of nanomembranes

Nanomembranes separate organic based pollutants, inorganic ions and virus-based pollutants from waste water. Though nanomembranes show higher efficiency at the research laboratory level, upscaling to lower cost and making them perfect on an industrial scale remains a challenge. To overcome this problem towards productive upscaling, first of all, collaborative work between manufacturing companies and research institution is required. Secondly, there is a need to improve the resistivity of nanomembrane to avoid decomposition of membrane. Zwitter ionic surface grafting based polymers may serve as a prominent candidate for the advancement of new generation nanomembranes. Thirdly, there is a need to focus on the development of multi-functional membranes for better application at an industrial scale. At last, improvement in the sensitivity of polyamide membranes towards oxidants like chlorine and ozone is also required.

Table 4: Pros, cons and uses of nanomembranes.

Types of nanomembrnes	Pros	Cons	Uses	References
Aquaporin based membranes	Better permeability and improved ionic selectivity.	Reduced mechanical stability.	Less pressure desalination.	Tang et al. 2015
Nanocomposite membranes	Improved water permeability, high foul resistance, high hydrophilicity, good mechanical and thermal stability.	Less resistance	Used as flame retardant, in solar cells.	Esfahani et al. 2019
Nanofiber membranes	Bactericidal, excellent porosity.	Pore blocking, conceivable discharge of nanoparticles.	Used as filter cartridge, in ultrafiltration.	Feng et al. 2010
Self-assembling membrane	Homogenous nonporous membrane.	Availability of laboratory scale only.	Ultrafiltration.	Cornwell et al. 2015

1.2.4 *Nanosorbents*

Nanosorbents hold various assets like high sorption capability that makes them more powerful and appropriate for water treatment (Montemagno et al. 2004). Though there are scarcities of nanosorbents for commercial use, many works are under process to produce nanosorbents in a larger amount (Salim and Ho 2015). Most commonly used nanosorbents are carbon black, graphene oxide, polymeric and metal oxide nanosorbents. Polymeric nanoadsorbents like dendimers are very useful in eliminating heavy metals and organic pollutants from waste water (Fuwad et al. 2019). For example, with the help of dendimer-ultrafiltration, copper ions were removed (Shen et al. 2014). Furthermore, the percentage of organic pollutants and dyes removed is almost 99% (Peng et al. 2017). Zeolites are other classes of nanosorbents, on which various ions like silver and copper can be incorporated (Giwa et al. 2017). Zeolites also possess anti-microbial properties (Diallo 2009). Magnetic nanosorbents are also being used for removing pollutants from waste water (Sahebi et al. 2019). Table 5 shows a list of various nanosorbents and their functions.

1.2.4.1 Some pros, cons, and future prospective of nanosorbents

Nanosorbents have received huge interest in dealing with ecological issues related to refinement of waste water. To treat contaminant in a shorter time, a sorbent should have a huge surface area and an exceptional adsorption rate. Challenges that need to be addressed entirely for commercialization purposes are production scalability, selectivity and stability of the material. However, there is a need to come up with a method to treat wastewater and develop some novel nanosorbents which could be applied to separate toxic ions and contaminants from waste water (Khajeh et al. 2013). Future prospects of nanosorbents are commendable as various research groups throughout the world are working on improving adsorption mechanism. They are concerned about waste water pollutants that induce health risks and are dedicated to finding a proper solution with the application of nanosorbents.

Table 5: Various nanosorbents and their utilities.

Nanosorbents	Utilities	References
Nanosorbents based on carbon	To treat water contaminated with nickel ions.	Rodovalho et al. 2016
Graphite oxide	Removal of dyes.	Sun et al. 2015
Nano-Aerogels	Elimination of uranium from water.	Krstic and Pesovski 2018
Nanoclay	Removal of dyes and hydrocarbons.	Kyzas and Matis 2015
Nano iron oxides	To remove toxic pharmaceutical materials and hormones from water.	Unuabonah and Taubert 2014
Nano metal oxides	Removal of various heavy metals.	Wang et al. 2016a
Polymer fibres	Elimination of arsenic and toxic metals.	Yadav et al. 2019
Regenerable polymeric nanosorbents	Elimination of organic and inorganic pollutants.	Lee et al. 2012

1.2.5 *Nano and micromotors*

Nano and micromotors convert energy from various sources to machine-driven force. These motors are motorized by using fuels, magnetic field or electric field and they possess many significant applications (Jurado-Sánchez et al. 2018). There is two major mechanistic pathway involved in cleaning polluted water by nano and micromotors. They either absorb or degrade contaminants by various routes: sodium borohydride assisted, hydrogen peroxide assisted, biocatalytic, or photocatalytic.

The adsorption mechanism involves a multi-step modus operandi that preliminarily involves the transfer of target contaminants from the liquid phase to the surface of porous solid particles and, secondly, diffusion of these solid particles into the interior. Adsorption efficiency of various nano/micromotors, which was recently developed, was studied for pollutants like metal ions, oil droplets, and organic compounds (Eskandarloo et al. 2017).

Degradation mechanism: Many nano and micro motors are also being developed for removing organic pollutants from water which works on the principle of degradation mechanism that is in contrast to adsorption mechanism. Several strategies including sodium borohydride assisted, hydrogen peroxide assisted, biocatalytic or photocatalytic degradation are in use (Eskandarloo et al. 2017). Out of these several strategies, sodium borohydride ($NaBH_4$) assisted strategy is summarized as follows:

In 2015, Srivastava et al. reported palladium catalysed degradation of organic pollutants in the presence of $NaBH_4$ as reductant. In their work, they demonstrated waste water-mediated synthetic micromotors for removal of model organic pollutant 4-Nitrophenol (4-NP) in presence of $NaBH_4$. These micromotors were prepared by growing palladim particles on thin films of titanium/iron/chromium. This was then rolled up to self-propelled microtubular motors. When the surface of the micromotor was coated with palladium particles in the presence of reductant $NaBH_4$ and pollutant 4-NP, an autonomous movement began through a bubble production mechanism.

Table 6: Record of nano and micromotors for refinement of water in presence of fuel.

Kind of Nanomotor	Target contaminants	Mechanism of operation	Fuel	References
Three dimensional printed motors	Droplets of oil	Adsorption	-	Wang et al. 2019b
Hybrid microrobots of CobaltNickel-Bismuth (III) oxide/ Bismuth oxychloride	Rhodamine B	Photocatalytic decay	Ultra violet radiation	Zhang et al. 2019
DNA-functionalized Gold/Platinum microtubes	Mercuric cation	Adsorption	Hydrogen peroxide	Wang et al. 2016b
Nanomotor of PEDOT/Platinum bilayer	Organophosphorous agents	Neutralization by oxidation	Hydrogen peroxide	Soler and Sánchez 2014
Silver incorporated zeolite	-	Adsorptive detoxification	Hydrogen peroxide	Li et al. 2014
CobaltNickel@ Platinum nanorods	Methylene blue, Rhodamine B and *para*-nitrophenol	Degradation	Borohydride	García-Torres et al. 2017
Gold nanoparticles/ Titanium dioxide/ Platinum nanomotor	Super organic mixture	Photocatalytic decay	Hydrogen peroxide	Liu et al. 2020
Polystyrene-Zinc-Iron coreshell microparticles	Rhodamine B	Fenton reaction	Hydrogen peroxide	Gao et al. 2013

This led to the consumption of target pollutant 4-NP as fuel, thereby creating a relatively non-toxic 4-Aminophenol. Table 6 shows a record of various nano and micromotors in presence of various fuels.

1.2.5.1 Advantages, disadvantages and future scopes of nano/micromotors

Nano/micromotors are reactive nano-based materials that possess essential properties to alter toxic pollutants to toxicity free compounds. Nanomachines open up a new avenue which decreases the entire clear out time and cost of the procedure (Safdar et al. 2017). The constant movement of nanomaterials through polluted sources can be used for transferring reactive nanoscale substances for water purification and for communication of proper mixing throughout the refinement process (Fu et al. 2014). To fulfil these demands, existing technologies are not sufficient. Therefore, more efforts are required and some issues need further research and development for commercial applications. Nano/micromotor is still a very immature technique as compared to other traditional approaches. New materials like graphene may be linked with nano/micromotors to treat waste water, as these materials are more efficient in waste water treatment. Furthermore, new mechanisms should be functionalized in nano motors for better results.

1.2.6 Disinfection and microbial management

Several decontamination methods are being employed in the purification of water which controls microbial pathogens. But some of these methods lead to the development of unsafe by-products. Oxidants like chloramines, ozone and free chlorine which are constituents of disinfectants, employed for purification of drinking water, react with natural water to produce by-products such as trihalomethanes, aldehydes and haloacetic acids. Most of these are carcinogenic. A number of pathogens are chemical disinfectants resistant and therefore need tremendously high doses of disinfectants, thereby leading to more formation of by-products. As a result, new alternate methods are required to be employed which will be more effective and at the same time will avoid the formation of harmful by-products. Herein application of anti-microbial nanomaterials comes into the picture because they are neither strong oxidizing agents nor they are reactive to water, so they don't produce any harmful by-products. Chitosan and peptides, titanium dioxide nanoparticles, fullerene nanoparticles and carbon nanotubes are some nanomaterials that possess anti-microbial properties. Some of their disinfecting mechanisms are reviewed and discussed as under:

A. Chitosan and peptides

Nanoparticles synthesized from peptides and chitosan possess many applications in water disinfection systems which are really cheap and affordable. Osmotic collapse of the bacterial cell membrane by formation of nanoscale channels is the antibacterial mechanism of natural peptides. There are various theories to explain the antibacterial mechanisms of chitosan nanoparticles. One of the theories proposed that when positively charged chitosan nanoparticles interact with the negatively charged cell membrane, an augmentation in membrane permeability eventually happens which leads to bursting and leakage of intracellular components. Another theory discussed a mechanism where chitosan particles penetrate the cell wall, binds with DNA and slow down the synthesis of RNA within the cell. Some of the uses of nanoscale peptides and chitosan are application in the surface coating of water storage tanks, application as anti-microbial agents in sponges and membranes.

B. Silver nanoparticles

When silver nanoparticles interact with bacterial cells, they liberate huge amount of silver ions (Ag^+). These ions are extremely reactive and they react with thiol groups in the enzymes within the cells to form reactive oxygen species (ROS). Formation of these species makes respiratory enzymes inactive which leads to cell death. Silver ions hoard within the cell membrane by forming pits, thereby compromising the permeability and integrity of cell membrane structure. In the presence of ultraviolet light, these ions also exhibit photocatalytic activities and this property is used for the disinfection of microbes. Several recent water disinfection and purification systems use membranes that contain silver nanoparticles impregnated into them.

C. Titanium dioxide (TiO_2) nanoparticles

In the presence of ultra violet radiation, titanium dioxide nanoparticles show excellent photocatalytic properties. Titanium dioxide nanoparticles exhibit anti-bacterial activities because it produces reactive oxygen species (ROS) like hydroxyl free radicals and peroxides which get formed inside the cell in the presence of ultra-violet radiation utilizing an oxidative and reductive sequence of chemical reactions. Titanium dioxide nanoparticles also show the photocatalytic property in the presence of visible light, and this is one of the added advantages of using titanium dioxide nanoparticles in waste water disinfection. These nanoparticles are stable in water, less toxic and can be used as thin films coated on the reactor surface.

D. Carbon nanotubes (CNTs)

Carbon nanotubes (CNTs) exhibit anti-microbial properties mainly by two means, i.e., by chemical reactions with pathogens and by physically restraining their passage through filters. There are reports which have shown that carbon nanotubes (CNTs) lead to the destruction of bacterial cells *via* oxidative reactions, which compromise cell membrane integrity and ultimately lead to its degradation. The anti-microbial property of carbon nanotubes (CNTs) through chemical means requires direct contact of carbon nanotubes (CNTs) with pathogens. This process is tedious and requires uniform and stable carbon nanotube suspension in water. Therefore, the application of carbon nanotubes in waste water treatment employing chemical interaction is limited. Carbon nanotubes are very efficient in filtering bacteria and other microbes by inhibiting their flow across the membrane. Single-walled carbon nanotubes (SWCNTs) possess a diameter in the range of 2 to 5 nanometres and effectively filter out all known pathogens.

E. Other methods

Besides waste water treatment and purification, nanomaterials also play a vital role in ground water remediation. Present method which is being used for ground water remediation is the Pump and Treat method (Savageand Diallo 2005). In this method, the ground water from downstream of the contamination source is pumped into the treatment plant, treated by various chemical separation methods and then again pumped back upstream of ground water source. Nanomaterials with a large surface area are well suited for these applications because they can reduce the requirement of a large volume of separation media, reusability and capability to be functionalized to target specific pollutants. Certain nanomaterials like Self Assembled Monolayers on Mesoporous Silica (SAMMS) and Single Enzyme Nanoparticles (SENs) are used effectively in waste water treatment as discussed below.

i. SAMMS

They are hexagonal honeycomb-like nanoporous ceramics. SAMMS can change the exposed functional groups of the monolayer and because of this, they can potentially bind with a broader range of molecules and with contaminants and pollutants in case of ground water remediation (Watlington 2005).

Syntheses of SAMMS can be carried out in three stages:

- Formation of micelle templates with an ordered liquid crystalline structure.
- Formation of a mesoporous anchor by precipitation of oxides on the surface of the micelle.
- Removal of surfactants by calcination of organic oxide materials and addition of bi-functional silanes.

ii. Single Enzyme Nanoparticles (SENs)

Enzymes carry out lots of applications like biosensing, chemical conversions and bioremediation. They are much more effective than synthetic catalysts because of their specificity and targeted effectiveness. But enzymes are very unstable under extreme conditions and are very short-lived, so their direct application in water purification is very limited. The above discussed problems can be overcome with the help of nanotechnology by synthesizing single enzyme nanoparticles (SENs), which contain enzymes that are surrounded by a protective cage (Katherine 2005).

SENs can be synthesized by following three steps:

- Covalent surface modification of enzyme by the creation of the vinyl group and dissolution in a non-polar solvent.
- Formation of polymers consisting of the vinyl group with free trimethoxysilane groups attached to the enzyme surface.
- Formation of a few nanometers thick silicate shell by hydrolysis of trimethoxysilane groups.

This shell protects enzymes from harsh conditions and thereby increases its life span, but it possesses few active sites through which it can bind to target molecules. SENs possess advantage that various enzymes are available for specific pollutants and no by-products are formed as there are no microorganisms involved.

1.3 Challenges and toxicity issues

Emerging nanotechnological processes possess specific challenges in waste water treatment (Dimapilis et al. 2018, Sultana et al. 2015). One of the significant challenges related to nanomaterial application is the lack of information on how nanomaterials are discharged in the environment, how they move into the water, and how they start existing in water (Bushra et al. 2014). Some studies have shown that nano silver, carbon nanotubes, and titanium dioxide nanoparticles have toxicological effects on human beings. Most nanoparticles hinder the cellular function of the immune system, lung and are also carcinogenic. (Dhakras 2011). A challenge linked to the usage of membrane technology is that their pores may get blocked after filtration, and their efficiency may start decreasing.

Nanotechnological research should not be focussed exclusively on the improvement of nanomaterials used in waste water treatment. Rather there is a need to focus on their future impact on the ecosystem and human health. The largest source of nanomaterial in the soil is sludge which is generated from waste water treatment plants. After treatment of waste water, total pollution including nanoparticles accumulates in the sludge. This sludge is then handled and transported in several ways. One of the major problems in the treatment phase is that several polyelectrolytes are added to waste water for flocculation and sedimentation of particles. These affect various qualities of nanoparticles like reactivity, mobility and dispersion. Sludge is then applied for several purposes and then ends up in the environment, where it either gets deposited as landfills or is used as a soil improver in agriculture. This area of waste water treatment has been completely neglected till now and there is a need for further developments in this field following the guidelines of environment protection.

1.4 Conclusions and future prospectives

In conclusion, we have seen that nanotechnological approaches are a boost to modern-day waste water treatment procedures. Its high efficiency for removing bacteria, viruses, and other organic contaminants compared to traditional waste water treatment methods is highly admirable. Applying this form of technology as an alternative to harmful disinfectants for purifying water is very demanding in the modern world. Although there are many such advantages of using nanotechnology in waste water treatment, at the same time, there are certain places where research and development are required. There is no digital monitoring technique that offers reliable real-time measurement on the dominance of nanoparticles that are existing in minor quantity in water. To diminish health risk, international research communities and research groups should plan proper guidelines. It is also important to take into consideration the cost effectiveness of this technology for waste water treatment.

References

Baruah, S., Khan, M.N. and Dutta, J. 2016 Perspectives and applications of nanotechnology in water treatment. Environ. Chem. Lett. 14: 1–14.

Bayoumi, T.A. and Saleh, H.M. 2018 Characterization of biological waste stabilized by cement during immersion in aqueous media to develop disposal strategies for phytomediated radioactive waste. Prog. Nucl. Energy. 107: 83–89.

Bhat, A.H., Rehman, W.U., Khan, I.U., Ahmad, S., Ayoub, M. and Usmani, M.A. 2018. Nanocomposite membrane for environmental remediation. pp. 407–440. *In*: Jawaid, M. and Khan, M.M. [eds.]. Polymer-Based Nanocomposites for Energy and Environmental Applications, Woodhead Publishing: Cambridge, UK.

Bhatkhande, D., Pangarkar, V. and Beenackers, A. 2002. Photocatalytic degradation for environmental applications-a review. J. Chem. Technol. and Biotechnol. 77: 102–116.

Bora, T. and Dutta, J. 2014. Applications of nanotechnology in wastewater treatment—A review. Journal of Nanoscience and Nanotechnology. 14: 613–626.

Bellon, T., Polezhaev, P., Vobecká, L. and Slouka, Z. 2019. Fouling of a heterogeneous anion-exchange membrane and single anion-exchange resin particle by ssDNA manifests differently. J. Membr. Sci. 572: 619–631.

Bushra, R.M., Shahadat, Ahmad, A., Nabi, S.A., Umar, K., Muneer, M., Raeissia, A.S. and Owais, M. 2014. Synthesis, characterization, antimicrobial activity and applications of polyanilineTi(IV) arsenophosphate adsorbent for the analysis of organic and inorganic pollutants. J. Hazard. Mater. 264: 481–489.

Ciambelli, P., La Guardia, G. and Vitale, L. 2019. Nanotechnology for green materials and processes. Stud. Surf. Sci. Catal. 179: 97–116.

Cornwell, D.J. and Smith, D.K. 2015. Expanding the scope of gels–Combining polymers with low-MOLECULAR-Weight gelators to yield modified self-Assembling smart materials with high-Tech applications. Mater. Horiz. 2: 279–293.

Dhakras, P.A. 2011. Nanotechnology applications in water purification and waste water treatment: A review," International Conference on Nanoscience, Engineering and Technology (ICONSET 2011), 285–291. doi: 10.1109/ICONSET.2011.6167965.

Diallo, M.S. 2009. Water Treatment by Dendrimer-Enhanced Filtration. U.S. Patent No. 2009/0223896, 10 September 2009.

Dimapilis, E.A.S., Hsu, C.S., Mendoza, R.M.O. and Lu, M.C. 2018. Zinc oxide nanoparticles for water disinfection. Sustain. Environ. 28: 47–56.

Esfahani, M.R., Aktij, S.A., Dabaghian, Z., Firouzjaei, M.D., Rahimpour, A., Eke, J. et al. 2019. Nanocomposite membranes for water separation and purification: Fabrication, modification, and applications. Sep. Purif. Technol. 213: 465–499.

Eskandarloo, H., Kierulf, A. and Abbaspourrad, A. 2017. Nano- and micromotors for cleaning polluted waters: focused review on pollutant removal Mechanisms. 9: 13850–13863.

Feng, C., Khulbe, K.C. and Matsuura, T. 2010. Recent progress in the preparation, characterization, and applications of nanofibers and nanofiber membranes via electrospinning/interfacial polymerization. J. Appl. Polym. Sci. 115: 756–776.

Fu, P.P., Xia, Q., Hwang, H.M., Ray, P.C. and Yu, H. 2014. Mechanisms of nanotoxicity: Generation of reactive oxygen species. J. Food Drug Anal. 22: 64–75.

Fuwad, A., Ryu, H., Malmstadt, N., Kim, S.M. and Jeon, T.J. 2019. Biomimetic membranes as potential tools for water purification: Preceding and future avenues. Desalination. 458: 97–115.

Gao, W.M. D'Agostino, Garcia-Gradilla, V., Orozco, J. and Wang, J. 2013. Multi-fuel driven janus micromotors. Small. 9: 467–471.

García-Torres, J., Serrà, A., Tierno, P., Alcobé, X. and Vallés, E. 2017. Magnetic propulsion of recyclable catalytic nanocleaners for pollutant degradation. ACS Appl. Mater. Interfac. 9: 23859–23868.

Giwa, A., Hasan, S.W., Yousaf, A., Chakraborty, S., Johnson, D.J. and Hilal, N. 2017. Biomimetic membranes: A critical review of recent progress. Desalination. 420: 403–424.

Gopalakrishnan, I., Samuel, S.R. and Sridharan, K. 2018 Nanomaterials-Based adsorbents for water and waste water treatment. Emerg. Nanotechnol. Environ. Sustain. 6: 89–98.

Hogen-Esch, T., Pirbazari, M., Ravindran, V., Yurdacan, H.M. and Kim, W. 2019. High Performance Membranes for Water Reclamation Using Polymeric and Nanomaterials. U.S. Patent No. 20160038885A, 29 October 2019.

Jurado-Sánchez, B. and Wang, J. 2018. Micromotors for environmental applications: A review. Environ. Sci. Nano. 5: 1530–1544.

Khajeh, M., Laurent, S. and Dastafkan, K. 2013. Nanoadsorbents: Classification, preparation, and applications (with emphasis on aqueous media). Chem. Rev. 113: 7728–7768.

Kochkodan, V. and Hilal, N. 2015. A comprehensive review on surface modified polymer membranes for biofouling mitigation. Desalination. 356: 187–207.

Krstic, V. and Pesovski, T.U.B. 2018. A review on adsorbents for treatment of water and wastewaters containing copper ions. Chem. Eng. Sci. 192: 273–287.

Kyzas, G.Z. and Matis, K.A. 2015. Nanoadsorbents for pollutants removal: A review. J. Mol. Liq. 203: 159–168.

Lee, X.J., Foo, L.P.Y., Tan, K.W., Hassell, D.G. and Lee, L.Y. 2012. Evaluation of carbon-Based nanosorbents synthesized by ethylene decomposition on stainless steel substrates as potential sequestrating materials for nickel ions in aqueous solution. J. Environ. Sci. 24: 1559–1568.

Lens, P.N.L., Virkutye, J., Jegatheesan, V., Kim, S.h. and Al-Abed, S. 2013. Nanotechnology for water and wastewater treatment, IWA Publishing.

Li, J., Singh, V.V., Sattayasamitsathit, S., Orozco, J., Kaufmann, K., Dong, R. et al. 2014. Water-Driven micromotors for rapid photocatalytic degradation of biological and chemical warfare agents. ACS Nano. 8: 11118–11125.

Liu, C., Hong, T., Li, H. and Wang, L. 2018. From club convergence of per capita industrial pollutant emissions to industrial transfer effects: An empirical study across 285 cities in China. Energy Policy. 121: 300–313.

Liu, J., Hong, C., Shi, X., Nawar, S., Werner, J., Huang, G. et al. 2020. Hydrogel Microcapsules with Photocatalytic Nanoparticles for Removal of Organic Pollutants. Environ. Sci. Nano. 7: 656–664.

Longwane, G.H., Sekoai, P.T., Meyyappan, M. and Moothi, K. 2019. Simultaneous removal of pollutants from water using nanoparticles: A shift from single pollutant control to multiple pollutant control. Sci. Total Environ. 656: 808–833.

Ma, S., Li, R., Lv, C., Xu, W. and Gou, X. 2011. Facile synthesis of ZnO nanorod arrays and hierarchical nanostructures for photocatalysis and gas sensor applications. J. Hazard Mater. 192: 730–740.

Ma, H., Guo, Y., Qin, Y. and Li, Y.Y. 2018. Review Nutrient recovery technologies integrated with energy recovery by waste biomass anaerobic digestion. Bioresour. Technol. 269: 520–531.

Mahmoodi, N.M. and Arami, M. 2009. Degradation and toxicity reduction of textile wastewater using immobilized titania nanophotocatalysis. J. Photoch. Photobio. B. 94: 20–24.
Mohan, R., Krishnamoorthy, K. and Kim, S. 2012. Enhanced photocatalytic activity of Cu-doped ZnO nanorods. Solid State Commun. 152: 375–380.
Montemagno, C., Schmidt, J. and Tozzi, S. 2004. Biomimetic Membranes. U.S. Patent No. 20040049230, 11 March 2004.
Muntha, S.T., Kausar, A. and Siddiq, M. 2017. Advances in polymeric nanofiltration membrane: A review. Polym.-Plast Technol. Eng. 56: 841–856.
Pang, Y.L. and Abdullah, A. 2013. Fe^{3+}doped TiO_2 nanotubes for combined adsorption-sonocatalytic degradation of real textile wastewater. Appl. Catal B: Environmental. 129: 473–481.
Pendergast, M.T.M. and Hoek, E.M.V. 2011. A review of water treatment membrane nanotechnologies, Journal: Energy and Environmental Science. 4: 1946–1971.
Peng, F., Xu, T., Wu, F., Ma, C.X., Liu, Y., Li, J. et al. 2017. Novel biomimetic enzyme for sensitive detection of superoxide anions. Talanta. 174: 82–91.
Qiu, X., Li, G., Sun, X., Li, L. and Fu, X. 2008. Doping effects of Co^{2+}ions on ZnO nanorods and their photocatalytic properties. Nanotechnology. 19: 215703.
Rajabi, H.R., Shahrezaei, F. and Farsi, M. 2016. Zinc sulfide quantum dots as powerful and efficient nanophotocatalysts for the removal of industrial pollutant. J. Mater. Sci. Mater. Electron. 27: 9297–9305.
Rajasulochana, P. and Preethy, V. 2016. Comparison on efficiency of various techniques in treatment of waste and sewage water-A comprehensive review. Resour. Efficience. Technol. 4: 175–184.
Rasolevandi, T., Naseri, S., Azarpira, H. and Mahvi, A.H. 2019. Photo-Degradation of dexamethasone phosphate using UV/Iodide process: Kinetics, intermediates, and transformation pathways. J. Mol. Liq. 295: 111703–111710.
Rayaroth, M.P., Aravind, U.K. and Aravindakumar, C.T. 2018. Photocatalytic degradation of lignocaine in aqueous suspension of TiO_2 nanoparticles: Mechanism of degradation and mineralization. J. Environ. Chem. Eng. 6: 3556–3564.
Richter, K.E. and Ayers, J.M. 2018. An approach to predicting sediment microbial fuel cell performance in shallow and deep water. Appl. Sci. 8: 2628.
Rodovalho, F.L., Capistrano, G., Gomes, J.A., Sodre, F.F., Chaker, J.A., Campos, A.F.C. et al. 2016. Elaboration of magneto-Thermally recyclable nanosorbents for remote removal of toluene in contaminated water using magnetic hyperthermia. Chem. Eng. J. 15: 725–732.
Safdar, M., Simmchen, J. and Jänis, J. 2017. Correction: Light-Driven micro-and nanomotors for environmental remediation. Environ.: Sci. Nano. 4: 2235.
Sahebi, S., Sheikhi, M. and Ramavandi, B. 2019. A new biomimetic aquaporin thin film composite membrane for forward osmosis: Characterization and performance assessment. Desalin. Water Treat. 148: 42–50.
Saleh, A., Parthasarathy, P. and Irfan, M. 2019. Advanced functional polymer nanocomposites and their use in water ultra-purification. Trends Environ. Anal. 24: 67–78.
Salim, W. and Ho, W.S.W. 2015. Recent developments on nanostructured polymer-Based membranes. Curr. Opin. Chem. Eng. 8: 76–82.
Savage, N. and Diallo, M.S. 2005. Nanomaterials and water purification: Opportunities and challenges. J. Nanoparticle Res. 7: 331–342.
Shen, Y.X., Saboe, P.O., Sines, I.T., Erbakan, M. and Kumar, M. 2014. Biomimetic membranes: A review. J. Membr. Sci. 454: 359–381.
Sizmur, T., Fresno, T., Akgül, G., Frost, H. and Moreno-Jiménez, E. 2017. Biochar modification to enhance sorption of inorganics from water. Bioresour. Technol. 246: 34–47.
Soler, L. and Sánchez, S. 2014. Catalytic nanomotors for environmental monitoring and water remediation. Nanoscale. 6: 7175–7182.
Srivastava, S.K., Guix, M. and Schmidt, O.G. 2016. Wastewater mediated activation of micromotors for efficient water cleaning. Nano Lett. 16: 817–821.
Sultana, S., Rafiuddin, Khan, M.Z., Umar, K., Ahmed, A.S. and Shahadat, M. 2015. SnO_2-SrO based nanocomposites and their photocatalytic activity for the treatment of organic pollutants. J. Mol. Struct. 1098: 393–399.

Sun, X., Liu, Z., Zhang, G., Qiu, G., Zhong, N., Wu, L., Cai, D. and Wu, Z. 2015. Reducing the pollution risk of pesticide using nano networks induced by irradiation and hydrothermal treatment. J. Environ. Sci. Health C. 50: 901–907.

Tahir, M.B., Kiran, H. and Iqbal, T. 2019. The detoxification of heavy metals from aqueous environment using nano-photocatalysis approach: A review. Environ. Sci. Pollut. Res. 26: 10515–10528.

Tang, C., Wang, Z., Petrini´c, I., Fane, A.G. and Hélix-Nielsen, C. 2015. Biomimetic aquaporin membranes coming of age. Desalination. 368: 89–105.

Tang, K., Gong, C. and Wang, D. 2016. Reduction potential, shadow prices, and pollution costs of agricultural pollutants in China. Sci. Total Environ. 541: 42–50.

Tchobanoglous, G., Stensel, H.D., Tsuchihashi, R. and Burton, F. 2014. Wastewater engineering: Treatment and resource recovery, Fifth Edition, New York, McGraw-Hill.

Tyagi, P.K., Singh, R., Vats, S., Kumar, D. and Tyagi, S. 2012. Nanomaterials use in wastewater treatment, International conference on nanotechnology and Chemical Engineering, Bangkok, December.

Umar, K., Dar, A.A., Haque, M.M., Mir, N.A. and Muneer, M. 2012. Photocatalysed decolourization of two textile dye derivatives, Martius Yellow and Acid Blue 129 in UV-irradiated aqueous suspensions of Titania. Desal. Water Treat. 46: 205–214.

Umar, K., Haque, M.M., Mir, N.A. and Muneer, M. 2013. Titanium dioxide-Mediated photocatalyzed mineralization of Two Selected organic pollutants in aqueous suspensions. J. Adv. Oxid. Technol. 16: 252–260.

Umar, K., Parveen, T., Khan, M.A., Ibrahim, M.N.M., Ahmad, A. and Rafatullah, M. 2019. Degradation of organic pollutants using metal-Doped TiO_2 photocatalysts under visible light: A comparative study. Desal. Water Treat. 161: 275–282.

Unuabonah, E.I. and Taubert, A. 2014. Clay-Polymer nanocomposites (CPNs): Adsorbents of the future for water treatment. Appl. Clay Sci. 99: 83–92.

Wang, H., Khezri, B. and Pumera, M. 2016b. Catalytic DNA-Functionalized self-Propelled micromachines for environmental remediation. Chem. 1: 473–481.

Wang, J., Wang, Z., Vieira, C.L.Z., Wolfson, J.M. and Pingtian, G. 2019a. Review on the treatment of organic pollutants in water by ultrasonic technology. Ultrasonics Sonochemistry. 55: 273–278.

Wang, L., Song, H., Yuan, L., Li, Z., Zhang, P., Gibson, J.K. et al. 2019b. Effective Removal of Anionic Re (VII) by Surface-Modified Ti_2CT_x MXene Nanocomposites: Implications for Tc (VII) Sequestration. Environ. Sci. Technol. 53: 3739–3747.

Wang, Y., Zhang, Y., Hou, C. and Liu, M. 2016a. Mussel-Inspired synthesis of magnetic polydopamine–Chitosan nanoparticles as bio sorbent for dyes and metals removal. J. Taiwan Inst. Chem. E. 6: 292–298.

Watlington, K. 2005. Emerging Nanotechnologies for Site Remediation and Wastewater Treatment, for U.S. Environmental Protection Agency.

Weng, B., Qi, M.Y., Han, C., Tang, Z.R. and Xu, Y.J. 2019. Photocorrosion Inhibition of Semiconductor-Based Photocatalysts: Basic Principle, Current Development, and Future Perspective. ACS Catal. 9: 4642–4687.

Xie, M., Jing, L., Zhou, J., Lin, J. and Fu, H. 2010. Synthesis of nanocrystalline anatase TiO2 by one-pot twophase separated hydrolysis-solvothermal processes and its high activity for photocatalytic degradation of rhodamine B. J. Hazard Mater. 176: 139–145.

Yadav, V.B., Gadi, R. and Kalra, S. 2019. Clay based nanocomposites for removal of heavy metals from water: A review. J. Environ. Manag. 232: 803–817.

Zhai T., Xie, S., Zhao, Y., Sun, X., Lu, X., Yu, M., Xu, M., Xiao, F. and Tong, Y. 2012. Controllable synthesis of hierarchical ZnO nanodisks for highly photocatalytic activity. CrystEngComm. 14: 1850–1855.

Zhang, Y., Yuan, K. and Zhang, L. 2019. Micro/nanomachines: From functionalization to sensing and removal. Adv. Mater. Technol. 4: 1800636–1800658.

2

Wastewater and Reuse for Agriculture

Nadar Hussain Khokhar,[1,*] *Sallahuddin Panhwar,*[1] *Hareef Ahmed Keerio,*[2] *Asim Ali,*[3] *Sayeda Sara Hassan*[4] and *Salah Uddin*[1]

2.1 Introduction

Direct use of wastewater without any treatment is widely practiced around the world (Elahi et al. 2017). Wastewater is directly used in different applications such as urban and industrial uses, irrigation of crops, orchards, managed forests, lawns, and landscape, recharging wetlands for promoting and improving biodiversity, aquaculture, recreation, environment, and artificial groundwater recharge (Shahid et al. 2020, Asano et al. 2007). The scarcity of fresh water supplies has led to alternate options such as direct use of wastewater, its conservation, and expanding current water supplies (Shahid et al. 2020). Ideally, treated wastewater can be used for all purposes as an alternative to freshwater. Generally, the use of wastewater is limited to non-potable applications, or at most to indirect potable uses. In indirect potable reuse, municipal wastewater is highly treated and discharged directly into groundwater or surface water sources with the intent of augmenting drinking water supplies (Rodriguez et al. 2009). Wastewater treatment technologies such as microfiltration, nanofiltration, and reverse osmosis are employed to wastewater before releasing it into the surface water bodies or direct injection in to the aquifer (Drewes et al. 2002).

The use of municipal wastewater in agriculture is the most common practice where agricultural lands are located near urban areas (Pedrero et al. 2010).

[1] NUST Balochistan Campus, National University of Sciences and Technology, Pakistan.
[2] Department of Environment Engineering, QUEST Nawabshah.
[3] Department of Civil Engineering Technology, the Benazir Bhutto Shaheed University of Technology and Skill Development, Khairpur (Mir's).
[4] US-Pakistan Center for Advanced Studies in Water, at Mehran University of Engineering and Technology Jamshoro.
* Corresponding author: nadaragk@gmail.com

Approximately twenty-million-hectare land is irrigated with treated or untreated municipal wastewater around the world (Jiménez and Asano 2008). Wastewater use is the most prominent practice in lower-income countries; wastewater is also extensively used in high-income countries located in arid and semi-arid regions (Kookana et al. 2014). Besides water scarcity, other drivers for the increasing use of wastewater may include growing urban wastewater flows due to increasing population and overexploitation of freshwater resources, more importantly, agricultural activities by urban households that intensify the use of urban wastewater for irrigation.

Extensive agricultural activities for more production in both low-income and middle-income countries are mostly associated with the use of untreated wastewater directly or indirectly from polluted rivers and streams (Weldesilassie et al. 2011, Van Rooijen et al. 2010). In those countries, freshwater supplies are either scarce or expensive while wastewater treatment plants are overloaded due to urban growth, and urban farmers use highly polluted water for agriculture due to no alternative choice. Mostly, urban farmers belong to poor family backgrounds and agriculture is the only source of food, income, and self-employment (World Bank 2000).

Municipal wastewater use in agriculture poses a high risk to the health of farmers themselves, their close family members, the surrounding community, and to those who consume the food produced on wastewater (Keraita et al. 2008, Drechsel and Evans 2010). Disease-causing microorganisms present in untreated or partially treated wastewater are the biggest risk (Drechsel and Evans 2010). The disease can spread in people who are dealing with wastewater in agricultural fields for irrigation and those who are consuming food that was produced with wastewater application, especially raw or uncooked food.

The guidelines provided by the World Health Organization (WHO) on the safe application of municipal wastewater and human excreta in agriculture or aquaculture have been used as a reference since 1989 for regulating water standards (WHO 1989). Research carried out later suggested accommodating local conditions added with other health-related interventions, such as promotion of hygiene. New guidelines of WHO were released in 2006 in which the use of wastewater was addressed as only a strategy of integrated risk management (WHO 2006). Now upgraded guidelines of 2006 include health-oriented goals, which focus on disease caused by wastewater use in agriculture.

In developed nations, wastewater is treated as per environmental standards before irrigation of food, fiber, fodder, orchards, and vineyards crops (Feigin et al. 2012, Angelakis and Bontoux 2001). While most of the under developing nations have environmental standards including wastewater, those standards are not strictly practiced (Qadir et al. 2010). Untreated wastewater has been used for centuries in agriculture and aquaculture. Thus, at the same time, we can consider wastewater as a resource and a dilemma. Wastewater contains essential nutrients for plant growth that can be used for irrigation and other environmental services (Qadir et al. 2010). Reuse of wastewater can be beneficial to the farmers, local community, and municipalities. However, reuse of wastewater comes with negative effects on humans and eco-environmental systems, and those problems need to be identified and assessed. The objective of this chapter is to evaluate the characteristics of municipal wastewater and its reuse in agriculture.

2.1.1 Supplies and sources of wastewater

In general, municipal wastewater is the mixture of domestic, industrial, and stormwater (Negulescu 2011, Reemtsma et al. 2010). Domestic wastewater can be sub-categorized into effluent coming from residential buildings, institutions, and commercial buildings. Industrial wastewater may consist of single or multiple manufacturing and food processing industries. Stormwater runoff from urban areas may be loaded with hazardous pollutants from nonpoint sources that may include soil sediments, organic and inorganic matter, and an excessive amount of nutrients, heavy metals, and harmful pathogens, fertilizers from farmyard manure and other sources, and pesticides. In developed countries, all three systems are separated; however, in some parts of the world, the municipal sewage system is also used as a stormwater sewer. Due to leaks and cracks in the sewerage systems, groundwater also joins the sewage system to be disposed of.

2.1.2 Composition of wastewater and major treatment plants

The overall quantity of wastewater and the concentration of each individual pollutant in wastewater may differ from community to community due to the variation in the population, their source of income, and living standards. However, all wastewaters generally consist of the following constituents: (1) organic and inorganic matter, (2) nutrients (nitrogen, phosphorus, and potassium), (3) toxic chemicals, and (4) harmful pathogens. The health effects of these constituents are described in Table 1.

Approximately, 330 km^3/year of municipal wastewater is generated worldwide (Drewes et al. 2002). Millions of hectares can be irrigated and fertilized with this huge amount of water. It is estimated that around 60% of the total municipal wastewater

Table 1: Wastewater constituents, and parameters.

Pollutants/Constituents	Parameters	References
Excess plant food nutrients	N, P, K, etc.	Hussain et al. (2010)
Particulate matters	Volatile compounds, settleable, suspended, and colloidal impurities	Hussain et al. (2010), Xie et al. (2022)
Pathogens	Viruses, bacteria, helminth eggs, fecal coliforms, etc.	Hussain et al. (2010), Panhwar et al. (2021a,b)
Biodegradable organics	BOD, COD	Hussain et al. (2010)
Stable organics	Phenols, pesticides, chlorinated hydrocarbons	Hussain et al. (2010)
Dissolved inorganic substances	TDS, EC, Na, Ca, Mg, Cl, and B	Hussain et al. (2010)
Heavy metals	Cd, Pb, Ni, Zn, As, Hg, etc.	Hussain et al. (2010), Keeerio et al. (2021)
Hydrogen ion concentration	pH	Hussain et al. (2010)
Residual chlorine in tertiary treated wastewater	Both free and combined chlorine	Hussain et al. (2010)

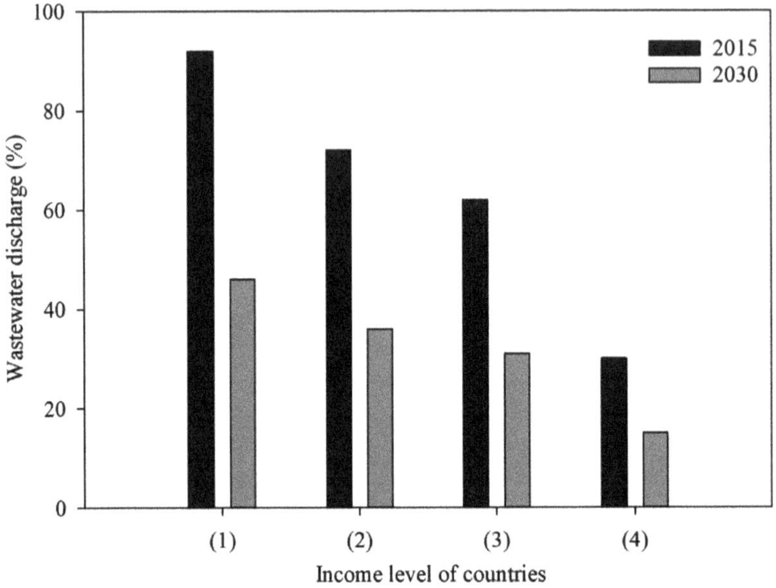

Fig. 1: Shows the percentage of untreated wastewater discharged into the environment by (1) Low-income countries, (2) Lower middle-income countries, (3) Middle higher-income countries, (4) Higher-income countries.

generated worldwide pass through treatment plants before its reuse. Figure 1 shows there is a great variation of untreated wastewater discharge between low- and high-income countries (Khalid et al. 2018). At present, 92% and 30% of untreated wastewater is released into the environment by low and high-income countries, respectively. About 80% of the total wastewater treatment plants in the world are present in high-income developed countries. In France alone, 17,000 wastewater treatment plants are functional along with 800,000 km long sewerage network. China has 3272 sewage treatment facilities with a treatment capacity of 140 million cubic meters per day. USA had 15,591 wastewater-treatment plants in 2000. In California, about 75% of wastewater is processed before irrigation (Khalid et al. 2018). In low-income countries, the scenario is different from high-income countries due to a lack of financial and technical resources. For example, about 30% of the total untreated wastewater generated in Pakistan is used for the irrigation of 32,500 ha, while 64% of the untreated wastewater is released into the environment. In India, only 24% of total industrial and municipal wastewater generated passes through treatment plants (Khalid et al. 2018).

2.1.3 *Wastewater sources and quantity*

In general, there is diurnal variation in the flow of domestic wastewater flow (Penn et al. 2012). Depending upon the usage of freshwater, usually flow of wastewater is minimal in early morning and maximal in the late morning followed by another maximal peak in the late evening after dinner (Friedler et al. 2013). However, peak flow rate and average flow may vary with the size of a city and the living standard of the community. The patron of peak flow may vary due to festive occasions, the timing of religious practices, business hours, sessional tourisms, large educational campuses, and so on (Henze and Comeau 2008, Penn et al. 2017).

Wastewater flow from industries follow the fixed pattern of minimal, maximal, and average flow depending upon the process involved, process timing, number of working hours, and water requirement for the specific process. Flow variation may occur during repair and maintenance, and shutdown during a specific season. Therefore, in industrial wastewater patron seasonal variation is more significant than daily variation. In municipal wastewater flow, if the contribution of industrial wastewater is significant, then variation in industrial wastewater has great importance in municipal wastewater management.

In developed countries, estimation of wastewater generation is determined by living standards, economic factors, and consistency in the water supply. However, in poor or under developing countries, economic factor is not considered, where water supplies are limited and uncertain. Estimation of wastewater generation in developing countries generally depends on the availability and minimal usage of freshwater.

2.2 Past and present of wastewater use in agriculture

Domestic wastewater in agriculture is historical. Angelakis et al. (2005) reported that nearly 4000 years ago, the Minoans started to collect and reuse wastewater at Knossos, Crete. Since ancient times, human excreta have been used in East Asia for soil and crops fertility. Countries like Japan, Korea, and China convert human excreta to fertilizer for agriculture production. The earliest reported sewage farms of the Scotland and Germany in the sixteenth and seventeenth centuries served as a hub of wastewater collection and utilization in agricultural lands (Shuval et al. 1986).

In the mid-nineteenth century, sewage farms largely served as disposal facilities of municipal wastewater mainly produced from fast-growing countries and cities/states of Europe and the United States. The use of sewage farms decreased with the advent of advanced wastewater treatment processes such as biological treatment of wastewater that includes trickling filters and activated sludge (Asano et al. 2007, Keerio et al. 2020a,b). However, treating wastewater with advance treatment technology can yield the fertilizer too and that can be used for agricultural purposes; this causes no need to dispose the municipal wastewater into fresh rivers (Asano et al. 2007) Scientific awareness of the possible transmission of water-borne disease through the food chain further limited the scope of sewage farms and complete abandonment in the western world (Shuval et al. 1986).

Pandemic disease and widespread wars bring significant change in society's behavior with both fresh and wastewater use. The effects of World War II brought advancement in scientific knowledge of wastewater treatment and reuse for agricultural production in both developed and under-developing nations, especially more arid regions. Advanced research in the United States was carried out to investigate microbial and health aspects of municipal wastewater reuse in agriculture. Health departments of the United States developed standards and established regulations for the reuse of municipal wastewater of irrigation. California was the first state that issued regulations in 1918 for wastewater reuse in agriculture; later on, those regulations were upgraded with more strict applications (Asano and Levine 1996). Those upgraded regulations were handed over to design engineers, health departments, and the farming communities in the United States. The regulations of California were replicated in other states of the United States and many countries including developing countries.

2.2.1 Fundamental issues in application of wastewater for agriculture

There is an important relationship between the income level of countries (such as low, middle, and high) and the problem it faces in wastewater use in agriculture. Insufficient water supply due to unplanned work, poor sanitation, and the use of untreated wastewater in urban areas is linked with poor economic development. Contrary to that, sufficient water supply, proper sanitation, and effective use of wastewater reflect the high level of economic development. Improved sanitation and proper wastewater treatment in urban areas is mainly linked with economic development (Scheierling et al. 2010).

2.2.1.1 Municipal wastewater treatment based on income level

Low-income countries: Generally, if the treatment plants exist in poor or low-income countries, they are poorly designed, work at the lowest capacity, and function poorly. For example, there are 70 wastewater and sludge treatment facilities in Ghana working at less than 10 percent efficiency (Murray and Drechsel 2011). In low-income countries, policymakers and end-users face challenges at the economic, institutional, and technical levels when they think about improvements or upgradation in existing sanitation or management of wastewater treatment. Affordability of initial and running cost is a crucial issue. Under this situation, cheap and unconventional treatment is considered a feasible alternative option, and these options are not availed in the richer neighborhood. In low-income countries, two main factors affect most wastewater treatment systems such as the high frequency of electricity load shading and minimal water availability per capita. Frequent electricity shortfall can affect electric appliances and the treatment efficiency of the plant.

Middle-income countries: Wastewater treatment and management are receiving prime attention and interest in middle-income countries, and more intense regulations are being established, along with the will of decision-makers. However, treatment of wastewater has significantly advanced, but operational difficulties due to some deficiencies, and issues in investments and fund transfer persist. The difference between installed capacity and the real capacity of wastewater plant is necessary, otherwise it can be misleading design information, as is illustrated in a survey conducted in Latin America and the Caribbean countries (Lawryshyn and Jaimungal 2010) and in another study reported from China (Xie 2009).

High-income countries: Generally, wastewater is treated in high-income countries, except in few cases. The developed countries such as countries in OECD have invested significant resources and funds to achieve a high level of treatment. In

North America, more than 90 percent of total wastewater passes through treatment plants and is mostly treated at secondary levels, and in some cases, it is treated at tertiary levels (WHO/UNICEF 2000). Progress in developing wastewater treatment facilities in the United States is generally subsidized by the government; for instance, in 1972–1989, federal construction approved US$56 billion to local governments to build secondary wastewater treatment facilities (Helmer and Hespanhol 1997).

2.2.2 *Increasing trends on use of wastewater*

Four main drivers are responsible for increasing the reuse of municipal wastewater for irrigation of crops around urban areas. These driving factors will be even more important in the future, the quality of treated wastewater would be the top priority.

Rising water demand: At present, many countries are facing the challenges of water scarcity. The population growth of water-scarce countries is projected to increase by 44 percent of the total population of the world by 2050 (Molden and de Fraiture 2010). Countries with enough water supplies will go under water stress due to unequal distribution of water supplies. For instance, the climate of Latin America falls under the category of humid tropics, still, there is 20 percent of the land that is arid or semiarid and receives only 5 percent of the total water supplies while 60 percent of the population is living in this arid zone (Bartone 1990).

Climate change has a great role in the increasing trend of water stress. According to the Intergovernmental Panel on Climate Change (2007), due to increasing global warming, unpredicted precipitation, melting glaciers, and frequent events of extreme droughts are expected in near future. The areas affected with drought are more inclined to face severe drought events with greater intensity and higher frequency. Specifically, drought-affected regions in the subtropics are likely to become significantly drier due to increased water stress (Meehl et al. 2007). Water quality will be affected in drier regions due to climate change and result in reduced assimilative capacity, and an increase in salinity (Sadoff and Muller 2009). Under these conditions, demand for the easily available, most reliable, and sustainable resource that is wastewater will increase.

Expanding urbanization: The biggest share of the world's population is living in cities. Specifically, the growth rate of urbanization in developing countries is very high. The urban population in urban areas is expected to increase by four billion by the end of 2030. Particularly, population in poor countries is projected to increase a maximum of 539 million which is double the current population (UNDESA 2008).

Rising urban wastewater: Increasing wastewater generation in urban areas is linked with the substantially increasing population, especially in under-developing countries. Economic development in such countries tends to shift the living standard of the population and as a result, wastewater generation increases. In recent years, sanitation services in urban areas have been improved, including sewerage. According to WHO/UNICEF (2008), some 779 million people in urban areas received advanced sanitation services from 1990 to 2006. However, despite all these attempts to improve, these efforts hardly kept pace with the growth rate of the urban population. As per

Millennium Development Goals, sanitation for all by 2025 requires a substantial pace of improvement. Nevertheless, wastewater generation will grow consistently and substantially, and therefore wastewater management requires time.

Increasing agricultural practices in and around urban areas: Increased wastewater generation due to the population growth in urban areas in developing countries is also associated with increased irrigation water demand to meet the food requirements of the growing urban population.

Very limited estimates of urban agriculture activities are available. According to UNDP (1996) report, around 800 million urban people were linked with agricultural activities and produce around 15 percent of the world's total food. Drechsel et al. (2006) reported that in West Africa, around 20 million urban people were involved in urban agriculture and produce 60 to 100 percent of vegetables. The contribution of high-income cities, for example Hong Kong and Singapore, in urban agriculture is fairly high.

2.3 Current methods for regulating the reuse of wastewater in agriculture

2.3.1 Guidelines for microbes contaminated wastewater reuse

Municipal wastewater is full of harmful pathogens such as fecal coliforms, bacteria, viruses, helminths eggs, and many others. These pathogens are capable of causing diseases if they come in contact with humans in sufficient amount. Intestinal nematodes can cause higher infection than bacteria. Viruses pose minimal risks to human health. In 1985, World Health Organization (WHO) issued standard guidelines of irrigation water for agriculture and water standards for aquaculture. The aim of WHO guidelines was to minimize the risk of transmission of infectious diseases without compromising resource conservation and reuse. The key notations of WHO guidelines were:

Controlled irrigation: Nematode eggs should not exceed more than one per liter.

Uncontrolled irrigation: Restricted irrigation criteria, plus fecal coliform should not exceed one thousand per hundred milliliters.

Restricted irrigation means the application of water to those crops that are not directly consumed by humans such as cotton, sunflower, and other crops. Crops pass through some process before consumption such as wheat, rice, barley, pulses, and other crops; these crops fall under Category A; crops in Category B are fodder, pastures, and fruit trees.

Unrestricted irrigation falls under Category C in which water is applied to those crops that are consumed directly including salads, carrot, cucumber, lettuce, etc., and also water is applied to lawns, public parks, sports grounds, and so on.

Guidelines for wastewater reuse for irrigation were not established in European countries except Germany and France. When European Union formulated the guidelines for wastewater use in agriculture, the guidelines aimed to cover the economic aspects, conservation of soil and groundwater, increasing crop yield, and improving sanitation to protect public health.

2.3.2 *Guidelines for chemical impurities of wastewater reuse*

Many water quality guidelines and standards of irrigation are mostly emphasized on microbial aspects rather than chemical aspects perhaps due to acute health effects. Chang et al. (1996) reported that a limited number of wastewater quality standards have been developed, especially application in agriculture. Moreover, the guidelines of the US EPA 1992 about wastewater reuse in agriculture do not address public health issues; rather they address transport of harmful and toxic pollutants from wastewater irrigation into soil, plant, and groundwater.

2.4 Cost-effective treatment systems of wastewater

The level of wastewater treatment depends on the type of crop, soil and environmental conditions, and regulatory obligations. The cost of a high polluted wastewater treatment is relatively higher. However, this cost can be sometimes acceptable as they cope with the value of crop yield, its significance, water use efficiency, and public concern (Schleich et al. 1996). If there are no obligations from environment quality control agencies, then cost optimization should be the prime objective of the wastewater treatment facility. However, previously reported studies indicated that improving wastewater quality is not compromised on the cost of the treatment facility (Schwarz and Mcconnell 1993).

The on-field situation in the developing world is different; mostly, wastewater is used directly without treatment mainly due to the initial and running cost of the wastewater treatment plant and loss of vital nutrients. However, in the opinion of general-public, wastewater treatment is essential for public health, aesthetic value, and religious beliefs (Mara 2000). In response to depleting freshwater sources, drought-tolerant crop farming, dealing with the hot climate, and most importantly, cost of available freshwater resources, advanced research and practical on-ground development have been carried out especially in arid or semi-arid regions of the world.

Treatment of municipal wastewater is an established technology; multiple processes are involved to treat wastewater effectively (Cheremisinoff 2001). If municipal wastewater is not included with a high concentration of industrial wastewater, then the conventional wastewater treatment process is applied, in which wastewater passes through primary sedimentation and secondary biological treatment. In many developing countries, the availability of a diverse range of technologies, along with the associated expenses of installation, operation, and upkeep, pose significant obstacles to the implementation of effective wastewater treatment systems. The lack of affordable and sustainable options further complicates the process of adapting wastewater treatment methods in these nations.

2.5 Framework for risk evaluation and management (WHO 2006 Guidelines)

It is very important to access the potential risk of harmful pathogens present in municipal wastewater used in agriculture. These pathogens can pose a significant threat to people engaged in agriculture activities and those who consume fruits, vegetables, and other food grown in wastewater.

The guidelines of WHO (2006) about the safe reuse of municipal wastewater are radically different from those guidelines of WHO 1973 and 1989 (WHO 1973, 1989). The main focus of WHO 2006 guidelines is risk analysis and its appropriate management; these guidelines are followed by Stockholm Framework approach (Bartram et al. 2001). This framework of risk management is also used for both fresh water and sanitation. However, in previous guidelines of WHO, required volume of treated wastewater was specified.

WHO 2006 approach for microbial risk of municipal wastewater reuse is (1) to describe the bearable maximum capacity of the extra burden of microbial disease, (2) to drive bearable risk of microbial infectious disease, (3) to ensure limits of pathogen reduction to control above factors 1 and 2, (4) to describe how to achieve those limits of pathogen reduction, (5) to describe a mechanism to manage the microbial risks of municipal wastewater (Table 2). This approach was adopted after Disability Adjusted Life Years (DALY) that was formed in 1993 by the WHO and with the cooperation of World Bank (World Bank 1993, Murray et al. 1996). Based on these methods and concepts, in 2006 National Guidelines of Australia for Water Recycling: Mitigating Health and Environmental Risks were developed. In 2007, Quantitative Microbial Risk Analysis (QMRA) was developed and used by (Mara et al. 2007).

2.5.1 *Associated challenges*

The combination of quantitative microbial risk assessment (QMRA) and QMRA-Monte Carlo risk simulations (QMRA-MC) makes it a very advanced tool to access the safety factors of wastewater reused in agriculture (Mara and Sleigh 2010). However, QMRA and QMRA-MC are associated with multiple challenges, especially those challenges that become even more critical in low-income and middle-income economies. Therefore, the following measures are suggested:

1. Training of resource persons and engineers in the sanitary, environment, and agriculture departments is necessary so that they become familiar with the application of QMRA under local conditions.
2. The regulators should be briefed about the application of QMRA; this will help them to establish new regulations for the reuse of wastewater in agriculture. Those regulations will address the health concerns of people who are in direct contact with wastewater in agriculture fields, and those who consume food grown in wastewater.

Table 2. A mechanism for the microbial risks management of municipal wastewater used in agriculture.

Steps	Activities and remarks
Step 1 Describe the bearable maximum capacity of the extra burden of microbial disease.	According to WHO 2006 guidelines, the default value of ≤ 10–6 DALY loss per person per year is consistent with acceptable health risks of completely treated drinking water. For low-income and middle-income countries, a more precise value is ≤ 10–4 DALY loss per person per year.
Step 2 Drive bearable risk of microbial infectious disease.	The bearable risk of a disease per person per year can be calculated using the equation: "tolerable DALY loss per person per year ÷ DALY loss per case of the disease", and the tolerable risk of infection per person per year from the equation: "tolerable disease risk ÷ disease/infection ratio".
Step 3 Quantify pathogenic reduction to control above steps 1 and 2.	QMRA-MC simulations are used to calculate the desired pathogenic reductions, such as, from untreated wastewater to consumption of treated water—such that the bearable risk of disease is calculated in Step 2, and therefore the bearable extra load of disease chosen in Step 1, do not exceed.
Step 4 Determine how to achieve the desired limit of pathogen reduction.	The reduction of the pathogen can be achieved through wastewater treatment, which is necessary for the health protection of those who directly deal with wastewater irrigation in the fields and the health of those who consume food produced on wastewater irrigation.
Step 5 Verification mechanism	The pathogenic reduction must be monitored and verified to ensure the proper functioning of a wastewater treatment system. A Hazard Analysis Critical Control Point (HACCP) system can be used to monitor the efficiency of the treatment system.

Sources: WHO Guidelines for the Safe Use of Wastewater, Excreta and Greywater. Vol. I: Policy and Regulatory Aspects. Vol. II: Wastewater Use in Agriculture. Vol. III: Wastewater and Excreta Use in Aquaculture. Vol. IV: Excreta and Greywater Use in Agriculture. Geneva: World Health Organization (2006).

3. Other government officials need to train in QMRA approaches to ensure the health risks-control through (a) approval of approach, (b) working with the farmers, and (c) to ensure the safety of foods consumed.

These measures can significantly help in the proper implementation of QMRA, and the reuse of municipal wastewater can be safe and cost-effective.

2.5.2 *Recommendations for ongoing practices*

If governments intend to implement regulations of reuse of wastewater in agriculture, it is strongly recommended to adopt QMRA-MC, so that the direct and indirect health risks of municipal wastewater can be addressed properly. Policymakers can make decisions based on QMRA-MC to correct the current practices and achieve the ultimate goal of reducing health risks of food producers such as field workers and farmers, and consumers such as the local public.

The QMRA-MC also addresses the cost-effectiveness of the wastewater used in agriculture, for example, to examine a question of what will be the health risk of people who consume more lettuce. The answer can be found by putting the correct range of lettuce being consumed in a computer program (QMRA-MC). Similar type of questions can be answered through QMRA-MC by providing accurate data. Policymakers, implementers, end-users, and consumers can benefit from this tool to reduce public health risks at all levels.

2.6 Improving wastewater quality and its use in agriculture

2.6.1 *Progressing towards planned wastewater irrigation*

To mitigate the health-hazard wastewater irrigation, every country needs to consider progressive approaches in response to their specific health priorities. These approaches can be different to achieve the ultimate goal to meet the internationally-established health-based standards such as set in guidelines of WHO 2006.

Middle-income countries: Generally, these countries have established some policies, approved legislation, developed institutes, and regulations. Still, they require a comprehensive approach for the planned framework of water resource management that includes both fresh water and wastewater. These countries also may not have comprehensively addressed the use of wastewater in agriculture. Affordability of treatment cost is a critical problem in these countries, and the strategy of stepwise upgrading and improvement of wastewater treatment systems should be adopted.

High-income countries: Standard regulations for fresh water and wastewater are implemented stringently. However, some of the countries have partially addressed wastewater reuse for irrigation of food crops. Many low-income countries are trying to focus on more stringent quality standards but cannot afford them. Therefore, strategic planning and step-by-step execution should be adapted.

2.6.2 *Stepwise up-gradation of wastewater treatment*

Shifting from unmanaged to managed reuse of wastewater for irrigation, the application of wastewater treatment should be the ultimate strategic goal for low-income countries. For such countries, a simpler and cheap treatment system should be adopted as an initial step, and gradually move toward the up-gradation of the existing system, expand its capacity, and improve overall efficiency. A crucial part of this initial step is to identify and fix the responsibilities of the community that generates wastewater and sanitation-service providers. Proper monitoring of the treatment system and verify the targets of treatment should be strictly followed.

When planning a proper wastewater treatment system, especially in developing countries, wastewater is treated in ways such as:

- *On-site treatment of municipal wastewater and its application:* Wastewater management is done at the wastewater generation point. These on-site systems may treat both black and gray water individually or mixed water. For example, greywater separately collected in storage tanks can be used for gardens at home, and composted human excreta can be used as soil and plant nutrients.

- *Communal treatment of wastewater and its application:* In this system, wastewater is collected from multiple residential buildings and diverted closer to treatment facility and then treated wastewater is used for small farming activities. An appropriate communal system may compose of condominial sewers, three tank systems, septic tanks, anaerobic ponds, storage ponds, and Imhoff tanks. The treated wastewater can be applied to communal plots for irrigation.
- *Decentralized treatment of wastewater and its application*: Wastewater is collected from separated medium-size populations or small portions of big cities. Collected wastewater is treated near the point of generation and treated water is generally used for larger agriculture fields in or near urban areas. This system may include waste settling ponds, anaerobic sludge reactors, storage reservoirs, or primary treatment with chemical application.
- *Centralized treatment of wastewater and its application:* Wastewater and stormwater are collected from big cities using an extensive pipe network and transported to a central treatment facility usually located near the conventional irrigation facilities. This system may compose of waste settling ponds. Wastewater may be treated with primary treatment with chemical application along with sand filtration (Libhaber 2007). High-income countries prefer to use membrane technologies (Bixio et al. 2006).

2.6.3 Planning to improve industrial effluent treatment

In cities with a large number of industries producing a huge amount of wastewater, pretreatment of effluent should be enforced for the successful operation of the treatment plant. Generally, industrial effluent is treated to the limits of wastewater quality standards and then discharged into municipal sewerage. It ensures that biological, chemical, organic, inorganic contaminants and heavy metals generated by industries do not enter the system to damage the pipe network and inhibit the treatment process; poorly treated industrial effluent should not be allowed to discharge into municipal sewers.

The important components of a successful pretreatment of industrial effluents are described by (Idelovitch and Ringskog 1997):

- The data collection system of effluent discharge.
- An industrial effluent discharging system into municipal sewers.
- Effluent tested from certified laboratories and reported by the concerned industry.
- Random examination by the wastewater quality control authorities.
- Penalties for noncompliance.
- Meterage and tariffs are based on volume and the pollutants load.
- Industrial participation in designing and implementation of standards.
- Need-based professional training and financial help to industries.

2.7 Conclusion

Increasing population and consequently rising food demand have compelled us to utilize all available resources to expand agricultural land and increase food production. Municipal wastewater is a resource that is either free or available at a low-cost in arid and low-income countries. Irrigation with municipal wastewater is an evolving priority and demand for wastewater will increase continuously. Specifically, four driving forces are responsible for increasing demand: inadequate water supplies in various countries and regions of the world; expanding cities in low-income countries; increasing wastewater flows due to enlargement of efficient water supply and proper sewerage system, and more people in urban areas involved in agricultural pursuits. Wastewater is a reliable source for irrigation, nutrients rich source, can increase yield, and reduces pollution at the point of generation; managed municipal wastewater can be used for livelihood for people living in urban areas, and more importantly, it provides food security. However, municipal wastewater is one of the best available resources for irrigation; contrary to that, if wastewater is not properly treated and managed, it may create health risks for both food producers and food consumers; it may deteriorate the environment as well.

Municipal wastewater offers both risks and benefits. The fundamental objectives of irrigation with municipal wastewater should be based on: (1) to reduce health-risk of public; (2) to reduce risk of environmental deterioration; (3) to promote livelihoods using wastewater; and (4) to incorporate wastewater into important water resource management. These objectives can be pursued based on the economic condition of any specific country such as low-income countries may focus on minimizing health risk through microbial disease and improve livelihood, while middle and high-income countries may focus on reduction of chemical risks and environmental deterioration.

References

Angelakis, A. and Bontoux, L. 2001. Wastewater reclamation and reuse in Eureau countries. Water Policy. 3(1): 47–59.

Angelakis, A., Koutsoyiannis, D. and Tchobanoglous, G. 2005. Urban wastewater and stormwater technologies in ancient Greece. Water Research. 39(1): 210–220.

Asano, T. and Levine, A.D. 1996. Wastewater reclamation, recycling and reuse: past, present, and future. Water Science and Technology. 33(10-11): 1–14.

Asano, T., Burton, F. and Leverenz, H. 2007. Water reuse: issues, technologies, and applications. McGraw-Hill Education.

Bartone, C.R. 1990. Water quality and urbanization in Latin America. Water International. 15(1): 3–14.

Bartram, J., Fewtrell, L. and Stenström, T.-A. 2001. Harmonised assessment of risk and risk management for water-related infectious disease: an overview. IWA Publishing, London.

Bixio, D., Thoeye, C., Wintgens, T., Hochstrat, R., Melin, T., Chikurel, H. et al. 2006. Wastewater reclamation and reuse in the European Union and Israel: Status quo and future prospects. International Review for Environmental Strategies. 6(2): 251–268.

Chang, A.C., Page, A.L., Asano, T. and Hespanhol, I. 1996. Developing human health-related chemical guidelines for reclaimed wastewater irrigation. Water Science and Technology, 33(10-11): 463–472.

Cheremisinoff, N.P. 2001. Handbook of water and wastewater treatment technologies. Butterworth-Heinemann.

Drechsel, P. and Evans, A.E. 2010. Wastewater use in irrigated agriculture. Irrigation and Drainage Systems. 24(1): 1–3.

Drechsel, P., Graefe, S., Sonou, M. and Cofie, O.O. 2006. Informal irrigation in urban West Africa: An Overview.

Drewes, J., Heberer, T. and Reddersen, K. 2002. Fate of pharmaceuticals during indirect potable reuse. Water Science and Technology. 46(3): 73–80.

Elahi, E., Zhang, L., Abid, M., Javed, M.T. and Xinru, H. 2017. Direct and indirect effects of wastewater use and herd environment on the occurrence of animal diseases and animal health in Pakistan. Environmental Science and Pollution Research. 24(7): 6819–6832.

Feigin, A., Ravina, I. and Shalhevet, J. 2012. Irrigation with treated sewage effluent: management for environmental protection. Springer Science and Business Media.

Friedler, E., Butler, D. and Alfiya, Y. 2013. Wastewater composition. Source Separation and Decentralization for Wastewater Management, 241–257.

Helmer, R. and Hespanhol, I. 1997. Water pollution control: a guide to the use of water quality management principles. CRC Press.

Henze, M. and Comeau, Y. 2008. Wastewater characterization. Biological wastewater treatment: Principles Modelling and Design, 33–52.

Idelovitch, E. and Ringskog, K. 1997. Wastewater treatment in Latin America: old and new options. World Bank Publications.

Intergovernmental Panel on Climate Change, 2007. Climate Change 2007: Synthesis Report. Geneva, Switzerland: IPCC.

Jiménez, B. and Asano, T. 2008. Water reuse: An international survey of current practice, issues and needs. IWA London.

Keerio, G.S., Keerio, H.A., Ibuphoto, K.A., Laghari, M., Panhwar, S. and Talpur, M.A. 2021. Arsenic removal through bio sand filter using different bio-adsorbents. Journal of Water and Land Development.

Keerio, H.A., Bae, W., Park, J. and Kim, M. 2020a. Substrate uptake, loss, and reserve in ammonia-oxidizing bacteria (AOB) under different substrate availabilities. Process Biochemistry. 91: 303–310.

Keerio, H.A. and Bae, W. 2020b. Experimental Investigation of Substrate Shock and Environmental Ammonium Concentration on the Stability of Ammonia-Oxidizing Bacteria (AOB). Water. 12(1): 223.

Keraita, B., Drechsel, P. and Konradsen, F. 2008. Perceptions of farmers on health risks and risk reduction measures in wastewater-irrigated urban vegetable farming in Ghana. Journal of Risk Research. 11(8): 1047–1061.

Khalid, S., Shahid, M., Bibi, I., Sarwar, T., Shah, A.H. and Niazi, N.K. 2018. A review of environmental contamination and health risk assessment of wastewater use for crop irrigation with a focus on low and high-income countries. International Journal of Environmental Research and Public Health. 15(5): 895.

Kookana, R.S., Williams, M., Boxall, A.B., Larsson, D.J., Gaw, S., Choi, K. et al. 2014. Potential ecological footprints of active pharmaceutical ingredients: an examination of risk factors in low-, middle-and high-income countries. Philosophical Transactions of the Royal Society B: Biological Sciences. 369(1656): 20130586.

Lawryshyn, Y. and Jaimungal, S. 2010. Optimization of a municipal wastewater plant expansion: A real options approach. Journal of Applied Operational Research. 2(1): 33–42.

Libhaber, M. 2007. Appropriate technology for wastewater treatment and reuse in developing countries. Proceedings of the international symposium on water supply and sanitation for all: obligation of the water professionals for our common future, pp. 87–130.

Mara, D. 2000. The production of microbiologically safe effluents for wastewater reuse in the Middle East and North Africa. Water, Air, and Soil Pollution. 123(1): 595–603.

Mara, D., Sleigh, P., Blumenthal, U. and Carr, R. 2007. Health risks in wastewater irrigation: comparing estimates from quantitative microbial risk analyses and epidemiological studies. Journal of Water and Health. 5(1): 39–50.

Mara, D. and Sleigh, A. 2010. Quantitative Microbial Risk Analysis: The 2006 WHO Guidelines and Beyond– How to Use the QMRA-Monte Carlo Computer Programs in Wastewater-use Planning. Leeds: School of Civil Engineering, University of Leeds.

Meehl, G.A., Stocker, T.F., Collins, W.D., Friedlingstein, P., Gaye, A.T., Gregory, J.M. et al. 2007. Global climate projections. Chapter 10.

Molden, D. and de Fraiture, C. 2010. Comprehensive assessment of water management in agriculture. Agricultural water management (Print). 97(4).

Murray, A. and Drechsel, P. 2011. Why do some wastewater treatment facilities work when the majority fail? Case study from the sanitation sector in Ghana. Waterlines. 135–149.

Murray, C.J., Lopez, A.D. and Organization,W.H. 1996. The global burden of disease: a comprehensive assessment of mortality and disability from diseases, injuries, and risk factors in 1990 and projected to 2020: summary. World Health Organization.Negulescu, M. 2011. Municipal waste water treatment. Elsevier.

Panhwar, S., Aftab, A., Keerio, H.A., Sarmadivaleh, M. and Tamer, U. 2021a. A novel approach for real-time enumeration of *Escherichia coli* ATCC 47076 in Water through high multi-functional engineered nano-dispersible electrode. Journal of The Electrochemical Society. 168(3): 037514.

Panhwar, S., Aftab, A., Muqeet, M., Keerio, H.A., Solangi, G.S., Suludere, Z. et al. 2021b. High-performance *E. coli* antibody-conjugated gold nanorods for the selective electrochemical detection of pathogens in drinking water. Journal of Electronic Materials, 1–7.

Pedrero, F., Kalavrouziotis, I., Alarcón, J.J., Koukoulakis, P. and Asano, T. 2010. Use of treated municipal wastewater in irrigated agriculture—Review of some practices in Spain and Greece. Agricultural Water Management. 97(9): 1233–1241.

Penn, R., Hadari, M. and Friedler, E. 2012. Evaluation of the effects of greywater reuse on domestic wastewater quality and quantity. Urban Water Journal. 9(3): 137-148.

Penn, R., Schütze, M., Gorfine, M. and Friedler, E. 2017. Simulation method for stochastic generation of domestic wastewater discharges and the effect of greywater reuse on gross solid transport. Urban Water Journal. 14(8): 846–852.

Qadir, M., Wichelns, D., Raschid-Sally, L., McCornick, P.G., Drechsel, P., Bahri, A. et al. 2010. The challenges of wastewater irrigation in developing countries. Agricultural Water Management. 97(4): 561–568.

Reemtsma, T., Miehe, U., Duennbier, U. and Jekel, M. 2010. Polar pollutants in municipal wastewater and the water cycle: occurrence and removal of benzotriazoles. Water Research. 44(2): 596–604.

Rodriguez, C., Van Buynder, P., Lugg, R., Blair, P., Devine, B., Cook, A. et al. 2009. Indirect potable reuse: a sustainable water supply alternative. International Journal of Environmental Research and Public Health. 6(3): 1174–1203.

Sadoff, C. and Muller, M. 2009. Water management, water security and climate change adaptation: early impacts and essential responses. Global Water Partnership Stockholm.

Schleich, J., White, D. and Stephenson, K. 1996. Cost implications in achieving alternative water quality targets. Water Resources Research. 32(9): 2879–2884.

Schwarz, G.E. and Mcconnell, V.D. 1993. Local choice and wastewater treatment plant performance. Water Resources Research. 29(6): 1589–1600.

Shahid, M., Khalid, S., Murtaza, B., Anwar, H., Shah, A.H., Sardar, A. et al. 2020. A critical analysis of wastewater use in agriculture and associated health risks in Pakistan. Environmental Geochemistry and Health, 1–20.

Shuval, H.I., Adin, A., Fattal, B., Rawitz, E. and Yekutiel, P. 1986. Wastewater irrigation in developing countries: Health effects and technical solutions.

UNDESA (United Nations Department of Economic and Social Affairs). 2008. Creating an Inclusive Society: Practical Strategies to Promote Social Integration. Final Report of the Expert Group Meeting. Available at: http://www.un.org/esa/socdev/egms/docs/2008/Paris-report.pdf (accessed on 29 August 2014).

UNDP (United Nations Development Program). 1996. Urban Agriculture: Food, Jobs and Sustainable Cities. UNDP Publication Series for Habitat II, Volume One. New York: UNDP.

Van Rooijen, D.J., Biggs, T.W., Smout, I. and Drechsel, P. 2010. Urban growth, wastewater production and use in irrigated agriculture: a comparative study of Accra, Addis Ababa and Hyderabad. Irrigation and Drainage Systems. 24(1): 53–64.

Weldesilassie, A.B., Boelee, E., Drechsel, P. and Dabbert, S. 2011. Wastewater use in crop production in peri-urban areas of Addis Ababa: impacts on health in farm households. Environment and Development Economics. 16(1): 25–49.

WHO. 1973. Reuse of Effluents: Methods of Wastewater Treatment and Health Safeguards.
Technical Report Series No. 517. Geneva: World Health Organization.

WHO. 1989. Health Guidelines for the Use of Wastewater in Agriculture and Aquaculture. Technical Report Series No. 778. Geneva: World Health Organization.

WHO. 2006. WHO Guidelines for the Safe Use of Wastewater, Excreta and Greywater. Vol. I: Policy and Regulatory Aspects. Vol. II: Wastewater Use in Agriculture. Vol. III: Watewater and Excreta Use in Aquaculture. Vol. IV: Excreta and Greywater Use in Agriculture. Geneva: World Health Organization.

WHO/UNICEF. 2008. Progress on drinking water and sanitation: special focus on sanitation. World Health Organization.

World Bank. 1993. World Development Report 1993: Investing in Health, Volume1. The World Bank.

World Bank. 2000. Project Appraisal Document: Tehran Sewerage Project. Report No. 20154 IRN. Washington, DC: World Bank.

Xie, J. 2009. Addressing China's water scarcity: recommendations for selected water resource management issues.

Xie, X., Yuan, K., Yao, Y., Sun, J., Lin, L., Huang, Y. et al. 2022. Identification of suspended particulate matters as the hotspot of polycyclic aromatic hydrocarbon degradation-related bacteria and genes in the Pearl River Estuary using metagenomic approaches. Chemosphere. 286: 131668.

3

Photocatalytic Wastewater Remediation using Nanoparticles

Present Challenges and Future Outlook

Adil Shafi,[1,*] *Nafees Ahmad*,[1,3,*] *Shah Imtiaz*,[2] *Sabeeha Jabeen*,[3] *Ashaq Hussain Bhat*,[4] *Suhail Sabir*[1] and *Mohammad Zain Khan*[1,*]

3.1 Introduction

The contamination and scarcity of water has emerged as a major and growing concern which the whole world is facing. Water has been rendered unfit due to the addition of pharmaceuticals, personal care products, antibiotics, xenobiotics, steroid hormones, organic dyes, inorganic wastes and volatile organic compounds (Obare et al. 2004, Mahmoodi et al. 2007, Das et al. 2015, Pawar et al. 2015). These pollutants are titled as emerging contaminants due to their potential harmful effects on both terrestrial and aquatic environments. The water containing these effluents, when utilised for human consumption, poses several detrimental problems to human health (Sha et al. 2019, Guerra et al. 2018, Byrne et al. 2018). This problem has gained attention from all over the globe and vast numbers of scientists and researchers are working hard to eliminate or detoxify these contaminants of emerging concern. From time to time, several techniques have been utilized for decontamination of wastewater, but conventional techniques are not able to remove emerging contaminants and their metabolites efficiently from water. Hence, development of advanced wastewater treatment technologies is required intending to enhance the ultimate quality or biodegradability of the water. Among various advanced technologies, photocatalytic

[1] Environmental Research Lab, Department of Chemistry, Aligarh Muslim University, Aligarh-202002.
[2] Department of Chemistry, Aligarh Muslim University, Aligarh-202002.
[3] Department of Chemistry, Integral University Lucknow-226026.
[4] Department of Higher education, GDC Women Anantnag-192132.
* Corresponding authors: aadilshafi@gmail.com; Siddiquenafees123@gmail.com; Zn.khan1@gmail.com

degradation of pollutants using nanomaterials as catalysts correspond to innovative development for competitive advantage. The use of nanomaterials as photocatalysts in wastewater decontamination has attracted increased attention because of their peculiar features like cost-effectiveness, excellent electronic properties, high surface to volume ratio, quantum confinement and better catalytic properties (Byrne et al. 2018, Gehrke et al. 2015, Guerra et al. 2018, Ikram 2016). Nano-photocatalysis is a promising technique to decontaminate wastewater and has shown good prospects in energy storage as well. Photocatalytic water splitting has been the focus of cutting-edge research to generate clean and safe drinking water using solar radiations. The photocatalytic degradation of organic dyes, pesticides, chloro-compounds have been considered as an emerging and efficient technology for the wastewater remediation. Nano-photocatalysis is considered as a progressive oxidation process which produces the extremely reactive hydroxyl and superoxide radicals on absorption of photonic light. These hydroxyl and superoxide act as powerful, non-selective reactive species and initiate a series of oxidative degradation reactions of organic compounds present in wastewater (Ganie et al. 2021, Colmenares et al. 2009, Patil et al. 2016). From past few years, a lot of research has been carried on nano-photocatysis for environmental remediation and wastewater decontamination. In this chapter, we highlight the prospects of nanophotocataysts in degradation of organic, inorganic and biological pollutants in water. Moreover, different photocatalysts along with governing principles and reaction mechanism have been elaborated. The prospect of photo catalytic technology in sustaining the quality of water in future has also been discussed.

3.2 Nano-photo catalysis for waste water remediation

The process in which solar light is utilized to enhance the speed of reaction is called photocatalysis. If the same process occurs by use of nano-particles in the presence of light, it is known as nano-photocatalysis. Nano-photocatalysis is an oxido-reductive reaction occurring on the surface of nanocatalyst in the presence of sunlight (Ata et al. 2019, Venkateswarlu et al. 2019, Chen et al. 2019). A large number of nano-photocatalysts in the form of particles, rods, wires, sheets, cylinders and spheres have been developed to purify water and decontaminate pollutants. Nano-photocatalysts have found their use in purification of wastewater, as they enhance the catalyst reactivity because of their large surface area and few shapes dependent features. Nanoparticles show different response as compared to bulk materials because of different surface properties and distinct quantum effects. An increase in optical, magnetic, mechanical, electrical properties, etc., of nano photocatalysts has been observed as compared to bulk materials (Ong et al. 2018, Gomez-Pastora et al. 2017). Due to large surface area of nano photocatalyst, they show an increased oxidation ability and generation of more oxidation species. The generation of reactive species leads to efficient degradation of organic hazards, complete conversion of pollutants and simultaneous detoxification of multiple contaminants without forming the harmful intermediates (Wang et al. 2019). Therefore, nano-photocatalysts can be used in degradation of pollutants especially in polluted water. Nano-photocatalysis is broadly classified as heterogeneous and homogenous catalysis.

3.2.1 Heterogeneous photocatalysis

In the present time, a lot of attention has been seen towards heterogeneous photocatalysts because of their wide application in wastewater decontamination. Heterogeneous photocatalysis requires the catalyst that exists in different phase than that of water. Among various heterogeneous photocatalysts, semiconductor devices have shown excellent and promising applications in wastewater remediation. Fujishima et al. in 1972 provided us with a procedure that was found to be a great promising tool in water treatment and remediation (Fujishima et al. 1972). This approach was found to be cost effective approach for heavy metal detoxification in water. The technology has been found promising because of solar energy utilization and showing dependence on the surface area of nanomaterials. In another technology based on advanced oxidation process, effluents are split into water and carbon dioxide via photocatalysis. Carbon dioxide can be then easily removed from the water sources (Rizzo et al. 2019, Maniakova et al. 2021). This process has shown wide applications in aquatic environment for removal of effluents preferably from soil and air in the environment.

3.2.1.1 Detoxification or removal of pollutants from aqueous phase

Photocatalysis utilizing nanostructures of a metal oxide semiconductor such as titania (TiO_2), zinc oxide (ZnO), zinc stannate (Zn_2SnO), tungsten oxide (WO_3), etc., has proved a promising route of water purification, as it removes chemical as well as biological contaminants (Ali et al. 2019, Bhanvase et al. 2017). Several other semiconductor nanoparticles (CeO_2, In_2O_3, BiOBr, Co_3O_4), heterostructures, bimetallics, nanocomposites, etc., have found their application in degrading pollutants like chlorpyrifos, azo dyes, pesticides, etc. (Loeb et al. 2018, Yaqoob et al. 2020, Ahmad et al. 2015). Some recent reports have shown the use of nanotubes based on TiO_2 in removing pollutants like Congo red, azo dyes, chlorinated compounds and aromatic pollutants from the waste water. Of all these semiconductor photocatalysts, TiO_2 is excellent because of its chemical stability, easy availability, non-toxicity and low cost. Further, it exists naturally in three different states, i.e., rutile, brookite and anatase, of which latter one is considered as good nano-photocatalyst. The anatase form of TiO_2 possesses remarkable features such as low cost, good performance and high stability. It has the band gap of 3.2 eV and capability of absorbing in ultraviolet region (< 387 nm) (Kafizas et al. 2016, Chen et al. 2018, Yaqoob et al. 2021). One of the studies has used dimethyl sulfoxide as reference material with CdS/TiO_2 to evaluate its photocatalytic performance in waste water treatment. The nanomaterials doped with iron can be used and recycled easily because of its property of ferromagnetism (Makama et al. 2016). Lin et al. in one of the experiments studied decomposition of effluents of paper making, obtained from flocculation process and sequential batch reactor (SBR) in wastewater using a reactor with nano-TiO_2 colloids followed by other experimental techniques. The results obtained showed that removal rates of total chemical oxygen demand (COD) for SBR-photocatalysis and flocculation-photocatalysis were found to be 65.8% and 90.8%, respectively (Lin et al. 2021). A lot of research on metal oxide nano-photocatalysts is being carried out to improve the quality of water and sustainable remediation of environment. But, these MO semiconductor photocatalysts suffer some drawback in showcasing excellent potential. To overcome this limitation, one of the efforts taken by researchers is to add other elements like metals, non-metals and their ions, and dye sensitizers which can increase their photocatalytic performance in presence of visible light (Umar et al. 2016, Malik et al. 2014).

3.2.1.2 *Physical parameters and principle governing heterogeneous photocatalysis*

Physical parameters are the assumptions which determine the performance of photocatalysts in performing a reaction. For catalysis, there are five physical parameters which govern a photocatalytic reaction (Gnanaprakasam et al. 2015).

They can be enumerated as:

(1) Mass of photocatalyst sometimes called catalyst loading
(2) pH of the reaction medium
(3) Temperature of the experimental procedure
(4) Wavelength absorbed by the photocatalyst, also called band gap determination.
(5) Morphology of the photocatalyst

The physical parameters give us an idea about the conditions and performance of a photocatalyst in executing a degradation process. The amount of catalyst, pH and temperature are prerequisites and determine the optimal conditions of the photo catalytic reaction. Photocatalysts with wide band gap tend to absorb light in UV region and by reducing the band gap, the absorption can be shifted to visible region. The morphological studies have a profound effect on the activity and operation of photocatalyst. Photocatalysts with porous morphology are more efficient than those who have rough or smooth morphology (Li et al. 2020).

The principle governing heterogeneous photocatalysis involves the generation of reactive species on absorption of light. This process includes accumulation of solar light, separation and relocation of charge, followed by charge recombination and redox phenomenon. Glaze et al. in one of the research projects in 1987 provided the first evidence of advanced oxidation process where the lifetime of hydroxyl radicals was used as function of decontamination for an adequate amount of water. The technology of heterogeneous photocatalysis generally utilizes the semiconductor materials having light absorbing capability in UV or visible region. When a semiconductor is irradiated with light, the electron is drifted from the valence band towards conduction band. This excitation process abandons a hole, i.e., photogenerated opening (h^+) in the valence band and electron (e^-) in the conduction band. The relocation of electron hole pair takes place on surface of material after taking part in redox process. These electron/hole pairs on combining with oxygen and water produces reactive species, i.e., hydroxyl radical, superoxide radical, which on interaction with pollutants results in its degradation (Umar et al. 2016).

3.2.2 Homogenous photocatalysis

In this process, the nano-photocatalyst used for waste water treatment is soluble in water. These are very costly and are difficult to separate from water after treatment, hence showing self-toxicity. Therefore, they have found less use in waste water treatment as compared to heterogeneous catalysts. Transition metal complexes are examples of homogeneous photocatalysts because of their suitable band gap and stability to perform photocatalysis. The frequently adopted processes by homogeneous photocatalysts involve ozonation and photo-Fenton process.

3.2.2.1 Photo-Fenton and Fenton processes

Photo-Fenton process involves reduction of ferric ions to ferrous ions along with hydroxyl radical using Fenton's reagent in combination with UV light (Perez et al. 2002).

$$Fe\ (III)\ OH^{2+} + hv \rightarrow Fe^{2+} + OH$$

Photo-Fenton process has better prospects in removing organic contaminants present in the waste-water through reverse osmosis (RO). This process was found more efficient as compared to UV/H_2O_2 system in the deprivation of effluent components and has resulted in enhancement of both the quantity of water and reduction of total organic carbon (TOC).

The process of homogenous catalysis is governed by photo-Fenton principle. In this process, M^{2+} (M is mostly Fe) is produced from photolysis of M^{3+} by utilizing light, which further attacks the hydrogen peroxide to produce hydroxyl radicals (Barrera-Salgado et al. 2016, Pignatello et al. 1992). During this conversion, the M^{2+} again gets converted to M^{3+}. This back-and-forth reaction from M^{2+} to M^{3+} is known as Fenton reaction. In the next step, the hydroxyl radical produced recombines with organic water pollutants again and again to oxidize them, followed by mineralization to water molecules, carbon dioxide and other simpler molecules. A simple diagram showing Fenton's process is shown in Fig. 1 (Zhang et al. 2019).

Fenton process was reported for maleic acid oxidation by Fenton in 1884. Amongst all AOPs, Fenton process proved to be the easiest and most efficient in terms of removal rate. Fenton's reagent targets the organic pollutants such as polyphenols, soluble substances and aromatic nucleophilic compounds According to Harber and Weiss (1934), Fenton reaction is based on the electron transfer between H_2O_2 and Fe^{2+}. The complex system along with electron transfer acts as catalyst producing hydroxyl radical which in turn degrades organic compounds:

$$Fe^{2+} + H_2O_2 \rightarrow Fe^{3+} + OH^- + OH$$

The factors affecting the extent of Fenton processes are: pH of the solution, concentration of H_2O_2, number of ferrous ions, initial concentration of contaminants and the presence of other ions. Furthermore, Fenton reagent action can be enhanced once subjected to UV radiation. This process proved beneficial in terms of removal of pesticides, insecticides and pollutants from refinery, rubber, and plastic industries (Zhang et al. 2019).

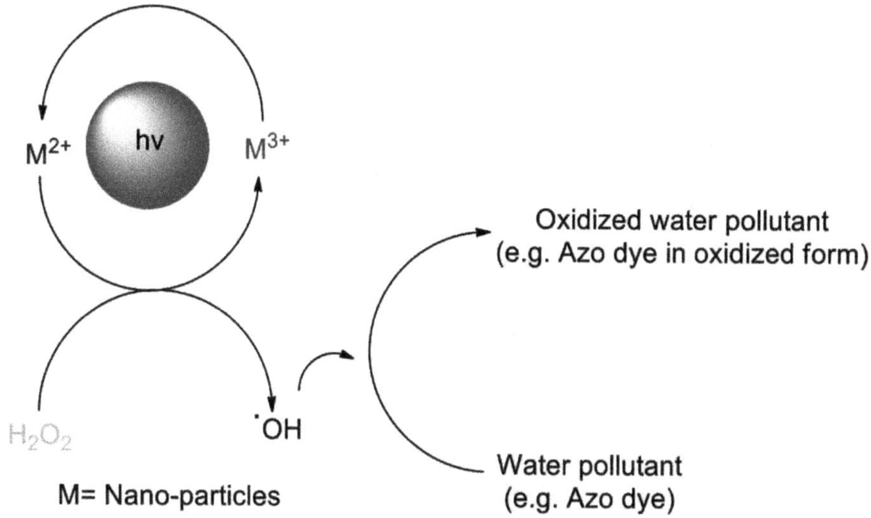

Fig. 1: Illustration of Fenton's process on surface of nano-photocatalyst.

3.2.2.2 Ozonisation

Ozone (O_3) is a strong oxidizing agent for water and wastewater with an oxidation potential of 2.07 V. Once ozone is dissolved in water, it undergoes a complex series of reactions in two different ways: by direct oxidation as molecular ozone or through formation of hydroxyl radicals and superoxide radicals (M'arimi et al. 2020).

$$O_3 + OH^- \rightarrow O_2^{*-} + HO_2^*$$

$$O_3 + O_2^{*-} \rightarrow O_3^{*-} + O_2$$

$$O_3^{*-} \rightarrow HO_3^{*-}$$

$$HO_3^* \rightarrow OH^* + O_2$$

Ozonation can be carried out at varied pH levels to elucidate the optimum conditions for reaction systems out of which, the best results are shown at alkaline pH, which can be ascribed to the reaction of organic and inorganic compounds with the molecular ozone and the radicals. Another significant parameter is the dose of ozone which affects the extent of oxidation. A 10 time's increase of ozone dosage enhances the rate of discoloration of treated water. Ozonation has revealed a strong oxidizing power with short reaction time whcih is helpful for the treatment of wastewater (Hodges et al. 2018).

3.2.2.3 Application in pollutant removal and remediation of wastewater

The waste water treatment for inorganic pollutants is very difficult because of their solubility in aqueous medium (Dahari et al. 2021). Further, coexistence of more than one contaminants, such as As(III) and Cr(VI) in drainage system, makes wastewater treatment more difficult. To overcome this, Liu et al. developed a strategy in which redox conversion and removal of both Cr(VI) and As(III) was done simultaneously in aqueous solution. In this strategy, Fe oxide was encapsulated in a carbon sphere denoted as HCS@Fe_3O_4 to reduce Cr(VI) to Cr(III) and oxidise As(III) to As(V). These converted Cr(III) and As(V) were easily removed in forms $Cr(OH)_3$ and $FeAsO_4$ precipitates, respectively (Liu et al. 2019). These findings provide an effectively synergistic conversion and removal strategy for applications in multiple-contaminated water treatment.

Jaihindh et al. (2019), in another study, developed the heterostructured nanomaterials that were used in detection as well as degradation of hazardous toxic chemicals from water. They reported nano structure of bismuth vanadate ($BiVO_4$) and used it as bifunctional catalyst in photocatalytic degradation as well electrochemical detection (Cr(VI)) (Jaihindh et al. 2019).

Fullerenes have found their use in waste water remediation by degrading various toxic metals because of its cage molecular structure. Moreover, because of their physicochemical properties, they have been found suitable for extraction of different species from the aqueous media (Zhang et al. 2013, Alekseeva et al. 2016). Carbon nano clusters, because of their low aggregation tendency, defective structure, and large surface area, have been used in adsorption of heavy metals from water. Further, carbon nanotubes have been used in degradation of Cd^{2+}, Cu^{2+}, Pb^{2+}, and Hg^{2+} during process of waste water treatment (Kaneko et al. 1993, Anitha et al. 2015, Zazouli et al. 2014, Gupta et al. 2015).

3.3 Strategies for improving photocatalytic materials for wastewater remediation

Residual and recalcitrant pollutants in water restrict the traditional treatment techniques to decontaminate and purify water as most of the pollutants have stable structure and are not easily removed. To remove these, advanced oxidation techniques are needed. Photocatalysis is one such technique to remove and eradicate these pollutants from waste water. Photocatalysis is an advance technique for purification and remediation of waste water coming from different resources. Photocatalysis by use of nanomaterial in form of oxides, sulphides, and hydroxides has the ability to absorb light and produce oxygen containing compound, which can prove helpful in eradication of wastes (Ahmad et al. 2017, 2019, 2020, 2021, Fernandez-Castro et al. 2015, Nosaka et al. 2017, Fanourakis et al. 2020). The application of photocatalysts in various reactions has been restricted by their large band gap and debauched recombination of electron-hole pairs. To overcome this phenomenon, various modifications have been done to alter the band gap of photocatalysts such as metal-ion doping, dye-sensitization, formation of nanocomposites, functionalization by suitable agents, capping, etc. (Patil et al. 2016, Bates et al. 2016). The modification of photocatalyst with suitable functional entity can impede the electron–hole pair recombination and accelerate the separation of photogenerated species, thus enhancing the photocatalytic activity.

3.3.1 Formation of nanocomposites

Since recent past, the fabrication of nanocomposites has attained the interest of scientific research. Compared to naked oxide photocatalysts, nanocomposites show improved and enhanced remediation activity. The enhanced activity can be ascribed to transfer of electron from one interface to other interface resulting in better charge separation and reduction in charge recombination (Opoku et al. 2017, Camargo et al. 2009, Shafi et al. 2019). Till date, several nanocomposites have been fabricated among metallic oxides and carbon-based materials. Carbon based metallic nanocomposites have proved efficient materials in various applications including wastewater decontamination. Graphene, a carbon-based material, has a unique property having good electric conductivity and large surface area (Li et al. 2020). Due to these effective properties, graphene/TiO_2-based photocatalyst has been prepared for wastewater decontamination. This composite has shown excellent photocatalytic activity due to formation of chemical bond and transfer of electrons from the photo catalyst to graphene, thus reducing the probability of recombination. Composites based on TiO_2/graphene attract scientific attention because of the multifunctional approach as highly efficient photo catalysts and adsorbent with high adsorption capacity (Wang et al. 2021, Ali et al. 2018).

However, TiO_2 decorated with carbon-based materials cannot serve as an effective photo catalyst because of the agglomeration of carbon layers, which undesirably affects their adsorption and photocatalytic properties. Modern researchers have overcome this difficulty by utilization of graphene-based aerogel as an auxiliary material for TiO_2 (Ton et al. 2018, Lettieri et al. 2018). Carbon nanotubes (CNTs) possess unique physicochemical, thermal, electrical and adsorption properties that make them intriguing for utilization in diverse applications as adsorbents, electrochemical sensors, nano-membranes, adsorbents and catalysts. CNTs can be modified according to the intended application through specific functionalization or fabrication processes. The modified CNTs can efficiently remove the contaminants in water and can expedite nanomaterial recovery and regeneration (Sarkar et al. 2018, Shahidi et al. 2018, Kausar et al. 2016). Since recent years, a vast number of carbon-based materials have been used to treat multiple contaminants of organic, inorganic, and biological origin. The applicability of CNTs in diverse fields demonstrate their robust prospective in practical applications, including wastewater remediation and purification. However, CNT-based materials have not penetrated the commercial market because of their reactive nature (Ates et al. 2017, Sireesha et al. 2018).

3.3.2 Functional nanoadsorbents

Nanoadsorbents are nano-dimension particles of inorganic or organic nature that possess the reactive surface and are capable of absorbing other substances. The potential applications of nanoadsorbents are mostly seen in the area of wastewater remediation and water purification. Nanosorbents are considered efficient materials as compared to traditional old sorbents and have shown better properties like large surface area and high substance specificity for the decontamination of water (Guerra et al. 2017, Vaseashta et al. 2007, Tratnyek et al. 2006).

Nanosorbents have excellent sorption capability which makes them more accurate and effective for water treatment. The commonly used nanoadsorbents are based on carbon, metal organic fragments, polymer nanocomposites, metal/metal oxide and metal hydroxides (Yu et al. 2017). The aggregate compound of different material like silver/polyaniline, silver/carbon, and carbon/TiO_2 have shown synergistic properties and have moderated the effect of toxic contaminants in the wastewater. CNTs hold measurable adsorption sites because of large surface area and hydrophobic surface properties and show good prospects in removing persistent pollutants and heavy metals from wastewater (Fuwad et al. 2019). Another important form of nanosorbents are zeolites, which have porous morphologies on which several nanoparticles like copper, zinc, nickel and silver ions can be implanted (Lettieri et al. 2018). The significant advantage of zeolites is controlled decoration of the amounts of metals and it can also serve as anti-bacterial agent. Furthermore, magnetic nanosorbents show efficient role in waste water remediation for the eradication of organic waste from water as compared to magnetic filtration (Tratnyek et al. 2006, Yu et al. 2017, Fuwad et al. 2019). The peculiar applications of nanosorbents are summarised below:

- Groundwater or soil remediation by novel colloidal composite of carbo-iron
- Nanoclays for adsorbing toxic phosphorus and other organic contaminants
- Nano-aerogels for removing toxic and carcinogenic uranium from groundwater
- Nano-iron oxides for the adsorption and simultaneous removal of pesticides, herbicides and pharmaceuticals from waste water
- Dendrimers, nano-metal oxides and polymer nanofibers for detoxifying and simultaneous removal of heavy metals and arsenic
- Carbon-based nano adsorbents eradicate excessive nickel ions present in water.

3.3.3 Nano-bioremediation

The eradication of environmental impurities such as heavy metals, organic and inorganic noxious wastes from water with the help of nanoparticles/nanomaterial synthesized from plants, fungi and bacteria is called nano-bioremediation (NBR) (Rizwan et al. 2014, Sharma et al. 2020). NBR is the favourable technique for the eradication and detoxification of several pollutants for environmental clean-up. Bioremediation is a kind of environmentally friendly and economically affordable option to eradicate waste from the water (Sharma et al. 2020). There are three main approaches of bioremediation, which include use of microbes, use of plants and use of enzymatic remediation.

Recent research has shown that hospital wastage (organic) such as atrazine, paracetomol, and chlorpyrifos can be removed with enzyme-based bioremediation in combination with phytoremediation (Sharma et al. 2020). The organic wastes can be rapidly degraded into simpler compounds by nanoencapuslated enzymes by the cooperative effect of plants and microbes.

3.4 Photocatalytic Reactor Design and System Evaluation

The efficiency of a photocatalyst in wastewater decontamination depends upon the configuration and design of the reactor used. The photocatalytic reactor used for wastewater treatment generally falls into two categories: (1) slurry type (2) reactor with immobilized photocatalysts (Hossain 2018). The slurry type reactor is mostly used where catalyst is suspended in liquid water. Although this type of reactor involves close contact between pollutants and photocatalysts and more exposed surface, they suffer some major drawbacks. Low light efficiency and hard separate ion of catalyst from the solution are the main limitations in operation of the reactor. These drawbacks can be eliminated by the immobilization of nanocatalyst in the reactor to avoid post-recovery problem of photocatalysts (Passalia et al. 2020).

A proper photoreactor is the one that helps carry out the photoexcitation, photoabsorption and phototransfer, radical generation and increased surface area to the best of its capabilities and brings about demineralisation of the organic pollutants.

3.5 Conclusion

In summary, waste-water treatment using emerging treatment methods like advanced oxidation processes and functionalized nanoadsorbents proved efficient remediation techniques for wastewater treatment. Apart from this, photocatalytic techniques based on nanomaterials can prove beneficial in reducing the clean-up time and can treat persistent pollutants by both oxido-reductive and catalytic processes. Photocatalytic wastewater decontamination is an effectual technique for the treatment of different recalcitrant and toxic pollutants.

The conventional techniques are not able to remove emerging contaminants and their metabolites from the water, and are thus considered less efficient as compared to modern and emerging treatment methods. The emerging treatment methods as discussed above require less manpower and are cost-effective and facile.

In order to further encourage the practicality of photocatalytic technology, several technical limitations need to be addressed. Few among them are potential catalyst development for efficient photo-efficiency, strategic catalyst immobilization, optimization of photo catalytic reaction parameters, development of innovative coupling system for enhanced photomineralization and effective design of photocatalytic reactor system for higher utilization of solar energy.

Acknowledgements

Authors are highly thankful to Aligarh Muslim University, Aligarh and University Sophistication Instrumentation facility, AMU for providing necessary and requisite research facilities. Nafees Ahmad is grateful for the PhD split-site scholarship awarded by the Commonwealth Scholarship Commission, United Kingdom (Award number INCN-2019-35).

References

Ahmad, A., Mohd-Setapar, S.H., Chuong, C.S., Khatoon, A., Wani, W.A., Kumar, R. et al. 2015. Recent advances in new generation dye removal technologies: novel search for approaches to reprocess wastewater. RSC Advances. 5(39): 30801–30818.

Ahmad, N., Sultana, S., Azam, A., Sabir, S. and Khan, M.Z. 2017. Novel bio-nanocomposite materials for enhanced biodegradability and photocatalytic activity. New Journal of Chemistry. 41(18): 10198–10207.

Ahmad, N., Sultana, S., Faisal, S.M., Ahmed, A., Sabir, S. and Khan, M.Z. 2019. Zinc oxide-decorated polypyrrole/chitosan bionanocomposites with enhanced photocatalytic, antibacterial and anticancer performance. RSC Advances. 9(70): 41135–41150.

Ahmad, N., Sultana, S., Sabir, S. and Khan, M.Z. 2020. Exploring the visible light driven photocatalysis by reduced graphene oxide supported Ppy/CdS nanocomposites for the degradation of organic pollutants. Journal of Photochemistry and Photobiology A: Chemistry. 386: 112129.

Ahmad, N., Anae, J., Khan, M.Z., Sabir, S., Yang, X.J., Thakur, V.K. et al. 2021. Visible light-conducting polymer nanocomposites as efficient photocatalysts for the treatment of organic pollutants in wastewater. Journal of Environmental Management. 295: 113362.

Alekseeva, O.V., Bagrovskaya, N.A. and Noskov, A.V. 2016. Sorption of heavy metal ions by fullerene and polystyrene/fullerene film compositions. Protection of Metals and Physical Chemistry of Surfaces. 52(3): 443–447.

Ali, I., AlGhamdi, K. and Al-Wadaani, F.T. 2019. Advances in iridium nano catalyst preparation, characterization and applications. Journal of Molecular Liquids. 280: 274–284.

Ali, M.H., Al-Afify, A.D. and Goher, M.E. 2018. Preparation and characterization of graphene–TiO2 nanocomposite for enhanced photodegradation of Rhodamine-B dye. The Egyptian Journal of Aquatic Research. 44(4): 263–270.

Anitha, K., Namsani, S. and Singh, J.K. 2015. Removal of heavy metal ions using a functionalized single-walled carbon nanotube: a molecular dynamics study. The Journal of Physical Chemistry A. 119(30): 8349–8358.

Ata, S., Tabassum, A., Bibi, I., Majid, F., Sultan, M., Ghafoor, S. et al. 2019. Lead remediation using smart materials. A review. Zeitschrift für Physikalische Chemie. 233(10): 1377–1409.

Ates, M., Eker, A.A. and Eker, B. 2017. Carbon nanotube-based nanocomposites and their applications. Journal of Adhesion Science and Technology. 31(18): 1977–1997.

Barrera-Salgado, K.E., Ramírez-Robledo, G., Álvarez-Gallegos, A., Pineda-Arellano, C.A., Sierra-Espinosa, F.Z., Hernández-Pérez, J.A. et al. 2016. Fenton process coupled to ultrasound and UV light irradiation for the oxidation of a model pollutant. Journal of Chemistry.

Bates, M.E., Grieger, K.D., Trump, B.D., Keisler, J.M., Plourde, K.J. and Linkov, I. 2016. Emerging technologies for environmental remediation: integrating data and judgment. Environmental Science & Technology. 50(1): 349–358.

Bhanvase, B.A., Shende, T.P. and Sonawane, S.H. 2017. A review on graphene–TiO2 and doped graphene–TiO2 nanocomposite photocatalyst for water and wastewater treatment. Environmental Technology Reviews. 6(1): 1–14.

Byrne, C., Subramanian, G. and Pillai, S.C. 2018. Recent advances in photocatalysis for environmental applications. Journal of Environmental Chemical Engineering. 6(3): 3531–3555.

Camargo, P.H.C., Satyanarayana, K.G. and Wypych, F. 2009. Nanocomposites: synthesis, structure, properties and new application opportunities. Materials Research. 12(1): 1–39.

Chen, S., Wang, Y., Li, J., Hu, Z., Zhao, H., Xie, W. et al. 2018. Synthesis of black TiO2 with efficient visible-light photocatalytic activity by ultraviolet light irradiation and low temperature annealing. Materials Research Bulletin. 98: 280–287.

Chen, W., Liu, Q., Tian, S. and Zhao, X. 2019. Exposed facet dependent stability of ZnO micro/nano crystals as a photocatalyst. Applied Surface Science. 470: 807–816.

Colmenares, J.C., Luque, R., Campelo, J.M., Colmenares, F., Karpiński, Z. and Romero, A.A. 2009. Nanostructured photocatalysts and their applications in the photocatalytic transformation of lignocellulosic biomass: an overview. Materials. 2(4): 2228–2258.

Dahari, N., Muda, K., Latif, M.T. and Hussein, N. 2021. Studies of atmospheric PM2. 5 and its inorganic water soluble ions and trace elements around Southeast Asia: a review. Asia-Pacific Journal of Atmospheric Sciences. 57(2): 361–385.

Das, S., Sen, B. and Debnath, N. 2015. Recent trends in nanomaterials applications in environmental monitoring and remediation. Environmental Science and Pollution Research. 22(23): 18333–18344.

Fanourakis, S.K., Peña-Bahamonde, J., Bandara, P.C. and Rodrigues, D.F. 2020. Nano-based adsorbent and photocatalyst use for pharmaceutical contaminant removal during indirect potable water reuse. NPJ Clean Water. 3(1): 1–15.

Fernández-Castro, P., Vallejo, M., San Román, M.F. and Ortiz, I. 2015. Insight on the fundamentals of advanced oxidation processes. Role and review of the determination methods of reactive oxygen species. Journal of Chemical Technology & Biotechnology. 90(5): 796–820.

Fujishima, A. and Honda, K. 1972. Electrochemical photolysis of water at a semiconductor electrode. Nature. 238(5358): 37–38.

Fuwad, A., Ryu, H., Malmstadt, N., Kim, S.M. and Jeon, T.J. 2019. Biomimetic membranes as potential tools for water purification: Preceding and future avenues. Desalination. 458: 97–115.

Ganie, A.S., Bano, S., Khan, N., Sultana, S., Rehman, Z., Rahman, M.M. and Khan, M.Z. 2021. Nanoremediation technologies for sustainable remediation of contaminated environments: Recent advances and challenges. Chemosphere. 275: 130065.

Gehrke, I., Geiser, A. and Somborn-Schulz, A. 2015. Innovations in nanotechnology for water treatment. Nanotechnology, Science and Applications. 8: 1.

Glaze, W.H., Kang, J.W. and Chapin, D.H. 1987. The chemistry of water treatment processes involving ozone, hydrogen peroxide and ultraviolet radiation.

Gnanaprakasam, A., Sivakumar, V.M. and Thirumarimurugan, M. 2015. Influencing parameters in the photocatalytic degradation of organic effluent via nanometal oxide catalyst: a review. Indian Journal of Materials Science.

Gómez-Pastora, J., Dominguez, S., Bringas, E., Rivero, M.J., Ortiz, I. and Dionysiou, D.D. 2017. Review and perspectives on the use of magnetic nanophotocatalysts (MNPCs) in water treatment. Chemical Engineering Journal. 310: 407–427.

Guerra, F.D., Campbell, M.L., Whitehead, D.C. and Alexis, F. 2017. Cover picture: tunable properties of functional nanoparticles for efficient capture of VOCs (Chemistry Select 31/2017). Chemistry Select. 2(31): 9888–9888.

Guerra, F.D., Attia, M.F., Whitehead, D.C. and Alexis, F. 2018. Nanotechnology for environmental remediation: materials and applications. Molecules. 23(7): 1760.

Gupta, S., Bhatiya, D. and Murthy, C.N. 2015. Metal removal studies by composite membrane of polysulfone and functionalized single-walled carbon nanotubes. Separation Science and Technology. 50(3): 421–429.

Hodges, B.C., Cates, E.L. and Kim, J.H. 2018. Challenges and prospects of advanced oxidation water treatment processes using catalytic nanomaterials. Nature Nanotechnology. 13(8): 642–650.

Hossain, M.F. 2019. Sustainable Design and Build. Building, Energy. Roads, Bridges, Water and Sewer Systems.

Ikram, S. 2016. Role of Nanomaterials and their Applications as Photo-catalyst and Senors: A Review. Nano Res. 2(10).

Jaihindh, D.P., Thirumalraj, B., Chen, S.M., Balasubramanian, P. and Fu, Y.P. 2019. Facile synthesis of hierarchically nanostructured bismuth vanadate: an efficient photocatalyst for degradation and detection of hexavalent chromium. Journal of Hazardous Materials, 367: 647–657.

Kafizas, A., Wang, X., Pendlebury, S.R., Barnes, P., Ling, M., Sotelo-Vazquez, C. et al. 2016. Where do photogenerated holes go in anatase: rutile TiO2? A transient absorption spectroscopy study of charge transfer and lifetime. The Journal of Physical Chemistry A. 120(5): 715–723.

Kaneko, K., Ishii, C., Arai, T. and Suematsu, H. 1993. Defect-associated microporous nature of fullerene C60 crystals. The Journal of Physical Chemistry. 97(26): 6764–6766.

Kausar, A., Rafique, I. and Muhammad, B. 2016. Review of applications of polymer/carbon nanotubes and epoxy/CNT composites. Polymer-Plastics Technology and Engineering. 55(11): 1167–1191.

Lettieri, S., Gargiulo, V., Pallotti, D.K., Vitiello, G., Maddalena, P., Alfè, M. et al. 2018. Evidencing opposite charge-transfer processes at TiO2/graphene-related materials interface through a combined EPR, photoluminescence and photocatalysis assessment. Catalysis Today. 315: 19–30.

Li, D., Song, H., Meng, X., Shen, T., Sun, J., Han, W. et al. 2020. Effects of particle size on the structure and photocatalytic performance by alkali-treated TiO2. Nanomaterials. 10(3): 546.
Lin, Z., Li, J., Shen, W., Corriou, J.P., Chen, X. and Xi, H. 2021. Different photocatalytic levels of organics in papermaking wastewater by flocculation-photocatalysis and SBR-photocatalysis: Degradation and GC–MS experiments, adsorption and photocatalysis simulations. Chemical Engineering Journal. 412: 128715.
Liu, F., Zhang, W., Tao, L., Hao, B. and Zhang, J. 2019. Simultaneous photocatalytic redox removal of chromium (vi) and arsenic (iii) by hydrothermal carbon-sphere@ nano-Fe_3O_4. Environmental Science: Nano. 6(3): 937–947.
Loeb, S.K., Alvarez, P.J., Brame, J.A., Cates, E.L., Choi, W., Crittenden, J. et al. 2018. The technology horizon for photocatalytic water treatment: sunrise or sunset?
Mahmoodi, N.M., Arami, M., Limaee, N.Y., Gharanjig, K. and Nourmohammadian, F. 2007. Nanophotocatalysis using immobilized titanium dioxide nanoparticle: degradation and mineralization of water containing organic pollutant: case study of Butachlor. Materials Research Bulletin. 42(5): 797–806.
Makama, A.B., Salmiaton, A., Saion, E.B., Choong, T.S.Y. and Abdullah, N. 2016. Synthesis of CdS sensitized TiO_2 photocatalysts: methylene blue adsorption and enhanced photocatalytic activities. International Journal of Photoenergy.
Malik, A., Hameed, S., Siddiqui, M.J., Haque, M.M., Umar, K., Khan, A. et al. 2014. Electrical and optical properties of nickel-and molybdenum-doped titanium dioxide nanoparticle: improved performance in dye-sensitized solar cells. Journal of Materials Engineering and Performance. 23(9): 3184–3192.
Maniakova, G., Salmerón, I., Polo-López, M.I., Oller, I., Rizzo, L. and Malato, S. 2021. Simultaneous removal of contaminants of emerging concern and pathogens from urban wastewater by homogeneous solar driven advanced oxidation processes. Science of The Total Environment. 766: 144320.
M'arimi, M.M., Mecha, C.A., Kiprop, A.K. and Ramkat, R. 2020. Recent trends in applications of advanced oxidation processes (AOPs) in bioenergy production. Renewable and Sustainable Energy Reviews. 121: 109669.
Nosaka, Y. and Nosaka, A.Y. 2017. Generation and detection of reactive oxygen species in photocatalysis. Chemical Reviews. 117(17): 11302–11336.
Obare, S.O. and Meyer, G.J. 2004. Nanostructured materials for environmental remediation of organic contaminants in water. Journal of Environmental Science and Health, Part A. 39(10): 2549–2582.
Ong, C.B., Ng, L.Y. and Mohammad, A.W. 2018. A review of ZnO nanoparticles as solar photocatalysts: Synthesis, mechanisms and applications. Renewable and Sustainable Energy Reviews. 81: 536–551.
Opoku, F., Kiarii, E.M., Govender, P.P. and Mamo, M.A. 2017. Metal oxide polymer nanocomposites in water treatments (Vol. 8, pp. 173–199). London (UK): IntechOpen.
Passalía, C., Flores, M., Santos, S.G., Paulista, L.O., Labas, M.D., Vilar, V.J. et al. 2020. Radiation modelling in the NETmix photocatalytic reactor: The concept of efficiencies in series. Journal of Environmental Chemical Engineering. 8(6): 104464.
Patil, S.S., Shedbalkar, U.U., Truskewycz, A., Chopade, B.A. and Ball, A.S. 2016. Nanoparticles for environmental clean-up: A review of potential risks and emerging solutions. Environmental Technology & Innovation. 5: 10–21.
Pawar, R.C. and Lee, C.S. 2015. Nanomaterial-Based Photocatalysis. Heterogeneous Nanocomposite-Photocatalysis for Water Purification. 25.
Pérez, M., Torrades, F., Domènech, X. and Peral, J. 2002. Fenton and photo-Fenton oxidation of textile effluents. Water Research. 36(11): 2703–2710.
Pignatello, J.J. 1992. Dark and photoassisted iron (3+)-catalyzed degradation of chlorophenoxy herbicides by hydrogen peroxide. Environmental Science & Technology. 26(5): 944–951.
Rizwan, M., Singh, M., Mitra, C.K. and Morve, R.K. 2014. Ecofriendly application of nanomaterials: nanobioremediation. Journal of Nanoparticles.
Rizzo, L., Agovino, T., Nahim-Granados, S., Castro-Alférez, M., Fernández-Ibáñez, P. and Polo-López, M.I. 2019. Tertiary treatment of urban wastewater by solar and UV-C driven advanced oxidation with peracetic acid: Effect on contaminants of emerging concern and antibiotic resistance. Water Research. 149: 272–281.

Sarkar, B., Mandal, S., Tsang, Y.F., Kumar, P., Kim, K.H. and Ok, Y.S. 2018. Designer carbon nanotubes for contaminant removal in water and wastewater: a critical review. Science of the Total Environment. 612: 561–581.

Shafi, A., Ahmad, N., Sultana, S., Sabir, S. and Khan, M.Z. 2019. Ag_2S-sensitized NiO–ZnO heterostructures with enhanced visible light photocatalytic activity and acetone sensing property. ACS omega. 4(7): 12905–12918.

Shahidi, S. and Moazzenchi, B. 2018. Carbon nanotube and its applications in textile industry–A review. The Journal of The Textile Institute. 109(12): 1653–1666.

Sharma, I. 2020. Bioremediation techniques for polluted environment: concept, advantages, limitations, and prospects. In Trace Metals in the Environment-New Approaches and Recent Advances. IntechOpen.

Sireesha, M., Jagadeesh Babu, V., Kranthi Kiran, A.S. and Ramakrishna, S. 2018. A review on carbon nanotubes in biosensor devices and their applications in medicine. Nanocomposites. 4(2): 36–57.

Ton, N.N.T., Dao, A.T.N., Kato, K., Ikenaga, T., Trinh, D.X. and Taniike, T. 2018. One-pot synthesis of TiO_2/graphene nanocomposites for excellent visible light photocatalysis based on chemical exfoliation method. Carbon. 133: 109–117.

Tratnyek, P.G. and Johnson, R.L. 2006. Nanotechnologies for environmental cleanup. Nano Today. 1(2): 44–48.

Umar, K., Aris, A., Ahmad, H., Parveen, T., Jaafar, J., Majid, Z.A. et al. 2016. Synthesis of visible light active doped TiO_2 for the degradation of organic pollutants—methylene blue and glyphosate. Journal of Analytical Science and Technology. 7(1): 1–8.

Vaseashta, A., Vaclavikova, M., Vaseashta, S., Gallios, G., Roy, P. and Pummakarnchana, O. 2007. Nanostructures in environmental pollution detection, monitoring, and remediation. Science and Technology of Advanced Materials. 8(1-2): 47.

Venkateswarlu, S., Kumar, B.N., Prathima, B., SubbaRao, Y. and Jyothi, N.V.V. 2019. A novel green synthesis of Fe_3O_4 magnetic nanorods using Punica Granatum rind extract and its application for removal of Pb (II) from aqueous environment. Arabian Journal of Chemistry. 12(4): 588–596.

Wang, J., Lin, W., Hu, H., Liu, C., Cai, Q., Zhou, S. et al. 2021. Engineering Z-system hybrids of 0D/2D F-TiO_2 quantum dots/gC_3N_4 heterostructures through chemical bonds with enhanced visible-light photocatalytic performance. New Journal of Chemistry. 45(6): 3067–3078.

Wang, Z., Li, C. and Domen, K. 2019. Recent developments in heterogeneous photocatalysts for solar-driven overall water splitting. Chemical Society Reviews. 48(7): 2109–2125.

Yaqoob, A.A., Parveen, T., Umar, K. and Mohamad Ibrahim, M.N. 2020. Role of nanomaterials in the treatment of wastewater: a review. Water. 12(2): 495.

Yaqoob, A.A., Ibrahim, M.N.M., Ahmad, A. and Vijaya Bhaskar Reddy, A. 2021. Toxicology and environmental application of carbon nanocomposite. In Environmental Remediation Through Carbon Based Nano Composites (pp. 1–18). Springer, Singapore.

Yu, L., Ruan, S., Xu, X., Zou, R. and Hu, J. 2017. One-dimensional nanomaterial-assembled macroscopic membranes for water treatment. Nano Today. 17: 79–95.

Zazouli, M.A., Yousefi, Z., Yazdani Cherati, J., Tabarinia, H., Tabarinia, F. and Akbari Adergani, B. 2014. Evaluation of L-Cysteine functionalized single-walled carbon nanotubes on mercury removal from aqueous solutions. Journal of Mazandaran University of Medical Sciences. 24(111): 10–21.

Zhang, B.T., Zheng, X., Li, H.F. and Lin, J.M. 2013. Application of carbon-based nanomaterials in sample preparation: A review. Analytica chimica acta. 784: 1–17.

Zhang, S., Gu, P., Ma, R., Luo, C., Wen, T., Zhao, G. et al. 2019. Recent developments in fabrication and structure regulation of visible-light-driven g-C_3N_4-based photocatalysts towards water purification: a critical review. Catalysis Today. 335: 65–77.

4

Resource Recovery from Greywater and its Reuse

Optimistic Approach

Deepanshu Kashyap and *Akshi K. Singh**

4.1 Introduction

Worldwide, most polluted water is untreated. It is a vital resource, but it is often considered a burden to be discarded. This view of wastewater needs to be changed.

Reuse is a way in which reclaimed water can be recycled for a variety of beneficial purposes, such as irrigation, agricultural irrigation, industrial processes, groundwater regeneration, direct/indirect water supply and groundwater filling (Sun et al. 2011). Wastewater recycling can be used for intangible urban programs such as toilet cleaning, street washing and fire protection. Because of different needs, ideas, and policies, different practices have sprung up worldwide. Newly installed wastewater treatment systems lead to the explicit use of recycled water by removing decaying nutrients, building materials and bacteria from contaminated water (Nathanson et al. 2022). Referring to Fig. 1, Nutrients and Energy can be extracted from wastewater and it also acts as an extra water source. Hence, wastewater is also considered as an important resource.

Wastewater recycling leads to a reliable and controlled local water source. In addition, it is associated with significant environmental benefits, such as the discharge of wastewater into sensitive natural areas, the reduction and prevention of pollution, and the creation, rehabilitation or improvement of wetlands (Rudolph et al. 2020). Reuse is a sustainable process and relatively inexpensive in the long run. In addition, water use should also consider public acceptance and safety. Advanced

Sustainability Cluster, Department of HSE and Sustainability and Civil Engineering, University of Petroleum and Energy Studies, Dehradun-248007.

* Corresponding author: akshiksingh@ddn.upes.ac.in

technology and medical workflow are designed and certified, declaring water quality, enjoyment, and safety for reuse applications.

Limitations of resource availability are driving an adjustment of current cultural creation frameworks, changing the concentration from residues treatment, like wastewater treatment, towards recovery of resources. Biotechnological processes offer a financial and flexible approach to focus and change resources from wastewater into significant items, which is essential for the innovative advancement of a support to-support bio-based economy (Puyol et al. 2017). The utilisation of aerobic wastewater treatment as a reductive medium is gaining interest because of its low activity and support costs. Likewise, it is not difficult to-acquire, with great adequacy and capacity for degrading toxins. Anaerobic wastewater treatment is a natural wastewater treatment without air or essential oxygen. Applications are coordinated toward the evacuation of natural poisons in wastewater and sludge. Complete supplanting of aerobic with anaerobic innovation isn't yet imaginable as the gushing nature of anaerobic treatment frameworks isn't acceptable. The anaerobic treatment is considered a pretreatment procedure applied in India, Brazil and Colombia, supplanting the more exorbitant enacted sludge processes (Dhote et al. 2012).

Sewage sludge, also known as bio-solids, is aging rapidly because of the ordinary increment of populace, metropolitan preparation, and modern improvements around the world (Tsai 2012). The sludge should be sufficiently treated and how to decrease the adverse consequences of its application should be ecologically figured out. Minimizing waste and embracing alternative waste management methods effectively reduce environmental impacts and promote sustainability by reducing

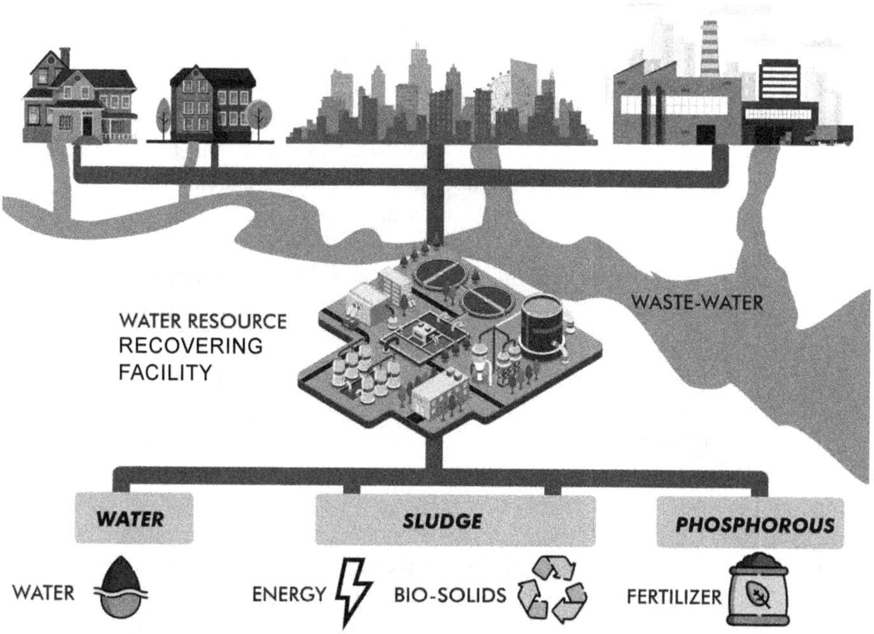

Fig. 1: Waste-water and Recovery (Wastewater? From Waste to Resource 2020).

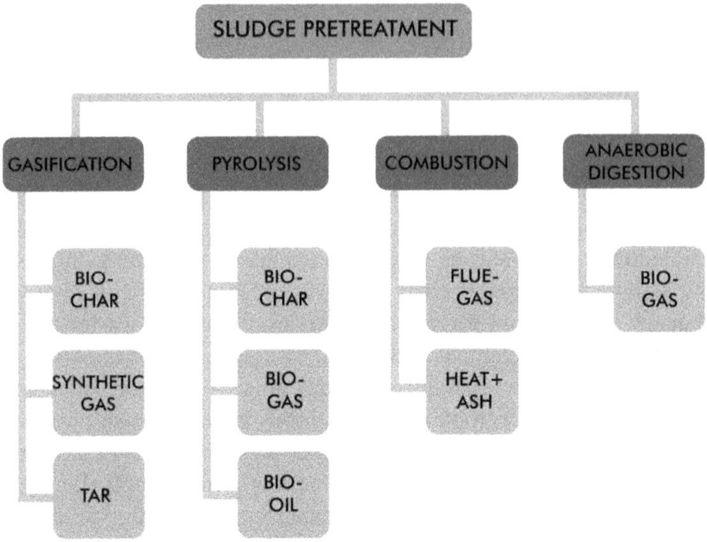

Fig. 2: Energy conversion techniques from sludge.

waste generation and reusing resources (Tsai 2012). Regardless, sludge of sewage which addresses energy, a source of material and the supplements, it is feasible to use it as an unrefined substance for any modern creation, energy generation & soil alteration (Kumar et al. 2017).

Energy recovery from waste management techniques plays a crucial role in minimizing storage space and maximizing the utilization of renewable energy sources. By converting waste into valuable energy, we not only reduce environmental impact but also foster a sustainable and efficient approach to resource management which can be seen in Fig. 2. From decantation and sedimentation, sludge is transferred to digesters where biogas is produced by O_2 free fermentation. Biogas from the digester can be additionally cleaned and moved up to deliver bio-methane that can be an immediate substitute for flammable gas or can be changed over to warm and electrical energy by cogeneration utilising warm reactors (Albalawneh and Chang 2015).

Anaerobic processes are broadly utilised for biological conversion due to their minimal expense and the capacity to utilise highly sticky natural waste without decreasing the high calorific value of the created biogas (methane and carbon dioxide). By anaerobic absorption, we mean a cycle during which the organic mixtures in the mud to be dealt with are changed over into methane and carbon dioxide in microbiological processes (Vatachi 2015). It happens in four phases: hydrolysis, acidogenesis, acetogenesis and methanogenesis. Anaerobic digestion of sludge and solid waste provides affordable energy production, surpassing the cost of natural gas, while also promoting sustainable waste management and reducing greenhouse gas emissions. Biodegradable mixtures are eliminated during anaerobic treatment, leaving few compounds in the effluent, for example, organic N compounds, ammonium, natural P mixtures, sulfides, and pathogens (Vatachi 2019, Stillwell et al. 2010).

To support heat recuperation at the part and the structure level, different nations can present WWHR (Waste-water Heat Recovery) in their construction regulations, and rules pointed toward further developing the energy proficiency of existing and new structures (Nagpal et al. 2021). An emphasis on lessening the environmental footprint is essential for traditional ammonium removal, but simultaneously it should also expand how much ammonium is for fertiliser creation. It has set off much research on recuperating ammonium from wastewater (Ye et al. 2018). Progressively rigid necessities on phosphorus (P) evacuation demand inventive answers for eliminating phosphorus from limited scope wastewater treatment works. Promising work has been performed to more readily comprehend metabolic pathways related to the extravagance of phosphorus-take-up by microscopic organisms, algae and bacteria, which give substitute which is suitable, feasible and minimal expense solutions for phosphorus evacuation at more limited sizes. Notwithstanding, accomplishing high degrees of phosphorus evacuation by employing such a system is regularly to the detriment of system simplicity, especially on biological treatment choices (Bunce et al. 2018, Chrispim et al. 2020).

4.1 Greywater and its reuse

All domestic wastewater, except for toilet waste, is known as greywater. It contains water for washing, bathing, and washing dishes and clothes. The amount of greywater varies significantly between families. Use in faulty areas is usually less than 15–20 litres per person per day, and in densely populated areas, it can produce ten times more value than high. Greywater is a source separated from private systems in natural cleaning, allowing controlled treatment systems rather than traditional wastewater treatment plants (Leal et al. 2010a). Greywater management involves both scientific methods and user involvement in maintaining the system. From the source to the final completion, every part of the program should be evaluated during the planning process. Greywater is often reused for other purposes and does not need to be disposed of in the sewage system as it does with other types of wastewater. It means that it is not wastewater but can be recycled for safe and profitable use, such as yard irrigation. Therefore, it works to install water for several purposes and therefore allows for conservation and reduces the amount of water sent to the sewage treatment system (Kurniawan et al. 2021).

Greywater that incorporates more toxic waste from clothes washing industries, dishwashers and, in some instances, kitchen sinks is called dark greywater. Some greywater sources and their constituents are introduced in Fig. 3.

The amount of greywater age relies upon the overall water utilisation, the expectation for everyday comforts, populace structures (i.e., age, orientation), occupant propensities, and water establishments of a given populace. Accordingly, greywater differs from half to 80% of the wastewater volume created by families and more than 90% on the off chance that vacuum latrines are introduced. The common volume of greywater differs from 90 to 120 l/p/d, but the volume of greywater in low pay nations that experience constant water deficiencies can be pretty much as low as 20–30 l/p/d (Jenssen and Vråle 2003). Greywater is produced because of the living habits of the people included. Subsequently, its attributes are exceptionally

Fig. 3: Greywater sources and their constituents (Birks and Hills 2007a).

factored and affected by way of life, the social way of behaving of the occupants, the accessibility of water, and the utilisation amount (Eriksson et al. 2002, Jefferson et al. 2004, Uddin et al. 2016). Greywater incorporates various scopes of natural matters, suspended solids, inorganic particles (Eriksson and Donner 2009) and *E. coli* (Winward 2007). The levels of these poisons/pollutants in greywater are lower in contrast to wastewater, and many researchers have noticed the inverse (Albalawneh and Chang 2015).

Fine qualities for greywater observing shift by country. Recovered greywater ought to fulfill four rules for reuse: aesthetics, hygienic safety, economic feasibility and environmental tolerance. The different reuse applications require different water quality specifications and request various medicines, shifting from straightforward process to further developed one.Not many re-use rules planned in light of greywater reusing (Li et al. 2009, Wichmann and Otterpohl 2009). Most nations apply similar principles to recovered civil wastewater as they do to greywater. A few nations have laid out particular principles for greywater reuse, like the UK, Germany, Jordan, Japan, and Australia (Gaulke 2006, Pidou and Memon 2007, Pidou et al. 2015).

4.2 Composition of greywater

Greywater contains high centralisations of effectively degradable natural material, like fat and oil from cooking, and tensides and different deposits from different cleansers. The greywater composition differs significantly, mirroring the occupants' way of life and the family's decision on synthetic substances for bathing, washing & laundry. Microorganisms possibly exist in the greywater division whenever polluted by excrement. The greywater climate is favourable for bacterial development, implying that it should be treated before reuse. Besides, untreated greywater effectively turns anaerobic (its oxygen is drained) and, along these lines, makes foul smell.

Greywater's composition varies, and it is generally a reflection of one's lifestyle and the type and choice of chemicals used in laundry, cleaning, and bathing. The qualities of greywater is also influenced by the quality of the water source and the type of distribution network (Abedin and Rakib 2013). There will be considerable differences in the composition of greywater from place to location and over time, owing to differences in water usage compared to the amount released. Chemical and biological degradation of some compounds in the transportation and storage network may potentially impact the composition. Greywater, in general, has significant quantities of readily biodegradable organic compounds and some essential elements, which are primarily produced in homes (Alsulaili and Hamoda 2015). Nutrients like nitrates and their derivatives, as well as phosphorus and its derivatives, as well as xenobiotic organic compounds (XOCs) and biological microorganisms such as faecal coliforms, salmonella, and general hydro-chemical components, are among them (Al-Jayyousi 2002, Dalahmeh et al. 2012). However, recent research has discovered pharmaceuticals, health and beauty products, aerosols, pigments, and hazardous heavy metals such as Pb, Ni Cd, Cu, Hg, and Cr in significant amounts in greywater (Arias et al. 2001, Argun et al. 2009). These pollutants in greywater indicate that greywater's composition's complexities are gradually increasing (Oteng-Peprah et al. 2018a).

4.3 Nutrients in greywater

The contents of nutrients in greywater are commonly less in contrast to normal mixed water (sewage). Content can be reduced to lower levels than in advanced treatments. There may be a high concentration of phosphorus (P) in some cases, but nitrogen levels remain constant.

Greywater ordinarily contains low degrees of supplements in contrast to typical wastewater from water-borne frameworks. Levels of nitrogen and other plant supplements are generally low; however, in some greywater, high convergences of phosphorous can be found. This phosphorous begins from washing and dishwashing powder, which is utilised to mellow the water. Washing and dishwashing powders without phosphorous are available. Generally, these are just about as modest and successful as those containing phosphorous, which makes picking sans phosphorus cleansers a reasonable choice. Assuming individuals utilise P-free cleansers, the phosphorus content of the greywater ought to be diminished to levels lower than ordinarily found in a high level treated wastewater. A few modern nations (for example, Norway) and a few urban communities in East Asia have prohibited washing powder containing phosphorous for water assurance. It makes sense why the degree of phosphorous in greywater in Norway is just 10–20% of the levels regularly found in Sweden (Boano et al. 2020).

Wastewater, agriculture, and nitrogen deposition from fossil fuel combustion are all significant sources of nutrient discharges (do Couto et al. 2013a, Imteaz et al. 2013). In the United States and the European Union, agricultural sources such as fertiliser leaching, runoff from cultivated fields, manure from concentrated livestock operations, and aquaculture are the primary sources of nutrient impairment in waterways. However in Asia and Africa, urban wastewater is the dominant source (Morales Aqualia 2014).

4.3.1 Primary components to consider in greywater management

The objectives of greywater management can be summarised as:

- Avoiding damage to buildings and areas adjacent to flooding and cold.
- Avoiding the formation of foul odours, stagnant water, and mosquito breeding grounds.
- Preventing drainage in sensitive areas.
- Preventing contamination of groundwater and drinking water ponds.
- Use of greywater as a source for crop growth, groundwater regeneration and land reform.

Successful greywater management incorporates technical features, such as the appropriate design and size of the various technological components involved. Equally important, however, are the 'soft' features of the system, such as the user's involvement in the operation and maintenance of the system. When planning a greywater system, all parts of the system, from their source to the recipients, should be considered (Shaikh and Ahammed 2020).

When the appropriate techniques and technology aren't in place or aren't adequately applied, greywater management goes from easy to exceedingly tricky. On the other hand, many industrialised countries have established various ways of handling, managing, and treating greywater, with some recycling greywater for potable and non-potable purposes. Before reuse or final disposal, treatment technologies have been utilised to lower the degree of contamination in greywater (do Couto et al. 2013b, Khan et al. 2018, Mohamed et al. 2018, Oteng-Peprah et al. 2018). They're all contaminant-specific and used in the same order as traditional wastewater treatment (pretreatment, primary, secondar, and tertiary). Each of these systems uses a physicochemical or biological treatment approach. Filtration, adsorption, and reverse osmosis, among other physical and chemical treatment processes, are used in physicochemical treatments. Activated sludge systems, trickling filters, waste stabilisation ponds, rotating biological contactors, and other biological treatment methods use a combination of microbes, sunlight, and oxygen manipulation; examples of such systems include activated sludge systems, trickling filters, waste stabilisation ponds, and rotating biological contactors, among others. Filtration, rotating biological contactors, membrane bioreactors, built wetlands, and up-flow anaerobic sludge blankets have been the most extensively utilised systems (UASBs). These devices have found use in managing the growing problem of greywater pollution in most developing countries. As a result, the performance of these systems is discussed in this review (do Couto et al. 2013a, Imteaz et al. 2013, Morales Aqualia 2014).

4.4 Other pollutants

Although generally low, metals and natural pollutants in the greywater can rise if environmental hazards are added to contaminated water. The metals come from the water itself, from the lining of the plumbing system and the pillars and the shampoos used in the house. Natural pollutants are found in many common household chemicals, such as shampoos, glues, protective substances and cleansers (Alsulaili and Hamoda 2015). In this case, too, people can significantly influence the greywater content. By using environmentally friendly chemicals and not exposing hazardous substances such as paint and solvents to the drain floor, metal levels and natural pollution in dark water can be kept low (Oteng-Peprah et al. 2018).

4.5 Treatment of greywater

Greywater is relatively harmless in terms of natural beauty and cleanliness. Problems are usually small and local. But if not properly controlled, greywater will be a powerful source of odour due to its high levels of nutrient depletion. When these chemicals are separated, natural processes can create anaerobic conditions within hours, creating a foul odour. Other therapeutic goals are to reduce the levels of natural pollutants, heavy metals, bacteria, and other microorganisms. A standard treatment procedure uses biofilm systems that eliminate water vapour by using bonds attached to attached bacteria to prevent biological damage under air conditions. With a good oxygen supply, a foul odour can be avoided. These systems range from comprehensive global systems to powerful and energy-intensive applications such as droplets and bio rotors (Zhou et al. 2020).

Compared to compact polluted water, greywater is harmless in terms of environment and sanitation. On the other hand, if not adequately controlled, greywater can be a vital odour source. Therefore, the main objective should be to remove high levels of easily degraded elements responsible for this. It should be done immediately because anaerobic and odour conditions will occur quickly (within hours if it is warm).

It should be noted that whenever white water is freely available to humans, it must first be treated to prevent BOD from causing anaerobic conditions. The second goal is to reduce bacteria and other microorganisms in the water. An additional desire for treatment and back treatment should be to reduce biological pollution and heavy metals. It is imperative when greywater is used to replenish groundwater and irrigation. It is usually a good measure to achieve the above objectives using aerobic techniques attached to biofilm. Often in these ways, biological degradation occurs in cooler climates (Albalawneh and Chang 2015, Wastewater? From Waste to Resource, 2020).

A few biological treatment frameworks have been applied for greywater treatment, including Sequencing Batch Reactor (SBR), Rotating Biological Contactor (RBC), Membrane Bioreactors (MBR), Up-flow Anaerobic Sludge Blanket (UASB) and Fluidized Bed Reactor (FBR) (Bani-Melhem et al. 2015, Khalid and Zakaria 2015). To eliminate sludge or biosolids biological frameworks are normally gone, and a sanitisation post-treatment stage by chlorination or UV to eliminate microorganisms

(Rakesh et al. 2020, Subramanian Srirangarayan 2020). Aerobic biological processes can accomplish phenomenal natural and turbidity evacuation rates. After aerobic greywater treatment processes, the vast majority of the biodegradable natural substances are taken out, and subsequently, microorganism re-development and scent issues are stayed away from, making the treated greywater steadier for capacity over more extended periods. Subsequently, medium to high strength greywater is recommended for treatment utilising natural cycles (Li et al. 2009).

Physical greywater treatment frameworks incorporate filtration and sedimentation. Filtration is generally utilised as a pretreatment technique (i.e., before natural or chemical treatment) or as a post-treatment strategy (i.e., preceding sanitisation). Filtration as a pretreatment strategy incorporates screen networks, nylon sock type filtration, sand bed filtration, rock filtration, metal sifters, and mulch tower framework (Birks and Hills 2007b). Exclusively utilising physical greywater treatment processes as the principal treatment technique is insufficient for greywater treatment since it doesn't ensure a satisfactory decrease of organics, supplements, and surfactants, besides in circumstances where the natural strength is meagre (Al-Husseini et al. 2021, Hassan and Al-Anbari 2021). The efficiency of the filtration methods relies upon the dispersion of greywater molecule size toxins and the filters' porosity; as a rule, the more modest the filters' porosity, the better the effluent quality (Ramon et al. 2004). Henceforth, coarse filters have a restricted impact on the evacuation of the poisons present in the greywater.

The conceptually applied biofilm techniques can be put in the process from a wide range of global application programs to robust systems such as deceptive filters and bio rotors (Fig. 4) (Heydari et al. 2021). Water systems such as lakes and wetlands, in some areas, will be found to be suitable for white water management (Khuntia et al. 2021). They are easy to build and operate, but they are not always the best solution to the problem of wastewater. In countries with cold winters and areas where water is scarce, lakes and wetlands should be consulted (Leal et al. 2010b, Zhang et al. 2021). On the other hand, in Introduction to Greywater Management, where the weather permits (with no cold and long dark periods), water systems can serve as an ideal solution for greywater treatment. It is especially true if aquatic biomass can be harvested and sold on the market (Chanakya and Khuntia 2014).

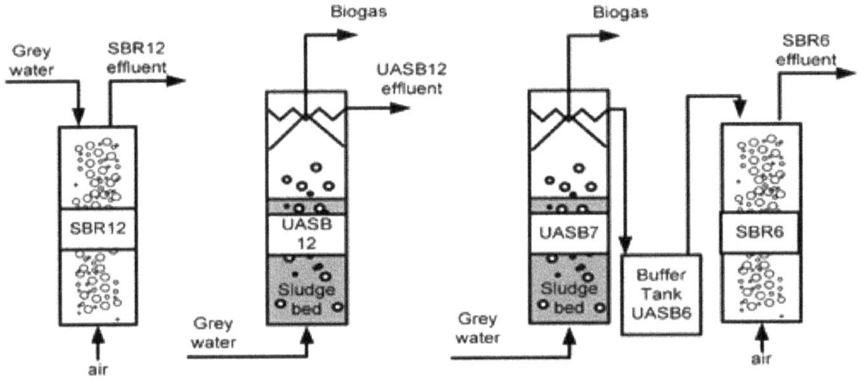

Fig. 4: Lab-scale reactors set up (Leal et al. 2010b).

4.6 Benefits of using greywater

Excess water loss is usually a natural and straightforward way to return water to the environment. After treatment, water is usually extracted from surface water, soaked in groundwater or used for irrigation. Typically, water can be extracted from open canals to be soaked in stormwater. Percolation in groundwater, Greywater returned to groundwater should be treated with proven and reliable methods. Water should be pumped into the ground at one meter or more depth. The subsoil should have sand (or small grain size). Safety areas around water sources need to be constructed. As a sixth rule, a maximum of one month in the entire space should be maintained before drainage (e.g., groundwater use). Use for irrigation, Water should be used in the ground or underground rather than sprayed. Plants where the leaves or stems can be eaten directly, such as fruit trees or bushes of berry, are best for irrigation. For the plants that are consumed raw, e.g., salad or spinach, it is recommended for at least one month between watering and harvesting. Some of the benefits are described below:

4.6.1 Freshwater conservation

Consumption of greywater reduces the amount of saltwater used in the home for other purposes. The reuse of greywater to wash toilets and water plants reduces the load of clean water needed, and if enough people make the same choice, water requirements can be significantly reduced to impact the environment positively. Reusing greywater to flush toilets and water plants saves freshwater, which is especially significant in dry conditions or droughts. It is imperative in regions with dry climates.

4.6.2 Reduced water wastage

The use of clean water for the various activities in the house produces much dirty water in sewage systems. This creates unnecessary wastage. The amount of wastewater can be reduced by half by reusing greywater in other activities before it is thrown into the wild. Greywater is wastewater that has been used for dishwashing, laundry, or bathing. Greywater reuse accomplishes two goals: it decreases the quantity of freshwater required to supply a community and minimises the quantities of wastewater entering the sewage or septic system (Morandi 2018).

Therefore, it means that fewer resources can be diverted to sanitation and that it saves even more litres of water that could be used before disposal, thus increasing water efficiency.

4.6.3 Reduced energy consumption

By reducing the need for clean water in each household, the energy needed to pump water into the house is reduced. Second, reuse of water reduces the burden on medical facilities to ensure that it is clean and thus reduces the total capacity of water distribution and sanitation companies. Reduced energy consumption means less electricity and, in turn, fuel consumption is also reduced and ultimately helps reduce greenhouse gas emissions. It means that energy can be diverted to more demanding resources or stored for later use. Having a greywater system at home allows you to contribute to the conservation of water and other energy resources.

An Eco-home is built using technologies to reduce energy usage and waste. One can reduce the amount of water they waste by up to 40,000 gallons per year by installing an Aqua2use greywater system. This method also permits the water that their plants can't take to replenish the groundwater table, which is fast depleting (Varzandeh 2011).

4.6.4 *Reducing the use of chemicals*

The reduction of water sent to wild plants also leads to reduced levels in the number of chemicals used to treat water. With less load, sewage treatment plants do not require more environmentally friendly chemicals as they reduce the cost of sanitation and the need for environmentally friendly chemical products. Long-term greywater disposal can lead to increases in soil chemical parameters, including pH, phosphate, and SAR, and pathogen indicators like *E. coli*, which, if allowed to persist, can have detrimental soil and human health consequences (Kumar et al. 2020).

4.6.5 *Beautiful plains*

The use of greywater in water sources provides another water source for plants, especially in areas in short supply. Regions that use reclaimed greywater thus provide an alternative to using non-spilt water. Drip lines connected to our greywater systems are a popular alternative to in-ground sprinkler systems. Greywater systems filter and distribute specific domestic water waste in the garden via a drip irrigation line (Andreadakis 2015).

Therefore, it creates beautiful scenery as plants retain their greenery and bloom, even in places where the climate is harsh. Combined with other technologies such as mulch and drip irrigation, the water provided can last a long time to ensure that the excellent condition of a person's yard is maintained even during the worst of seasons.

4.6.6 *Nutrition reuse*

Greywater restores nutrients that would otherwise be lost to the sewage system. The kitchen and bath water contains many organic matters that are not harmful to plants. It is not the same as dark water when natural levels are high. Dark water also contains many harmful bacteria like *E. coli*, which can cause disease if it pollutes the soil. Greywater provides plants with healthy nutrients that increase beauty.

4.6.7 *Natural water filtering*

Greywater used for irrigation plants is purified by plant and soil actions that counteract the use of chemicals that can, in excess, damage the soil and the organisms that live in it. Natural filters work to ensure environmental safety.

4.6.8 *Groundwater regeneration*

The water that is used most often keeps the soil filtered as it descends and ends up in the groundwater, where it fulfils its levels. It is a good thing as it ensures that water sources such as rivers that get their water from underground rivers are constantly supplied.

4.7 Conclusion

Of the 165 billion m^3 wastewater collected and processed annually worldwide, only 2% are currently used. The therapies we are proposing allow for the reuse of wastewater for various non-domestic purposes such as irrigation of agricultural land, irrigation and urban clean-up, industrial cleaning and water processing and new water tables. Greywater, also known as spelled greywater or sullage, is all wastewater produced from homes or office buildings from streams without sewage. Various greywater treatment technologies, such as the physiochemical, biological and advanced oxidation process (AOP), are used to remove natural contaminants and nutrients in grey water.

Recycling is an integral part of sustainable water resource management. Greywater can be an essential water source, especially in arid and tourist areas, where high water demand is expected during the dry season. Potential natural recycling benefits of black water include:

- Reduction of freshwater discharges into rivers and streams,
- A small environmental impact from septic tanks and water treatment plants,
- Reduced energy consumption and chemical pollution in water treatment,
- Groundwater recycling and nutrient recovery.

References

Abedin, S.B. and Rakib, Z. bin. 2013. Generation and quality analysis of greywater at dhaka city. Environmental Research, Engineering and Management. 64(2). doi: 10.5755/j01.erem.64.2.3992.

Albalawneh, A. and Chang, T.-K. 2015. Review of the greywater and proposed greywater recycling scheme for agricultural irrigation reuses. International Journal of Research -GRANTHAALAYAH. 3(12): 16–35. doi: 10.29121/granthaalayah.v3.i12.2015.2882.

Al-Husseini, T.H., Al-Anbari, R.H. and Al-Obaidy, A.H.M.J. 2021. Greywater environmental management: a review. IOP Conference Series: Earth and Environmental Science. 779(1). doi:10.1088/1755-1315/779/1/012100.

Al-Jayyousi, O. 2002. Focused environmental assessment of greywater reuse in Jordan. Environmental Engineering Policy. 3(1): 67–73. doi:10.1007/s100220100044.

Alsulaili, A.D. and Hamoda, M.F. 2015. Quantification and characterization of greywater from schools. Water Science and Technology. 72(11): 1973–1980. doi:10.2166/WST.2015.408.

Ambulkar, Archis and Nathanson, Jerry A. 2022. Wastewater treatment. Encyclopedia Britannica, 25 May. 2023, https://www.britannica.com/technology/wastewater-treatment. Accessed 22 June 2023.

Andreadakis, A. 2015. (2) (PDF) Grey Water Characterization and Treatment. Available at: https://www.researchgate.net/publication/283686808_Grey_Water_Characterization_and_Treatment?msclkid=e40b357caa0111ecb99ddfb195509395 (Accessed: March 22, 2022).

Argun, M.E., Dursun, S. and Karatas, M. 2009. Removal of Cd(II), Pb(II), Cu(II) and Ni(II) from water using modified pine bark. Desalination. 249(2): 519–527. doi:10.1016/j.desal.2009.01.020.

Arias, C.A., del Bubba, M. and Brix, H. 2001. Phosphorus removal by sands for use as media in subsurface flow constructed reed beds. Water Research. 35(5): 1159–1168. doi:10.1016/s0043-1354(00)00368-7.

Bani-Melhem, K., Al-Qodah, Z., Al-Shannag, M., Qasaimeh, A., Qtaishat, M.R. and Alkasrawi, M. 2015. On the performance of real grey water treatment using a submerged membrane bioreactor system. Journal of Membrane Science. 476: 40–49. doi:10.1016/J.MEMSCI.2014.11.010.

Birks, R. and Hills, S. 2007a. Characterisation of indicator organisms and pathogens in domestic greywater for recycling. Environmental Monitoring and Assessment. 129(1): 61–69. doi:10.1007/S10661-006-9427-Y.

Birks, R. and Hills, S. 2007b. Characterisation of indicator organisms and pathogens in domestic greywater for recycling. Environmental Monitoring and Assessment. 129(1–3): 61–69. doi:10.1007/S10661-006-9427-Y.

Boano, F., Caruso, A., Costamagna, E., Ridolfi, L., Fiore, S., Demichelis, F., Galvão, A., Pisoeiro, J., Rizzo, A. and Masi, F. 2020. A review of nature-based solutions for greywater treatment: Applications, hydraulic design, and environmental benefits. Science of the Total Environment. 711: 134731. doi:10.1016/J.SCITOTENV.2019.134731.

Bunce, J.T., Ndam, E., Ofiteru, I.D., Moore, A. and Graham, D.W. 2018. A review of phosphorus removal technologies and their applicability to small-scale domestic wastewater treatment systems. Frontiers in Environmental Science. 6, 8. doi:10.3389/fenvs.2018.00008.

Chanakya, H.N. and Khuntia, H.K. 2014. Treatment of gray water using anaerobic biofilms created on synthetic and natural fibers. Process Safety and Environmental Protection. 92(2): 186–192. doi:10.1016/J.PSEP.2012.12.004.

Chrispim, M.C., Scholz, M. and Nolasco, M.A. 2020. A framework for resource recovery from wastewater treatment plants in megacities of developing countries. Environmental Research. 188. doi:10.1016/j.envres.2020.109745.

de Castro Carvalho, I., Calijuri, M.L., Assemany, P.P., Machado, M.D.F., Neto, R.F.M., da Fonseca Santiago, A. and de Souza, M.H.B. 2013a. Sustainable airport environments: A review of water conservation practices in airports. Resources, Conservation and Recycling. 74: 27–36.

do Couto, E.D.A., Calijuri, M.L., Assemany, P.P., da Fonseca Santiago, A. and de Castro Carvalho, I. 2013. Greywater production in airports: Qualitative and quantitative assessment. Resources, Conservation and Recycling. 77: 44–51.

Dalahmeh, S.S., Pell, M., Vinnerås, B., Hylander, L.D., Öborn, I. and Jönsson, H. 2012. Efficiency of bark, activated charcoal, foam and sand filters in reducing pollutants from greywater. Water, Air, & Soil Pollution, 223(7): 3657–3671. doi:10.1007/S11270-012-1139-Z.

Dhote, J., Ingole, S. and Chavhan, A. 2012. Review on wastewater treatment technologies. International Journal of Engineering Research & Technology (IJERT), 1. Available at: www.ijert.org.

Energies | Free Full-Text | An Analysis of the Use of Biosludge as an Energy Source and Its Environmental Benefits in Taiwan (2012). Available at: https://www.mdpi.com/1996-1073/5/8/3064 (Accessed: March 22, 2022).

Eriksson, E. et al. 2002. Characteristics of grey wastewater. Urban Water. 4(1): 85–104. doi:10.1016/S1462-0758(01)00064-4.

Eriksson, E. and Donner, E. 2009. Metals in greywater: Sources, presence and removal efficiencies. Desalination. 248(1–3): 271–278. doi:10.1016/J.DESAL.2008.05.065.

Gaulke, L.S. 2006. On-site wastewater treatment and reuses in Japan. Proceedings of the Institution of Civil Engineers: Water Management. 159(2): 103–109. doi:10.1680/WAMA.2006.159.2.103.

Hassan, Eng. T. and Al-Anbari, R. 2021. (2) (PDF) Greywater Environmental Management: A Review. Available at: https://www.researchgate.net/publication/352915888_Greywater_Environmental_Management_A_Review (Accessed: March 22, 2022).

Heydari, Fariba, Shahriar Osfouri, Mohsen Abbasi, Mohammad Javad Dianat and Javad Khodaveisi. 2021. Treatment of highly polluted grey waters using Fenton, UV/H2O2process and UV/TiO2process. Membrane and Water Treatment. 12(3): 125–132. doi:10.12989/MWT.2021.12.3.125.

Imteaz, M.A., Ahsan, A. and Shanableh, A. 2013. Reliability analysis of rainwater tanks using daily water balance model: Variations within a large city. Resources, Conservation and Recycling. 77: 37–43. doi:10.1016/j.resconrec.2013.05.006.

Jefferson, B., Palmer, A., Jeffrey, P., Stuetz, R. and Judd, S. 2004. Grey water characterisation and its impact on the selection and operation of technologies for urban reuse. Water Science and Technology. 50(2): 157–164. doi:10.2166/WST.2004.0113.

Jenssen, P.D. and Vråle, L. 2003. Ecosan-closing the loop. GTZ.

Khalid, B.-M. and Zakaria, A.-Q. 2015. (2) (PDF) On the performance of real grey water treatment using a submerged membrane bioreactor system. Available at: https://www.researchgate.net/publication/270648111_On_the_performance_of_real_grey_water_treatment_using_a_submerged_membrane_bioreactor_system (Accessed: March 22, 2022).

Khan, Adnan Hossain, Mark Libby, Daniel Winnick, John Palmer, Mark Sumarah, Madhumita B. Ray and Sheila M. Macfie. 2018. Uptake and phytotoxic effect of benzalkonium chlorides in *Lepidium sativum* and *Lactuca sativa*. Journal of Environmental Management. 206: 490–497. doi:10.1016/j.jenvman.2017.10.077.

Khuntia, Himanshu Kumar, Sushmitha, M.B., Sadiya Hameed, Naveen Janardhana, Karthik, M.G., Madhuri, K.S. and Chanakya, H.N. 2021. Bench scale demonstration of greywater treatment in a 3-stage sequential process comprising anaerobic, aerobic, and vertical greenery system. Journal of Water Process Engineering. 43. doi: 10.1016/j.jwpe.2021.102246.

Kumar, Anshal, Himanshu Kumar Sadhya, Esar Ahmad and Shashivendra Dulawat. 2020. Application of bio-enzyme in wastewater (Greywater) treatment. International Research Journal of Engineering and Technology [Preprint]. Available at: www.irjet.net (Accessed: March 22, 2022).

Kumar, V., Chopra, A.K. and Kumar, A. 2017. A review on sewage sludge (Biosolids) a resource for sustainable agriculture. Archives of Agriculture and Environmental Science. 2(4): 340–347. doi:10.26832/24566632.2017.020417.

Kurniawan, Sigit, Novarini, Erna Yuliwati, Eko Ariyanto, Marlia Morsin, Rahmat Sanudin and Suratun Nafisah. 2021. Greywater treatment technologies for aquaculture safety: Review. Journal of King Saud University - Engineering Sciences [Preprint]. doi:10.1016/J.JKSUES.2021.03.014.

Leal, Lucía Hernández, Hardy Temmink, Grietje Zeeman and Cees J. Cees. 2010a. Comparison of three systems for biological greywater treatment. Water. 2(2): 155–169. doi:10.3390/W2020155.

Leal, Lucía Hernández, Hardy Temmink, Grietje Zeeman and Cees J. Cees. 2010b. Comparison of three systems for biological greywater treatment. Water. 2(2): 155–169. doi:10.3390/W2020155.

Li, F., Wichmann, K. and Otterpohl, R. 2009. Review of the technological approaches for grey water treatment and reuses. Science of the Total Environment. 407(11): 3439–3449. doi:10.1016/J.SCITOTENV.2009.02.004.

Mohamed, Radin, Adel Al-Gheethi, Adeleke Abdulrahman, Muhammad Shabery bin Sainudin, Siti Asmah Bakar and Amir Hashim Mohd Kassim. 2018. Optimization of ceramic waste filter for bathroom greywater treatment using central composite design (CCD). Journal of Environmental Chemical Engineering. 6(2): 1578–1588. doi:10.1016/j.jece.2018.02.006.

Morales Aqualia, N. 2014. Novel technologies for WWTP optimization in footprint, nutrients valorization, and energy. The Potential of Innovative Technologies to Improve Sustainability of Sewage Treatment Plants (Pioneer_STP) View project. Available at: https://www.researchgate.net/publication/263403844.

Morandi, C. 2018. (2) (PDF) How does greywater separation impact the operation of conventional wastewater treatment plants? Available at: https://www.researchgate.net/publication/329070462_How_does_greywater_separation_impact_the_operation_of_conventional_wastewater_treatment_plants?msclkid=09ec4a86aa0111ec8e7308d6fefb1529 (Accessed: March 22, 2022).

Nagpal, Himanshu, Jan Spriet, Madhu Krishna Murali and Aonghus McNabola. 2021. Heat recovery from wastewater—A review of available resource. Water (Switzerland). 13(9). doi:10.3390/w13091274.

Oteng-Peprah, M., Acheampong, M.A. and deVries, N.K. 2018. Greywater characteristics, treatment systems, reuse strategies and user perception—a review. Water, Air, and Soil Pollution. 229(8). doi:10.1007/s11270-018-3909-8.

Oteng-Peprah, M., de Vries, N.K. and Acheampong, M.A. 2018. Greywater characterization and generation rates in a peri urban municipality of a developing country. Journal of Environmental Management. 206: 498–506. doi:10.1016/J.JENVMAN.2017.10.068.

Pidou, Marc, Fayyaz Ali Mamon, Tom Stephenson, Bruce Jefferson and Paul Jeffrey. 2015. Greywater recycling: treatment options and applications. http://dx.doi.org/10.1680/ensu.2007.160.3.119, 160(3): 119–131. doi:10.1680/ENSU.2007.160.3.119.

Pidou, M. and Memon, F.A. 2007. Greywater recycling: treatment options and applications | Proceedings of the Institution of Civil Engineers - Engineering Sustainability. Available at: https://www.icevirtuallibrary.com/doi/abs/10.1680/ensu.2007.160.3.119 (Accessed: March 22, 2022).

Puyol, Daniel, Damien J. Batstone, Tim Hülsen, Sergi Astals, Miriam Peces and Jens O. Krömer. 2017. Resource recovery from wastewater by biological technologies: Opportunities, challenges, and prospects. Frontiers in Microbiology. Frontiers Media S.A. doi:10.3389/fmicb.2016.02106.

Rakesh, S.S., Ramesh, P.T., Murugaragavan, R., Avudainayagam, S. and Karthikeyan, S. 2020. Characterization and treatment of grey water: A review. International Journal of Chemical Studies. 8(1): 34–40. doi:10.22271/CHEMI.2020.V8.I1A.8316.

Ramona, G., Green, M., Semiat, R. and Dosoretz, C. 2004. Low strength graywater characterization and treatmentby direct membrane filtration. Desalination, 170(3): 241–250.

Rudolph, Karl Ulrich, Birte Boysen, Jens Hilbig, Faruq Shalizi, Keno Stroemer and Gabriele Walenzik. 2020. Drivers, challenges and solutions-Case studies for water reuse. doi:10.1016/bs.apmp.2020.07.009.

Shaikh, I.N. and Ahammed, M.M. 2020. Quantity and quality characteristics of greywater: A review. Journal of Environmental Management. 261. doi:10.1016/J.JENVMAN.2020.110266.

Stillwell, A.S., Hoppock, D.C. and Webber, M.E. 2010. Energy recovery from wastewater treatment plants in the United States: A case study of the energy-water nexus. Sustainability. 2(4): 945–962. doi:10.3390/su2040945.

Subramanian Srirangarayan, R. 2020. (2) (PDF) Characterization and treatment of grey water: A review. Available at: https://www.researchgate.net/publication/338488908_Characterization_and_treatment_of_grey_water_A_review (Accessed: March 22, 2022).

Sun, F., Chen, M. and Chen, J. 2011. Integrated management of source water quantity and quality for human health in a changing world. Encyclopedia of Environmental Health, pp. 254–265. doi:10.1016/B978-0-444-52272-6.00286-5.

Tsai, W.T. 2012. An analysis of the use of biosludge as an energy source and its environmental benefits in Taiwan. Energies. 5(8): 3064–3073. doi:10.3390/EN5083064.

Uddin, S.M.N., Li, Z., Adamowski, J.F., Ulbrich, T., Mang, H.P., Ryndin, R., Norvanchig, J., Lapegue, J., Wriege-Bechthold, A. and Cheng, S. 2016. Feasibility of a 'greenhouse system' for household greywater treatment in nomadic-cultured communities in peri-urban Ger areas of Ulaanbaatar, Mongolia: an approach to reduce greywater-borne hazards and vulnerability. Journal of Cleaner Production. 114: 431–442. doi:10.1016/J.JCLEPRO.2015.07.149.

Varzandeh, S. 2011 (2) (PDF) Turning wastewater (greywater) into electrical energy in buildings. Available at: https://www.researchgate.net/publication/327549546_TURNING_WASTEWATER_GREYWATER_INTO_ELECTRICAL_ENERGY_IN_BUILDINGS?msclkid=e40b02f8aa0111ec879755b9c24c1faa (Accessed: March 22, 2022).

Vatachi, N. 2015. Sewage Sludge Thermal Pretreatment Techniques, Internal Combustion Engines, Boilers And Turbines.

Vatachi, N. 2019. Wastewater sludge to energy production. A review, in IOP Conference Series: Materials Science and Engineering. Institute of Physics Publishing. doi:10.1088/1757-899X/595/1/012053.

Wastewater? From Waste to Resource (2020). Available at: https://www.worldbank.org/en/topic/water/publication/wastewater-initiative (Accessed: March 22, 2022).

Wichmann, F.L. and Otterpohl, R. 2009. (2) (PDF) Review of the technological approaches for grey water treatment and reuses. Available at: https://www.researchgate.net/publication/24145844_Review_of_the_technological_approaches_for_grey_water_treatment_and_reuses (Accessed: March 22, 2022).

Winward, G.P. 2007. (PDF) A study of the microbial quality of grey water and an evaluation of treatment technologies for reuse | Marc Pidou - Academia.edu. Available at: https://www.academia.edu/9117504/A_study_of_the_microbial_quality_of_grey_water_and_an_evaluation_of_treatment_technologies_for_reuse (Accessed: March 22, 2022).

Ye, Y., Ngo, H.H., Guo, W., Liu, Y., Chang, S.W., Nguyen, D.D., Liang, H. and Wang, J. 2018. A critical review on ammonium recovery from wastewater for sustainable wastewater management. Bioresource Technology. Elsevier Ltd, pp. 749–758. doi:10.1016/j.biortech.2018.07.111.

Zhang, Y., Tang, Q., Shi, P. and Katsumi, T. 2021. Influence of bio-clogging on permeability characteristics of soil. Geotextiles and Geomembranes. 49(3): 707–721. doi:10.1016/j.geotexmem.2020.11.010.

Zhou, Y., Li, R., Guo, B., Zhang, L., Zou, X., Xia, S. and Liu, Y. 2020. Greywater treatment using an oxygen-based membrane biofilm reactor: Formation of dynamic multifunctional biofilm for organics and nitrogen removal. Chemical engineering journal (Lausanne, Switzerland : 1996), 386. doi:10.1016/J.CEJ.2019.123989.

5

Resource Recovery from Wastewater

Harshal Deshmukh,[1] *Nikhil Nair,*[1] *Nilesh Wagh*[1,*] and *Jaya Lakkakula*[1,2,*]

5.1 Introduction

Water pollution is one of the most concerning issues which adversely affects the health of not only the humans but also of all the other life forms. In recent times, the world is developing, we are experiencing vast urbanization and a rapid increase in world population. All this development comes at the cost of natural resources, fossil fuels, etc., which have been used without any constraints. This development and industrialization have led to irrational depletion and consumption of resources. Outcome of this is generation of enormous wastes which, if not dealt with, would contribute to pollution. Water is especially vulnerable to pollution since it has the most dissolving ability compared to any other liquid. Anthropogenic activities contribute to the release of toxic and harmful substances from factories, cities, and farms, etc., resulting in the pollution of water and thereby generation of wastewater. This wastewater contains contaminants like pathogens, phosphorus, dyes, heavy metals, insecticides, in some cases oil and even radioactive materials, etc., which are harmful and need to be removed before it is discharged into water bodies. When the wastewater is not properly treated and is consumed, it may cause serious health issues like waterborne diseases, poisoning, accumulation of heavy metals, cancer, chronic asthma, etc. For example, when excess Zn (II) ions enter a person, it may cause neurological damage, insomnia, lethargy, liver, or kidney damage, etc. (Baby et al. 2011, Khlifi and Hamza-Chaffai 2010, Ernhart 1992). The contaminants present in the wastewater also cause algal blooms in water bodies which leads to eutrophication. The aquatic life is deprived of oxygen, eventually, the organisms

[1] Amity Institute of Biotechnology, Amity University, Mumbai Maharashtra, Mumbai-Pune Expressway, Bhatan, Panvel, Mumbai, Maharashtra, 410206, India.
[2] Centre for Computational Biology and Translational Research, Amity Institute of Biotechnology, Amity University, Mumbai, India.
* Corresponding authors: spencerjaya@gmail.com, waghnil@gmail.com

suffocate and die creating dead zones. The toxic heavy metals and chemicals from the industries are toxic to the aquatic ecosystem. They adversely affect their ability to reproduce and shorten their lifespan. These toxic substances enter the food chain with phytoplanktons and as the hierarchy of the food chain increases, the concentration of the toxins increase, this phenomenon is referred to as bioaccumulation. This accumulation occurs in living beings because the toxic compounds, when ingested, are taken up and stored by the body faster than they are metabolized or expelled (Sadegh et al. 2017).

Wastewater, if treated as a resource, it would be very beneficial in terms of environmental and economic aspects. These wastewaters contain heavy metals, precious metals, phosphorus, rare metals, dyes, and even oil which can be removed and recovered. The recovered material can be of use, making the wastewater treatment process more sustainable. The resource recovery approach has the potential to generate positive environmental outcomes, new sources of resources, and even profits. To recover a particular resource from wastewater, it must be first removed. There are several conventional techniques like precipitation, reduction, ion exchange, adsorption, and oxidation. However, among them, the adsorption process is the most effective one due to its great efficiency and economic feasibility (Zare et al. 2015, Sadegh et al. 2015, Madhava Rao et al. 2006).

Adsorption is a surface phenomenon in which the adsorbate aggregates on the adsorbent surface. When a liquid containing an absorbable solute meets a solid having a porous surface structure, intermolecular forces of attraction enable some of the solute molecules in the solution to be concentrated or deposited on the solid surface (Singh et al. 2011, Shen et al. 2009, Wang et al. 2012). In bulk materials, the constituent atoms of the material fulfil its bonding requirements (ionic, covalent, or metallic).

However, because the atoms on the adsorbent's surface are not surrounded by other adsorbent atoms, they can attract adsorbates (Wang et al. 2012, Tuzen and Soylak 2007, Moradi et al. 2016, Zhao et al. 2010, Apul et al. 2013). The exact nature of the bonding depends on the species involved, but the adsorption process is generally classified as physisorption (an adsorbate bound to the surface by weak Van der Waals forces), (an adsorbate bound to the surface by strong Van der Waals forces) (Sadegh et al. 2016) or due to electrostatic attraction (Biškup and Subotić 2005).

The quantity of solute adsorbed per unit weight of adsorbent as a function of the equilibrium concentration in the bulk solution at constant temperature is represented using adsorption isotherm models. Langmuir and Freundlich, Temkin, Harkin–Jura, and Dubinin–Radushkevich are examples of isotherm models (Apul et al. 2013, Kurniawan and Lo 2009, Ho and McKay 1998, Agarwal et al. 2016). Langmuir and Freundlich models are two of the most often utilized for describing adsorption (Wang et al. 2012, Ho and McKay 1998, Agarwal et al. 2016, Abdel Ghafar et al. 2015).

When the cost of material synthesis, processing, and production is high, desorption is essential. The capacity of loaded nanomaterials to regenerate is a critical component in increasing the efficiency of adsorption processes. The ability of adsorbents based on nanomaterials to remove the dye, phenols, heavy metals, metalloids, and organic contaminants from aqueous solution to regenerate throughout many adsorption/desorption cycles is critical to their economic feasibility. Through desorption, it is feasible to reduce the cost of the process as well as its reliance on a constant supply of nanomaterials. A successful desorption process demands the best mix of elutants, which is greatly dependent on the material type and adsorption mechanism (Das 2010). Elutants must be (1) non-destructive to nanomaterials, (2) inexpensive, (3) environmentally friendly, and (4) effective (Gautam and Chattopadhyaya 2016).

Activated carbon (AC), zeolites, biomaterials, and polymers are examples of adsorbents that have been utilized extensively for wastewater treatment (Zare et al. 2015, Sadegh et al. 2015, Madhava Rao et al. 2006, Biškup and Subotić, 2005, Ekmekyapar et al. 2006, Arias et al. 2002, Sublet et al. 2003). Nonetheless these materials have relatively low adsorption efficiency (Biškup and Subotić 2005). It was reported that nanomaterials like carbon nanotubes, nanofibres, nanosilica, nanospheres, etc., are very good adsorbent agents because of their high adsorption capacity even in dilute solutions, selectivity, and their economic feasibility. These nanoparticles are often surface modified with certain agents to further enhance their performance and selectivity (Sadegh et al. 2017).

In this chapter, an attempt has been made to highlight the importance and progress of various nanoparticles and their efficacy in the removal and recovery of various adsorbate molecules from the wastewater.

5.1.1 *Various nanoparticles subjected to resource recovery from wastewater*

5.1.1.1 *Carbon nanotubes (CNTs)*

Carbon nanotubes (CNTs) are cylinder-shaped molecules made up of rolled-up single-layer carbon atom sheets (graphene). The real space analysis of pictures of multiwall nanotubes revealed a wide variety of interlayer spacing (0.34 to 0.39 nm) (Ajayan and Ebbesen 1997). Their length might range from a few micrometres to millimetres. MWCNTs are always conducting and have at least the same conductivity as metals, but SWCNTs' conductivity is determined by their chiral vector: they can act as a metal and be electrically conducting, or they can act as a semiconductor and be non-conducting (Ali et al. 2014).

5.1.1.2 *Magnetic nanoparticles (MNPs)*

Magnetic nanoparticles exhibit phenomena such as high field irreversibility, high saturation field, superparamagnetism, additional anisotropy contributions, and shifted loops. Individual nanoparticles' magnetic behaviour is dominated by narrow and finite-size effects, as well as surface effects (Grancharov et al. 2005). For the synthesis of MNPs, numerous popular methods have been documented, including co-precipitation, microemulsion, thermal decomposition, solvothermal, sonochemical, microwave aided, chemical vapour deposition, combustion synthesis, carbon arc, and laser pyrolysis synthesis (Marathe and Doshi 2015).

5.1.1.3 *Graphite nanoplatelets (GNPs)*

Because of their superior physical and chemical qualities as well as their inexpensive cost, GNPs are a viable alternative to carbon nanotubes (Balandin et al. 2008, Kalaitzidou et al. 2007). The platelet thickness of GNPs ranges from 0.35 to

100 nm (Jang and Zhamu 2008). Graphene is a single-layered GNP with close-packed carbon atoms in a two-dimensional hexagonal lattice (Geng et al. 2009, Raza et al. 2011).

5.1.1.4 *nZVI*

To date, the most widely used technique in nano remediation initiatives has been nanoscale zero-valent iron particles (nZVI). The majority of nZVI examined are in the 10–100 nm size range as manufactured. The addition of another metal and the production of an noble metal alloy, such as Pd, Pt, Ag, Cu, Ni, is a well-documented approach of increasing the remediation properties (Yan et al. 2013). Small amounts of those metals put to the surface of nZVI increases the reactivity of the resulting particle and protects it from passivation (Geng et al. 2009). ZVI is a highly reactive reducing agent that reduces various organic molecules and modulates element oxidation states. This is what makes it useful in soil and groundwater clean-up (Stefaniuk et al. 2016).

5.1.1.5 *Nano silica*

Researchers are focusing their attention to silica nanoparticles (Si-NPs) because of its unique properties, including large surface area, high pore volume, variable pore size, great biocompatibility, capacity to encapsulate both hydrophilic and hydrophobic molecules, and scalable synthetic availability. The surface modification of silica can also be done easily using different molecular or polymeric moieties making them highly compatible with biological environments. Illustrating the huge potential of nanosilica in biomedical areas (Singh et al. 2019).

5.1.1.6 *Nanocrystal*

Because nanocrystals have a small surface area, they have a faster dissolution rate, higher saturation solubility, greater cell membrane/surface adhesiveness, and better oral bioavailability (Ige et al. 2013). As a result, nanocrystals have a quick beginning of action, few side effects, a high drug load, numerous delivery routes, low drug concentration, and overall improvement (Jahangir et al. 2020, Müller et al. 2011).

5.1.1.7 *Carbon nanosphere*

Carbon nanospheres (CNSs) are circular graphitic layers made up of pentagonal and hexagonal carbon rings paired together. They have an open dangling bond structure and range in size from 50 nm to 1 micrometre. This makes them suitable for a wide range of applications. Chemical decomposition, solution combustion, and hydrothermal synthesis procedures have all been used to create them (Kristianto et al. 2015).

5.2 Metallic nanomaterials for the recovery of resources from wastewater

This section highlights the role of various metallic nanomaterials for the recovery of resources from wastewater. These studies have been summarized in Table 1.

Table 1: Metallic nanomaterials used for the recovery of various resources from wastewater

Sr. No.	Type of nanoparticle	Recovered resource	Recovery rate	Adsorption capacity	Source of wastewater	References
1	Three-dimensionally ordered macroporous magnetic potential-responsive ion-exchange nanomaterials (3DOM MPiX)	Cesium (I) ions	98%		High-level radioactive wastewater	(Wang et al. 2017)
2	Titanate nanotubes (TiNTs)	Strontium (II) ions	77%	97 mg/g	Seawater	(Ryu et al. 2016)
3	Magnetic multi-walled carbon nanotubes @zeolite (MWCNT-Fe_3O_4@Zeo) nanocomposite	Lead and Thallium	97–99.3%	37.8 mg/g (Pb), 44.5 mg/g (Tl)	Not mentioned	(Gugushe et al. 2019)
4	Nanomagnetic hydroxyapatite	Cadmium (II)	40.20%	62.14 mg/g.	1000 mg/L $CdCl_2$	(Li et al. 2019)
5	Nanoscale zero-valent iron (nZVI)	Precious metals	>94–99%	250 mg/l1	Industrial wastewater	(Ling et al. 2018)
6	$ZnFe_2O_4$ nanoparticles	Molybdenum	97.60%	62.5 mg/g	Industrial wastewater	(Tu et al. 2016)
7	Palladium nanoparticles	Chromium (VI) and Cr (III)	85.75%	69.75 mg/g	Electroplating factory wastewater	(Tabatabaei et al. 2020)
8	Maghemite nanoparticles	Chromium (VI)	97.30%		Synthetic solution	(Hu et al. 2005)
9	Rhein-coated magnetic Fe_3O_4 nanoparticles	Pb (II) and Mg (II)	76.1–78.8% Pb (II) 82.2–84.0% Mg (II)	69.3 mg/g Mg (II) 64.9 mg/g Pb (II)	Model wastewater	(Sun et al. 2021)
10	Fe_3O_4@Me_6TREN nanoparticles	Heavy metals	95%	61.4 mg/g Cr(III) 245.0 mg/g Cu (II), 1136.2 mg/g Cd (II), 5.3 mg/g Pb (II)		(Jing and Shi 2020)

Table 1 contd. ...

...Table 1 contd.

Sr. No.	Type of nanoparticle	Recovered resource	Recovery rate	Adsorption capacity	Source of wastewater	References
11	Chitin nanofibrous	Noble metal ions		243.2mg/g Au (II), 136.5mg/g Ag (II), 217.8 mg/g Pt (II), 197.8mg/g Pd (II)	Oil and water emulsion	(Wang et al. 2019)
12	Surface-modified jacobsite ($MnFe_2O_4$) nanoparticles	Cr (VI)	98.90%	19.36 mg/g	Synthetic wastewater	(Hu et al. 2005)
13	Diatomite supported nanoscale zerovalent iron particles	Copper	99.90%		Acid mine wastewater	(Crane and Sapsford 2018)
14	Fe_3O_4 nanoadsorbents	Lead	90%	36 mg/g	Wastewater	(Nassar 2010)
15	Dendrimer-functionalized multi-walled carbon nanotubes	Heavy metal ions	Hg (II) (98.9%), Ca (II) (60.2%)		Industrial wastewater	(Iannazzo et al. 2017)
16	Orange peel modified magnetic nanoadsorbent	Cadmium (II)	82%		Aqueous solution	(Gupta and Nayak 2012)
17	Fe (II) oxide/hydroxide nanoparticles-based agglomerates suspension	Chromium (VI)	>95%			(Zelmanov and Semiat 2011)
18	PDMC functionalized nanoadsorbent	Au (II)	81.93%	127.3 mg/g		(Fu et al. 2017)
19	Imidazoline modified nanosilica	Pd (II)	86.79%	69.6 mg/g		(Zhang et al. 2018)
20	Activated carbon nanospheres	Ag (II)	almost 100%	152 mg/g		(Song et al. 2011)
21	Gum arabic modified magnetic nanoadsorbent	Cu (II)	93%	38.5 mg/g		(Banerjee and Chen 2007)
22	Nanoscale zerovalent iron (nZVI) and Au doped nZVI particles	Cd (II)	85%	78 mg/g - 188 mg/g		(Su et al. 2014)
23	Zerovalent Iron Nanoparticles	Zn (II), Cd (II), Pb (II),Ni (II),Cu (II), Ag (I)	>80%			(Li and Zhang 2007)

24	*S. cerevisiae* cells modified with nZVI	Ni (II)	77%	54.23 mg/g		(Guler and Ersan 2016)
25	Nanoscale Magnetic iron oxide	Neodymium	98.30%	24.88 mg/g		(Tu et al. 2015)
26	Thiol-functionalized chitosan nanoparticles	Hg (II)	>80%	1192 mg/g		(Nemati et al. 2019)
27	Graphene oxide nanocomposites	Hg (II)	99%	389 mg/g		(Hosseinzadeh et al. 2019)
28	Di-thio functionalized nanosilica	Hg (II)	>97%	538.9 mg/g		(Mehdinia et al. 2015)
29	Nanoscale zero-valent iron (nZVI)	Uranium	90%			(Ling and Zhang 2015)
30	Magnetically modified nanomaterials	Uranium	89%	310 mg/g	Wastewater	(El-Maghrabi et al. 2019)

Among the various nanomaterials used for the resource recovery, smart magnetic potential-responsive ion exchange nanomaterials (MPiX) with a three-dimensionally ordered macroporous (3DOM) structure were used for hazardous ion separation and recovery of c-Fe_2O_3 @nickel hexacyanoferrate (NiHCF). The open and linked porous channels, the large specific surface area of 3DOM, and the good affinity of NiHCF toward Cs (I) all contribute to the high ion exchange capacity, excellent selectivity, and quick ion insertion kinetics of such a 3DOM MPiX. Colloidal crystal templating was used to make 3DOM c- Fe_2O_3. Chemical co-precipitation was used to make 3DOM MPiX. Suitable quantity of 3DOM c-Fe_2O_3 was submerged in highly radioactive wastewater. Magnetic separation was used for 1 hour in $K_3[Fe(CN)_6]$ and 1 hour in $NiSO_46H_2O$. The 3DOM MPiX was dried for 12 hours at 80°C. For the adsorption testing, stable cesium (cesium-133) was employed. The three-electrode system was used to conduct the desorption test. The ability of 3DOM MPiX to absorb Cs (I) was investigated. The 3DOM MPiX nanomaterial produced has a 204 mg/g ion exchange capacity, great selectivity, and quick ion insertion kinetics. The combination of the 3DOM structure and coating MPiX materials with outstanding affinity for target ions results in such remarkable absorption performance. After 20 adsorption–regeneration cycles, the capacity of the 3DOM MPiX toward Cs (I) is practically invariable, and a high regeneration efficiency of more than 98 percent is maintained in each cycle. Cs (I) concentrations in the solution prior to and the DX-600 ion chromatograph were used to assess the results of the adsorption test (Wang et al. 2017).

The separation of strontium ions (Sr (II)) from seawater has received interest as a method for removing radioactive contaminants and recovering Sr (II). Titanate nanotubes (TiNTs) were manufactured using a simple hydrothermal process, their physicochemical properties were studied, and Sr (II) sorption behaviour was comprehensively examined under various reaction conditions corresponding to seawater environments. An analytical scanning electron microscope (SEM) with energy dispersive X-ray spectrometry (EDS) was utilized to investigate the morphology, and a high-resolution transmission electron microscope (TEM) was also employed to reveal the element distribution (Fig. 1). The standard alkaline hydrothermal technique was used to make TiNTs. TiNTs were disseminated at 1 g/L (0.02 g/0.02 L) in distilled water and sonicated for 30 seconds. The suspension was then given a desired amount of Sr (II) and a separate metal cation stock solution, and the pH was adjusted using an HCl or NaOH standard solution, which was then agitated for 30 minutes to allow equilibrium adsorption. During the 30-minute adsorption procedure, sample aliquots were periodically removed and filtered over a 0.2-μm cellulose acetate membrane. Inductively coupled plasma-atomic emission spectrometry was used to quantify the concentrations of Sr (II) and other metal cations. Inductively coupled plasma-atomic emission spectrometry was used to quantify the concentrations of Sr (II) and other metal cations (Ryu et al. 2016).

A one-pot solvothermal approach was used to make a magnetic multi-walled carbon nanotubes@zeolite (MWCNT-Fe_3O_4@Zeo) nanocomposite. Vibrating sampling magnotometry (VSM), X-ray diffraction (XRD), TEM, and SEM/EDS (energy dispersive X-ray spectroscopy) were used to describe the MWCNT- Fe_3O_4@Zeo. In ultrasonic-assisted magnetic solid phase extraction (UA-MSPE), zeolite-

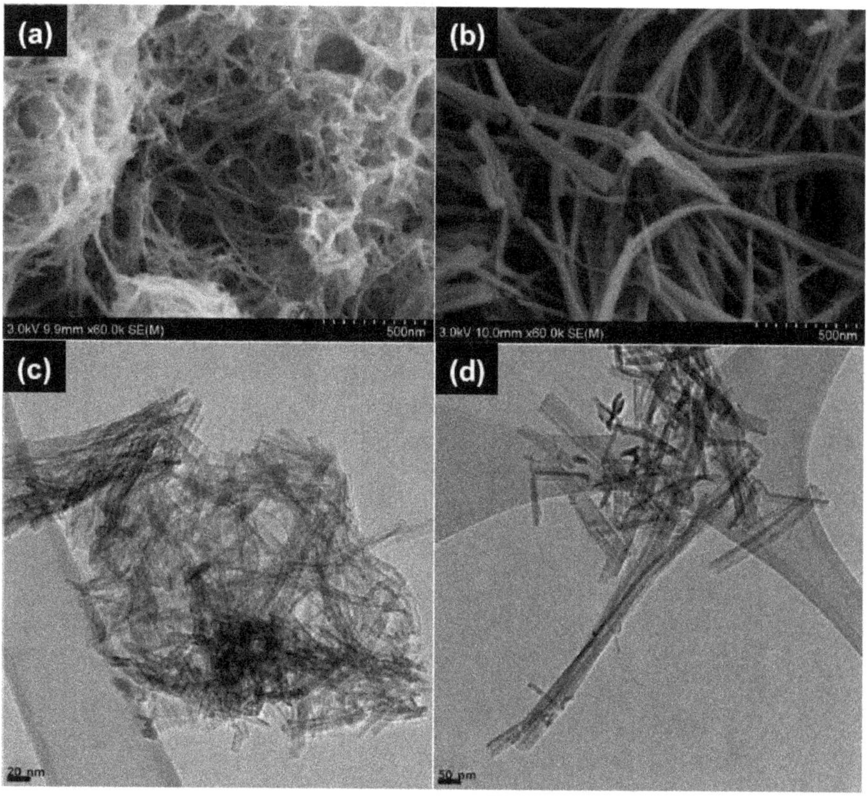

Fig. 1. SEM images of (a) TiNT-S and (b) TiNT-L and TEM images of (c) TiNT-S and (d) TiNT-L. Reproduced with permission from Ryu, J., Kim, S., Hong, H.J., Hong, J., Kim, M., Ryu, T., Park, I.S., Chung, K.S., Jang, J.S. and Kim, B.G. 2016. Strontium ion (Sr2+) separation from seawater by hydrothermally structured titanate nanotubes: Removal vs. recovery. Chemical Engineering Journal. 304: 503–510. https://doi.org/10.1016/j.cej.2016.06.131.

coated multi-walled carbon nanotubes–magnetite was utilized as an adsorbent for preconcentration of trace Lead (Pb) and Thallium (Tl). Inductively coupled plasma optical emission spectrometry (ICPOES) was used to identify and quantify the analytes. A multivariate strategy was used to optimize experimental factors impacting the preconcentration process. The method's accuracy was tested using standard reference material and a spiking recovery test, and satisfactory results (97–99.3%) were obtained. Pb and Tl had the highest adsorption capabilities of 37.8 mg/g and 44.5 mg/g, respectively. The developed UA-MSPE/ICP-OES technique was successfully used in environmental samples for simultaneous preconcentration and measurement of Pb and Tl (Gugushe et al. 2019).

In another such study, nanomagnetic hydroxyapatite (nHAP-Fe_3O_4) was used for the removal and recovery of cadmium ions (Cd (II)) from aqueous solutions. These nanoparticles are prepared by co-precipitation of ferrous salt, iron salt and hydroxyapatite (Wu et al. 2007). Rod-shaped nanoparticles were observed in TEM images, 35 nm thick and 170 nm long. It was observed that almost no nanoparticle was free, which means that nHAP bound tightly to Fe_2O_3. For adsorption, a $CdCl_2$

solution was prepared, and the nanoparticles were then added to 50 ml $CdCl_2$ solutions. The nanoparticles adsorbed the heavy metal present in the solutions and were separated with the help of a magnet. Atomic spectrophotometer was used to determine the amount of Cd (II) that was adsorbed. The kinetic study data revealed that adsorption equilibrium was achieved within 2 h. The as-prepared nanoparticles showed adsorption capacity of 62.14 mg/g according to the Langmuir model. The adsorption capacity increased from pH 2–7. The presence of common co-existing ions like Mg (II), Cu (II) and Na (I) decreased the adsorption capacity. Eluents like EDTA-Na_2 and HCl facilitated the regeneration of the nanoparticles (Li et al. 2019).

Nanoscale zero-valent iron (nZVI) has been proven effective for metal ion enrichment from wastewaters (Guler and Ersan 2016, Li and Zhang 2007, Su et al. 2014). nZVI is prepared by sodium borohydride reduction of ferrous or ferric salts. From TEM images, it was observed that nZVI have an iron core within a thin shell of iron oxides/hydroxides and are of size ranging from 20–100 nm. The oxide/hydroxide outer shell electrostatically attracts metal ions, and the core reduces and immobilises several heavy metals. nZVI is less to the pH changes and achieves removal efficiencies- > 94–99 percent in a wide pH range of 3–9. Enrichment of Au and Ag is done with help of nZVI. Firstly, Au and Ag are concentrated from a solution with the help of nZVI. A tiny galvanic cell is formed because of the deposition and reduction of gold seedlings. The iron acts as the anode and the gold seeds act as cathodes on which the gold ions get reduced and deposited and form pure Au nanoparticles. Ag nanoparticles were enriched in the same manner. nZVI is capable of enriching uranium from low-level sources like wastewater and seawater (Ling and Zhang 2015, Ling et al. 2018).

$ZnFe_2O_4$ nanoparticles were subjected for the removal and recovery of Molybdenum (Mo) from water. They are synthesized by the hydrothermally ferrite process (Tu et al. 2015). The XRD pattern of the prepared adsorbent confirms its purity and reveals the spinel structure of $ZnFe_2O_4$. TEM images showed that the size of the synthesized adsorbent nanoparticles ranged from 10–30 nm. Maximum removal efficiency-87 percent was achieved at pH 2.82 (Fig. 2). A drop in adsorption was observed above pH 4. Mo K-edge XANES (X-ray absorption near edge structure) was used to determine the oxidation state of adsorbed Mo, wherein it was found to be Mo (VI). It was also observed that reduction of Mo did not occur. The maximum adsorption capacity value of 62.5 mg/g was determined by the Langmuir isotherm model. Desorption studies revealed that adsorbed Mo could be easily desorbed from $ZnFe_2O_4$ surface and replaced by OH^- from NaOH. The results showed that 99.9 percent of Mo can be removed with a removal efficiency of 97.6 percent (at pH 3.02). The findings in this study prove the efficacy of $ZnFe_2O_4$ nanoparticles in industrial wastewater treatment for the removal and recovery of Mo (Tu et al. 2016).

Removal of Chromium from electroplating wastewater is done by electrochemical method (Cr (VI)) and adsorption. To enhance the efficiency of reduction of Cr (VI) in the electroplating process, Palladium nanoparticles were coated on the copper cathode wire. Pd NPs convert the H_2 into H radical which reduces Cr (VI) to Cr (III). From SEM images, the approximate diameter of Palladium nanoparticles on the copper substrate surface was found to be 26.8–55.5 nm. A cellulose sulphate

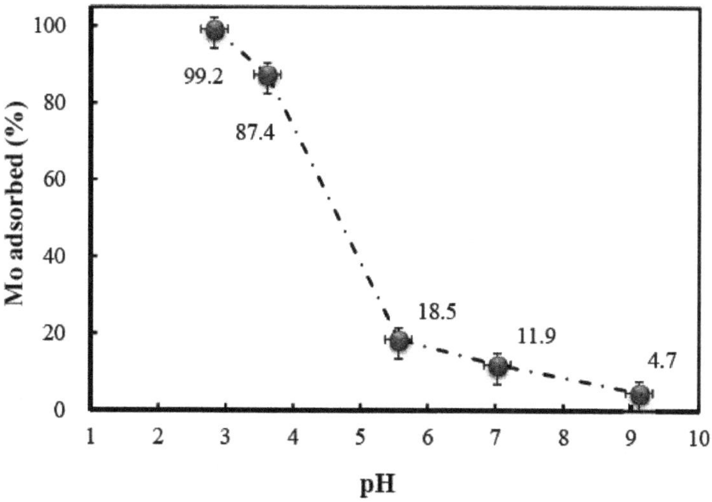

Fig. 2: Adsorption tendency of Mo onto ZnFe2O4 under different pHs. Conditions: solid/liquid = 2 g L^{-1}, T = 298 K, initial Mo concentration = 110 mg L^{-1}. Reproduced with permission from Tu, Y.J., Chan, T.S., Tu, H.W., Wang, S.L., You, C.F. and Chang, C.K. 2016. Rapid and efficient removal/recovery of molybdenum onto ZnFe2O4 nanoparticles. Chemosphere. 148: 452–458. https://doi.org/10.1016/j.chemosphere.2016.01.054.

adsorbent was prepared for the adsorption of the reduced Cr (III) ions. It was observed that the removal percentage increased when the initial concentration of Cr (VI) ions was increased. A semi-continuous system was employed to enhance Cr (VI) reduction efficiency, where they were found to be almost 100 percent efficient. The removal percentage of Cr (III) by cellulose sulphate was 85.75 percent. Certain experiments were performed which revealed the dual mode of action of cellulose sulphate for Cr (III) removal. Cellulose sulphate generates dense sludge, which makes it easier to drain and has a high sedimentation rate. It showed a maximum adsorption capacity of 69.75 mg/g. The results obtained from the experiments show that the use of Palladium nanocoating made the process of Cr (VI) adsorption more efficient (Tabatabaei et al. 2020).

In a similar study, Hu et al. synthesized and used maghemite (γ-Fe_3O_4) nanoparticles for the removal and recovery of Cr (VI) from wastewater. Sol-gel technique by was used for developing maghemite nanoparticles. The average diameter of the nanoparticles was found to be 10 nm from the TEM images. XRD data revealed that the maghemite nanoparticles were highly crystalline in nature. It was observed that about 90 percent of total Cr (VI) present in the solution was adsorbed in the first minute itself, this is attributed to the maghemite nanoparticles in which almost all the sites of adsorption are exteriorly present, hence easily accessible to Cr (VI). At an initial concentration of 50 mg/L, the removal efficiency was found to be 97.3 percent. It was found that at pH 2.5, common ions present in the solutions did not compete with Cr (VI) for adsorption. The adsorption of Cr (VI) was found to be exothermic. When the adsorption data was analysed, it was found that it correlated well with the Freundlich isotherm. The adsorption capacity of maghemite nanoparticles was determined to be 19.2 mg/g at pH 2-3 and room temperature. According to the regeneration studies, the maghemite nanoparticles can be effectively recovered to carry out re-adsorption. The above study indicates maghemite nanoparticles to be effective for Cr (VI) removal and recovery from wastewaters (Hu et al. 2005a).

A magnetic adsorbent Rhein-coated magnetic Fe_3O_4 nanoparticle (RMNP) for Pb (II) and Mg (II) was produced, and the adsorption process was examined using low-field NMR. Fe_3O_4 nanoparticles and Rhein were used to make RMNPs. The combination was sonicated for 90 minutes after adding 1 mg Rhein to a 100 mL suspension water system containing Fe_3O_4 nanoparticles (1 mg). Finally, RMNPs were produced by centrifuging for 10 minutes at 12,000 rpm and washing with ultrapure water numerous times. Using quercetin-coated Fe_3O_4 nanoparticles (QMNP) as a magnetic sensor/adsorbent, an experiment was devised to identify and remove Pb (II) or Mg (II) from water. Under an external magnetic field, 0.9 mg RMNP was added to 3 mL Pb (II) (25 mg/l) or Mg (II) (25 mg/l) aqueous solution, and the precipitate was removed after a period of standing under an external magnetic field. QMNP was used as a sensor to quantify the levels of Pb (II) or Mg (II) in the filtrate. Atomic absorption spectroscopy (AAS) was used to determine the original concentration of Pb (II) or Mg (II), as well as the concentration in the filtrate. The mixture was filtered after 90 minutes, and the concentration of Pb (II) or Mg (II) in the filtrate was determined by AAS. The results showed that RMNP removed 76.1–78.8 percent of Pb (II) and 82–2–84.0 percent of Mg (II) from the aqueous system, indicating that RMNP is a good material for removing Pb (II) and Mg (II) from aqueous systems (Sun et al. 2021).

A straightforward way to graft Me_6TREN ligands onto the surface of SiO_2-coated Fe_3O_4 to generate functionalized Fe_3O_4@Me_6TREN nanoparticles (NPs) to act as a new nano-adsorbent of heavy metals in contaminated water was analysed. As a potent nano-gripper, Fe_3O_4@Me_6TREN NPs can selectively concentrate various heavy metals to achieve categorized and simple separation. Furthermore, by washing the NPs with stronger chelators of heavy metal ions, the NPs can be regenerated. The Fe_3O_4 NPs were synthesized by the solvothermal method. Fe_3O_4 @Me_6TREN NPs were dispersed in 6 mL of aqueous Cr (III), Cu (II), Pb (II), and Cd (II) ions, each with a starting concentration (C_0) of 5 mg/mL and a different pH value at 30°C. The capacity of the NPs for four metal ions varies greatly, with Cr (III) (61.4 mg/g), Cu (II) (245.0 mg/g), and Cd (II) (1136.2 mg/g) being the strongest and Pb (II) (5.3 mg/g) being the weakest. The Fe_3O_4@Me_6TREN NPs' significant difference in adsorption capacity for the four metal ions chosen is advantageous for the classified recovery (Jing and Shi 2020).

A one-step green recovery of noble metal ions from surfactant-stabilized oil/water emulsions was reported using a multifunctional membrane made by filtration of partially deacetylated chitin nanofibers. For preparation of chitin nanofibrous membrane (CNFM), chitin powder was dispersed in 33 percent (w/w) KOH aq. solutions with reducing agent $NaBH_4$. To achieve partial deacetylation, the mixture was manually agitated. Centrifugation was used to separate the resulting dispersion. To eliminate alkali, the precipitate was rinsed with deionized water before being freeze-dried. In acetic acid aq. solutions, deacetylated chitin powder was disseminated. The suspension was manually agitated for 24 hours to thoroughly protonate it before being treated to ultrasonic treatment. The translucent chitin nanofibers suspension was produced after centrifugation and dialysis. The extraction efficiency of the CNFM was evaluated using four noble metal ions (Au (III), Ag (I), Pd (II), and Pt (IV)) as models. The oil/water emulsions with noble metal ions were made by mixing hexane with 0.2 wt. percent Tween 80 with water containing noble metal ions. At a pressure of 0.2 bar, the as-prepared oil/water emulsions containing noble metal ions were placed over the CNFM for filtering. The adsorbed noble metal ions on the CNFM were heated for 30 minutes after filtering to create noble metal nanoparticles laden with noble metals. When the CNFM's chelation and reduction abilities were combined, the CNFM was able to directly extract noble metal ions from an oil/water emulsion, with maximal absorption of Au (III), Ag (I), Pd (II), and Pt (IV) ions reaching 243.2, 136.5, 217.8, and 197.8 mg, respectively (Wang et al. 2019).

The efficiency of surface-modified jacobsite ($MnFe_2O_4$) nanoparticles for the removal and recovery of Cr (VI) from synthetic wastewater was examined in this study. Using a co-precipitation approach followed by a surface redox reaction, ten nanometer modified $MnFe_2O_4$ nanoparticles were generated as a novel adsorbent. Cr (VI) adsorption onto modified $MnFe_2O_4$ nanoparticles reached equilibrium in as little as 5 minutes, and the adsorption results suited the Langmuir model well. At pH 2, the highest absorption of 31.5 mg of Cr (VI)/g of modified $MnFe_2O_4$ was achieved, which was like other commonly used adsorbents including activated carbon and sawdust. In a pH range of 2–10, the effects of ligands (EDTA, SO_4^{2-}, NH_4^+) and ionic strength were investigated. Over the entire pH range tested, EDTA and SO_4^{2-} hindered Cr (VI) adsorption, whereas NH_4^+ improved Cr (VI) absorption at pH greater than 6.5. X-ray diffraction and X-ray photoelectron spectroscopy revealed that the mechanisms leading to Cr (VI) adsorption by modified $MnFe_2O_4$ nanoparticles are a mix of electrostatic interaction and ion exchange. Regeneration investigations revealed that the modified $MnFe_2O_4$ nanoparticles could be reused without compromising adsorption capacity, and that Cr (VI) could be recycled without affecting its valence (Hu et al. 2005b).

Similarly, a recent study proposes the notion of metals 'precision mining', which is described as a procedure for selective *in situ* absorption of a metal from a substance or medium, followed by retrieval and recovery of the target metal. To prove this hypothesis, nanoscale zerovalent iron (nZVI) was placed onto diatomaceous earth (DE) and examined for selective Cu absorption and subsequent release from acid mine drainage (AMD). The AMD and nZVI-DE were used in batch tests at concentrations of 4.0–16.0 g/L. A four-acid digest was used to prepare each sample for ICP-OES (Inductively coupled plasma—optical emission spectrometry) analysis. First, 0.01 g was put in a Polytetrafluoroethylene (PTFE)-lined microwave digest cell, followed by 3 mL of analytical grade 45.71% hydrofluoric acid (HF) and 12 hours of waiting. The container was then placed in a microwave digest oven and heated at 200°C for 30 minutes before cooling for 15 minutes with 6 mL of aqua regia solution (1:1 ratio of analytical grade 32 percent hydrochloric acid (HCl) and 70 percent nitric acid (HNO_3)). The resulting solution was then neutralized for 20 minutes at 150°C with 18 mL of analytical grade 4% boric acid (H_3BO_3) before being allowed to cool for 15 minutes. Despite the relatively low concentration of nZVI onto and within the DE (2.6 wt. percent), exposure of nZVI-DE concentrations 12.0 g/L to the AMD led to the quick and near entire recovery of Cu from solution (> 99.9 percent removal after 1 h), presumably by electrochemical cementation (Crane and Sapsford 2018)

Fe_3O_4 nano adsorbents were shown to be a viable alternative to traditional adsorbents for the rapid removal of metal ions from wastewater with good removal efficiency in one such study. The adsorption of Pb (II), a common metal ion found in wastewater, onto Fe_3O_4 nano adsorbents proved successful in removing Pb (II). Adsorption was extremely fast, and equilibrium was reached in under 30 minutes. Adsorption was also substantially influenced by the initial Pb (II) content, pH, and temperature. At pH 5.5, the most removal was seen. As the starting concentration of Pb (II) and temperature increased, so did adsorption. The adsorption isotherms were also calculated, and both the Langmuir and Freundlich models adequately represented them, with the Freundlich model fitting better than the Langmuir model. The endothermic character of the adsorption process, as well as its spontaneity and physisorption, were validated by the thermodynamics of Pb (II) adsorption onto Fe_3O_4 nano adsorbents. Nano adsorbents may be employed repeatedly without losing their adsorption capability, according to desorption and regeneration experiments. Fe_3O_4 nano adsorbents are recommended for the rapid removal and recovery of metal ions from wastewater effluents since they are fast, effective, and affordable (Nassar 2010).

Another study uses dendrimer-functionalized multi-walled carbon nanotubes (MWCNT) for the removal and recovery of toxic heavy metal ions from wastewaters. Carbon nanotubes (CNTs) have been used for the adsorption of removal of heavy metal ions because of their adsorption capacity. This adsorption by CNTs can be made more efficient with the help of chelating agents like triazole as they have coordination ability towards heavy metal ions. This study combines the adsorption activity of CNTs and the chelating activity of triazole based dendrimers to develop triazole dendrimer-functionalized carbon nanotubes (MWCNT-TD2) and a phosphonate (α-aminophosphonate) derivative MWCNT-TD2P. HRTEM (High-resolution transmission electron microscopy) images indicated that the functionalization of MWCNTs did not cause intrinsic damages to its peculiar structure. It was observed that the addition of triazole dendrimers to the CNT surface somewhat reduced the binding ability towards metal ions due to the steric hindrance of dendrimer moieties. A slight preference towards Pb ions was observed by the MWCNT-TD2 system. The addition of α-aminophosphonate resulted in chelating behaviour towards mercury ions (98.4%) indicating its application for selective removal of Hg (II). A study was done to investigate the competition for dendrimer sites amongst 4 metal ions (Hg, Ni, Pb and Ca). The results confirmed the excellent chelating activity of α-aminophosphonate dendrimers (MWCNT-TD2P) towards Hg (II) ions and overall, less affinity towards Ca (II) by both the MWCNT systems. The findings of this study suggest α-aminophosphonate dendrimer combined with MWCNTs (MWCNT-TD2P) for the selective removal and recovery of mercury ions from wastewaters (Iannazzo et al. 2017).

Gupta et al. synthesized a novel magnetic nano adsorbent by the surface modification of Fe_3O_4 NPs along with orange peel powder (MNP-OPP) for removal and recovery of cadmium ions from aqueous solutions. Magnetic nanoparticles are often considered for adsorption because of their high effective surface area, ease of recovery and they can be surface modified and functionalized. Orange peel is known to be a bioadsorbent having active carboxyl and hydroxyl functional groups (Feng et al. 2009, Sha et al. 2009). Orange peel powder is co-precipitated with magnetic Fe_3O_4 nanoparticles (MNPs) to obtain MNP-OPP. From the TEM images, the average diameter of MNP-OPP was found to be 32–35 nm. XRD analysis indicated that the nanoadsorbent was amorphous. It was found that an increase in initial Cd ion concentration increased the amount of Cd ions adsorbed. High adsorption levels were obtained at pH 6-8. Ion exchange and complexation were found to be the principal mechanisms behind the adsorption of Cd ions. The presence of humic acid at higher pH led to complex formation with Cd ions, which reduced the adsorption capacity of MNP-OPP. The thermodynamic study indicated that adsorption was spontaneous. It was revealed that the adsorption would be efficient at a higher temperature. The maximum adsorption capacity was found to be 76.92 mg/g. On desorption by HNO_3, 98 percent adsorbed Cd ions were desorbed. The desorption efficiency dropped by just 4.74 percent after 5 consecutive adsorption-desorption cycles, indicating its great reusability. The MNP-OPP were tested in electroplating industry wastewater, and 82 percent of Cd ions were adsorbed despite the presence of competitive metal ions like Ni and Zn. This study puts forward MNP-OPP as a great candidate for Cd adsorption from wastewaters, considering it is eco-friendly, cost-effective, efficient, and highly reusable (Gupta and Nayak 2012).

A similar study deals with the removal and recovery of Cr (VI) from aqueous solutions. Chromium ions are usually removed by reduction of hexavalent ion to trivalent ion followed by precipitation of the trivalent chromium (Hu et al. 2005a, b). The hexavalent chromium can be adsorbed onto the surface of Fe^0, which is followed by electron transfer, thus reducing Cr (VI) to Cr (III). This trivalent chromium obtained is precipitated by the addition of NaOH in the solution and separated from the solution in the form of chromium oxides/hydroxides. This study uses iron (Fe (III)) oxide/hydroxide nanoparticles-based agglomerates suspension (AggFe) as an adsorbent Cr (VI). The adsorbent was prepared by agglomeration of iron oxide/hydroxide-based nanoparticles (NanoFe). TEM images of NanoFe revealed that the average size was 55 nm. pH studies showed that increasing pH values reduced adsorbent activity. Adsorption studies indicate that > 95 percent of the removed Cr was adsorbed. After adsorption, the pH is raised above the zero-charge point, imparting a negative charge to the adsorbent surface, hence repelling the negatively charged. $BaCl_2.2H_2O$ is added in a stoichiometric amount to the concentrated chromium solution. $BaCrO_4$ particles were formed which were filtered out using a 0.45 μm filter. AggFe showed recovery efficiency of 95–97 percent for chromium. The iron nanoparticles were reusable after desorption. The findings of this study put forward AggFe as a very efficient and cost-effective option for Cr (VI) removal and recovery (Zelmanov and Semiat 2011).

Fu et al. developed a nano adsorbent by functionalising nano silica with poly methacryloxyethyltrimethyl ammonium chloride (PDMC) for selective removal and recovery of gold ions from aqueous solutions. 3-chloropropyltriethoxysilane (CTO) is added to nanosilica to form CTO-silica nanoparticles (CTO-SNP). PDMC is grafted on CTO-SNPs by atom transfer radical polymerization to yield PDMC-SNP. TEM images revealed the size of the nanoparticles to be 12 nm. The optimum pH value for Au (III) adsorption was found to be 3.0. It was observed that the adsorption of Au (III) increased when the concentration of Au (III) was increased. The maximum adsorption capacity was determined to be 127.3 mg/g. Au (III) ions were adsorbed by PDMC-SNP by monolayer adsorption. In the presence of competitive ions, PDMC exhibited relatively higher adsorption for Au (III) than other ions like Ba (II), Cu (II), Pb (II) and Mg (II). Regeneration of the adsorbent was done using an eluent constituting HCl and thiourea. The Au (III) removal rate was 81.93 percent which decreased by 5.6 percent in ten successive cycles. This data indicates that PDMC-SNP can maintain high adsorption efficiency even after 10 cycles. The findings of this study suggest PDMC functionalized silica nanoparticles for selective removal and recovery of Au (III) (Fu et al. 2017).

Another study discusses the removal of palladium (II) by adsorption using imidazole modified nanosilica, from an aqueous solution. The silica nanoparticles were functionalized with imidazoline groups to obtain ELD-silica nanoparticles (ELD-SNPs). The size of ELD-SNPs from TEM images was found to be 15 nm. FT-IR (Fourier Transform Infrared Spectroscopy) and XPS (X-ray photoelectron spectroscopy) analysis confirmed the introduction of imidazole groups onto silica nanoparticles. The adsorption increased from pH 1-4 and decreased on further increase in pH. Maximum adsorption was observed at pH 4. Adsorption kinetic study shows that the Pd (II) adsorption efficiency rapidly increases for the first 100 min and reaches equilibrium in 180 min. The adsorption data reveals that the pseudo-second-order model is followed. The adsorption capacity of Pd (II) increased when the initial concentration of Pd (II) was increased. The maximum adsorption capacity was 69.6 mg/g. In a mixture with competitive ions like Cu (II), Zn (II), Pb (II) and Ba (II), the Pb (II) adsorption was 86.79 percent. This indicates ELD-SNP's ability of selective adsorption of Pd (II). Thiourea and HCl are used for desorption. The adsorption after 5 consecutive cycles was found to be > 85 percent, indicating its high reusability. Thus, imidazole functionalized nanosilica is a promising candidate for selective adsorption and recovery of palladium from wastewaters (Zhang et al. 2018).

Song et al. synthesized activated carbon nanospheres for the removal and recovery of silver (II) ions from aqueous solutions by adsorption. Carbon nanospheres (CNS) were activated by NaOH treatment. The FESEM (Field Emission Scanning Electron Microscope) images indicate that the CNS had a uniform diameter of size 400–500 nm. It was observed that the NaOH treatment did not noticeably affect the CNS morphology. FTIR analysis showed richer surface hydroxyl and deprotonated carboxylic groups on the activated CNS (CNS/OH). These groups contributed to increased adsorption activity. Adsorption data well correlated with the Langmuir model. The maximum adsorption capacity was found to be 144.9 mg/g, very near to the calculated experimental value of 152.1 mg/g. It was observed that the adsorption of Ag (II) was very fast, and 100 percent removal efficiency was achieved in less than 6 min. The activated CNS efficiently adsorbed Ag ions even from dilute solutions. The adsorption followed the pseudo-second-order model of kinetics. Desorption experiments revealed that 96–99 percent of Ag (I) was recovered within just 5 min. The CNS were reactivated after desorption with NaOH. It was found that the adsorption capacity was almost the same as that of the original CNS/OH. This indicates the high regeneration capacity of CNS/OH. The findings of this study suggest that activated carbon nanospheres can be used for the removal and recovery of silver present in wastewaters even if present in traces (Song et al. 2011).

A novel magnetic nano-adsorbent was produced by treating Fe_3O_4 nanoparticles with gum arabic to remove copper ions from aqueous solutions. Gum arabic (GA) was linked to Fe_3O_4 via the interaction between the carboxylic groups of gum arabic and the surface hydroxyl groups of Fe_3O_4. The surface modification did not cause Fe_3O_4 to change phase, but it did result in the creation of secondary particles with diameters in the micron range. The shift of the isoelectric point from 6.78 to 3.6 nm and a wavelength range of 13–67 nm. In aqueous solutions at pH 2–6 and 300–340 K, the adsorption of copper ions by magnetic nano adsorbents (MNP or GA-MNP) was examined. By putting 25 mg of copper ions-loaded GA-MNP in 5 mL of HCl solutions, the desorption of copper ions was investigated. The GA-MNP was removed after 30 minutes of swirling at 200 rpm, and the concentration of copper ions in the liquid solution was measured to determine the amount of copper ions desorbed. The desorption percentages in the HCl solutions of pH 3.0, 2.0, and 1.5, respectively, were determined to be 81, 92, and 93 percent (Banerjee and Chen 2007).

The reactivity of nZVI to co-contaminants and by-product formation, as well as the use of different catalysts to reduce nitrite yield from nitrate and the transformation of nZVI after reaction were investigated in a recent study. The presence of Cd (II) and nitrate in groundwater has a significant impact on nZVI's ability to remove each other. Nitrate stimulates nZVI to remove more Cd (II), lowering Cd bioavailability. Furthermore, instead of ionic iron, nitrate promotes the formation of stable iron oxide (magnetite) from nZVI. The formation of magnetite prevents a significant increase in exchangeable iron concentration, which has been identified as a potential toxicant because of the use of nZVI in the environment. On the other hand, Cd (II) has both positive and negative nitrate-reduction effects: Cd (II) (50 mg/L) significantly improves nZVI reactivity with nitrate, but a significant amount of nitrite is produced. Doping nZVI with Au (1 percent) appeared to reduce the nitrite yield ratio significantly more than nZVI alone or doping nZVI with Cu or Ag. In the presence of nitrate, nZVIeAu has the same Cd (II) removal capacity as nZVI and 85 percent removal efficiency was observed at 100 mg-Cd (II)/L load (Su et al. 2014).

The ability of nanoscale zerovalent iron (nZVI) for the sequestration of Zn (II), Cd (II), Pb (II), Ni (II), Cu (II), and Ag in water is substantially higher than that of traditional materials. The iron nanoparticles feature a core-shell structure, which leads to outstanding capabilities for concurrent sorption and reductive precipitation of metal ions, according to studies using high-resolution X-ray photoelectron spectroscopy (HR-XPS). In nZVI reactions with metal cations, Cu, Ag, and Pb have extremely high removal effectiveness (> 99 percent). Cr and Zn removal effectiveness is very outstanding (92–97 percent). However, significant levels of Ni and Cd (39-64 percent) remain in the solution. Separate studies with single metal ions of Ni (II) or Cd (II) are also carried out, with substantially higher removal efficiencies (> 80 percent) for both (Li and Zhang 2007).

In an interesting study, investigation was planned to assess if *S. cerevisiae* cells treated with nZVI (magnetic bio sorbent) could be used to remove Ni (II) ions from aqueous solutions. FTIR, XRD, and SEM investigations were used to describe this composite, which was produced in ethanol utilizing a borohydride reduction process in atmospheric circumstances. The effects of initial solution pH, magnetic bio sorbent quantity, contact time, starting Ni (II) concentration, and temperature were studied in batch tests. For the Ni (II) solution, pH 5 was chosen as the best value. An increased Ni (II) removal was observed which was attributed to an increase in magnetic bio sorbent biosorptive and active sites, as shown by the influence of magnetic bio sorbent quantity on Ni (II) biosorption. The equilibrium data was subjected to Langmuir, Freundlich, Dubinin–Radushkevich, and Temkin isotherm models. XRD, FTIR, and SEM analyses demonstrated that the surface area of the magnetic bio sorbent has quite a rough surface, which is a good feature for biosorption. The maximum biosorption capacity was found to be 54.23 mg/g (pH 5.0, magnetic bio sorbent amount 3 g/L, and contact time 3 h). Biosorption kinetics were a good fit to the pseudo-second-order kinetic model. The Ni (II) biosorption onto the magnetic bio sorbent was found to be endothermic and spontaneous (Guler and Ersan 2016).

In a recent study, the recovery of spiked Nd from seawater media by magnetic iron oxide was investigated (Fe_3O_4). Batch tests were carried out to determine the adsorption processes and Nd mobilization equilibrium states. To test the reusability of the synthesized magnetic iron oxide, the adsorption–desorption operations were repeated three times. During the adsorption process, the pH and ionic strength of solutions were both considered important considerations. When the pH was higher than 7, the Nd recovery reached a maximum value of close to 100 percent in 30 minutes, according to the findings. Because the pH of seawater is around 8.2, this result also suggested that our manufactured magnetic iron oxide may be used in Nd recovery from seawater. Desorption experiments utilizing various concentrations of HNO_3 were studied following Nd adsorption to evaluate the desorption process and the feasibility of regenerating spent adsorbent. When the concentration of HNO_3 is greater than 0.05 N the reuse ability on the magnetic iron oxide for Nd recovery, 98.3 percent of Nd can be desorbed. Adsorption competition between H^+ ions and Nd (III) cations on the surface of the magnetic iron oxide suggest that at the examined circumstances, a 0.05 N HNO3 solution is sufficient for Nd desorption. The results clearly revealed that 99.9 percent of Nd can be recovered in the first trial, followed by 99.1 percent, 99.1 percent, and 97.8 percent, respectively, in subsequent cycles. About 5 mg/L Nd was introduced to the actual seawater to examine Nd recovery performance in various pH atmospheres to evaluate the viability of the magnetic iron oxide in real seawater. In genuine seawater media, Nd recovery might reach 99.3 percent, according to the findings (Tu et al. 2015).

In a recent method, thiol-functionalized chitosan (CS) based nanoparticles were used for adsorption of mercury (Hg (II)) from aqueous solutions. Chitosan is a biodegradable natural polymer that possesses functional groups as active sites, hence making it a good adsorbent. CS NPs were synthesized using the ionic gelation method, in which microfluidic (MF) and bulk mixing (BM) methods were employed separately. From the TEM images, the NPs obtained from the MF method were found to be more uniform than those from the BM method. NPs fabricated by MF were

18 ± 3 nm in size. The CS was functionalized to enhance adsorption. The highest adsorption was observed when the molar ratio of epichlorohydrin to cysteamine was 2.73. It was found that lower crystallinity of CS was associated with high adsorption capacity, hence nanoscale CS structure (amorphous phase) contributes to higher adsorption capacity. Magnetic (Fe_3O_4) NPs were added to aid the separation of Cs NPs after adsorption of Hg (II) from the solution. 0.5 mol/L HCl was used for adsorbent recovery. The findings reveal the mentioned nanoadsorbent as a promising candidate for Hg (II) removal and recovery from wastewaters (Nemati et al. 2019).

For the removal of mercury (Hg (II)) by adsorption from aqueous solutions, Oceanside et al. used magnetic graphene oxide nanocomposites. The reversible addition-fragmentation chain transfer (RAFT) method was used for carrying out controlled polymerisation to synthesize polymer-functionalized Fe_3O_4 and graphite oxide (GO) (Moad et al. 2013). Firstly, magnetic graphene oxide NPs are prepared, and a RAFT agent is added to form magnetic GO-RAFT. The polymer-coated magnetic GO (nanocomposite adsorbent) is prepared by copolymerization of AA (Acrylic acid) and AM (Acrylamide) monomers by the RAFT method. From TEM images, the size of Fe_3O_4 NPs was found to be 15–20 nm. The SEM images indicate a considerable improvement in the porosity of the nanocomposite after grafting. XRD analysis reveals that the nanocomposite has an amorphous structure which is the result of polymer chain grafting onto the surface of magnetic GO. It was found that the total surface area increased by the addition of GO substrate and acrylic monomers. The nanoadsorbent is capable of selective adsorption of Hg (II) as it can attract through electrostatic interactions with Hg (II). The incorporation of magnetic NPs further contributed to the adsorption of mercury. It was observed that at higher pH, the mercury adsorption capacity was high. The presence of Hg-Cl complexes resulted in the reduction of Hg removal. The Langmuir isotherm model correlated with the adsorption data, with maximum adsorption of Hg (II) of 389 mg/g. The magnetic nanoadsorbent (MNA) proved to be reusable and maintained 86 percent of initial adsorption capacity after 5 regeneration cycles. The findings of this study suggest MNA as a potential adsorbent for selective removal and recovery of Hg (II) from wastewater (Hosseinzadeh et al. 2019).

Similarly, magnetic di-thio functionalized mesoporous silica nanoparticles (DT-MCM-41) were used for the high-efficiency removal of mercury from water samples. MCM-41 is a silica-based mesoporous structure with hexagonal pores, high surface area, large volume pores, presence of many internal hydroxyl groups, high surface activity, etc. (Selvam et al. 2001). It is these features that MCM-41 have been researched upon for the development of functionalized mesoporous silica for adsorption (Shahbazi et al. 2011). The dithiocarbamate (DT) group is known to have a strong tendency of mercury ion interaction. Magnetic nanoparticles have been used for incorporation into mesoporous structures for magnetic separation. High-efficiency adsorbents are synthesized by grafting DT groups into the interior surface of the magnetic mesoporous silica nanoparticles. DLS (dynamic light scattering) analysis was done for size estimation which showed that the average size of nanoparticles was 187 nm. The adsorbent was able to completely remove Hg (II) from the aqueous sample with almost 100 percent efficiency. Adsorption at pH 6 was found to be the highest. The maximum adsorption capacity for the nanoadsorbent was found to be 538.9 mg/g, which was more than any other silica-based adsorbent for Hg removal. The contact time was short, and the efficiency was greater when compared to any other similar sorbents that have been reported. The presence of other heavy metal ions did not significantly affect Hg adsorption. For recovery of mercury by desorption, 1 M HNO_3 and 2 percent thiourea was used. The recovery value was close to 100 percent and the recovery efficiency of Hg (II) was > 98% after 3 times of use. The findings in this study establish the given nano adsorbent as highly efficient for removal and recovery of Hg from wastewaters (Mehdinia et al. 2015) magnetic di-thio functionalized mesoporous silica nanoparticles (DT-MCM-41.

In another study involving nanoscale zero-valent iron (nZVI), which is increasingly being used in hazardous waste treatment and environmental remediation and has been suggested as a potential agent for uranium immobilization, is used to demonstrate instant separation, enrichment, and encapsulation of low-level uranium from water. In batch experiments with uranium concentrations in the range of 2.32–882.68 μg/L, fast and nearly total removal of uranium from water was obtained. Under totally mixed conditions, the residual uranium concentration in water was decreased below 1 g/L in all trials within 2 minutes. Inductively coupled plasma mass spectrometry (ICPMS) was used to measure total dissolved uranium. With a standard potential (Eh) of less than 500 mV, the solution containing nZVI was highly reducing. Experiments with solution pH ranging from 3.5 to 8 revealed that pH had no effect on separation efficiency or rate. Repeated dosing of the nZVI solution with uranium at 50–300 mg/L was also used in the tests. The power of nZVI for uranium sequestration and enrichment is further demonstrated by XEDS (Energy-dispersive X-ray spectroscopy) quantification analysis: after 24 h, 38 percent of the total uranium in the nZVI particle deposited toward the centre (10 percent of volume), whereas the central 50 percent volume had roughly 87.3 percent of the uranium in the solid phase. In contrast, uranium deposited on the surface and within fractures of nZVI after 1 h is primarily (> 80 percent) on the outer shell (25 percent volume), and all uranium associated with the iron oxide is deposited on the surface and within fractures (Ling and Zhang 2015).

A recent study described a simple microwave-assisted technique for making magnetically modified hydroxyapatite nanoparticles (MNHA) and assessed their adsorption capacity for removing Uranium (VI) ions from wastewaters. In a small amount of water, iron nitrate and urea were well dissolved. A specific weight of hydroxyapatite nanopowder was carefully added, and the dispersion was then ultrasonicated at room temperature for 15 minutes. The resulting solution was microwaved for 3.0–5.0 minutes at 900 W and 2.45 GHz in a home microwave. Several approaches were used to completely describe the produced MNHA nanocomposite. The phases of the MNAP nanocomposite were identified using an X-ray diffractometer. The FTIR spectrometer was used to identify functional groups and bonds. The morphology and particle size of the produced MNHA particles were examined using a transmission electron microscope. A TEM apparatus with an Energy Dispersive X-ray spectroscope was used to determine the composition. The effect of initial uranyl ion concentration on the adsorption process was investigated by shaking 30 mg of the adsorbent for 120 minutes at pH 5.0 and 25°C with varied concentrations of uranyl ions ranging from 100 to 300 mg/L. When the starting concentrations of U (VI) ions were raised from 100 to 300 mg/L, the quantity of absorption by the produced MNHA rose from 140 to 310 mg/g. The desorption of U (VI) ions was investigated using several eluents, including HCl, NaOH, $NaHCO_3$, Na_2CO_3, and CH_3COONa. Because carbonates form exceptionally stable and soluble complexes with uranyl ions, Na_2CO_3 demonstrated the highest desorption efficiency of 89 percent (El-Maghrabi et al. 2019).

5.3 Other nanomaterials for the recovery of resources from wastewater

This section highlights the role of various other nanomaterials subjected to the recovery of resources from wastewater. These studies have been summarized in Table 2.

Bisphenol F (BPF) removal and recovery by multi walled carbon nanotubes (MWCNTs) were explored in this study. Transmission electron microscopy was used to examine the morphology of MWCNT. Fourier transform infrared spectroscopy was used to identify the functional groups on the MWCNT surface (FTIR). For each adsorption isotherm data point, duplicate samples were taken, and for each pH effect data point and the ionic strength trials, triplicate samples were taken. NaCl was used to regulate the ionic strength effect in the studies. The pH of bisphenol is used in adsorption isotherm investigations. At each temperature, the thermodynamic parameters for the adsorption process were obtained. The recovered BPF was determined by eluting the BPF absorbed by MWCNT with 10 mL ethanol/1.0 M sodium hydrate solution (7:3, v/v). Each adsorption isotherm data point received duplicate samples, while the pH impact data points and ionic strength tests received triplicate samples. The effluents were collected from a municipal wastewater treatment facility. To investigate the absorbability of BPF, three MWCNTs were used as sorbents. The adsorption parameters of each sorbent were tuned to make their BPF adsorption efficiency as high as possible. MWCNT adsorption efficiency of BPF was higher than MWCNT–COOH and MWCNT–OH. MWCNT has an adsorption efficiency of 98.88 percent, MWCNT–COOH has an adsorption efficiency of 87.11 percent, and MWCNT–OH has an adsorption efficiency of 71.09 percent. The total rate processes appeared to be influenced by both intraparticle diffusion and external mass transfer, but mostly driven by external mass transfer, according to the adsorption mechanism analysis (Zhang et al. 2013).

A recent study deals with Fe_3O_4 nanoparticles on C100 (polymeric cationic resin) for removal and recovery of phosphate from water. The Fe_3O_4 nanoparticles are immobilised on C100 (polymeric matrix) using a variation of the Intermatrix Synthesis (IMS) protocol paired with the co-precipitation method to obtain magnetic nanocomposites (NCs). SEM-EDS images showed the immobilisation of the Fe_3O_4 NPs on the NC surface. It was found that increased NP concentration on the NC leads to a decrease in the adsorption capacity. It was so because the surface/volume ratio of NPs decreased due to an excess of NPs. It was determined that 23.59 mg_{Fe}/g_{NC} is the optimum iron content for phosphate adsorption. The optimum pH for phosphate adsorption is pH 7 because of maximum adsorption capacity achieved and NC stability. It was found that NaOH (0.5 M) was the most effective for the

Table 2: Other nanomaterials for the recovery of resources from wastewater.

Sr. No.	Type of nanoparticle	Recovered resource	Recovery efficiency	Maximum adsorption capacity (mg/g)	Source of wastewater	References
1	Multi-walled carbon nanotubes (MWCNTs)	Bisphenol F (BPF)	95%	83.43	Municipal wastewater	(Zhang et al. 2013)
2	Nanocomposite of immobilized magnetite nanoparticles on cationic polymer	Phosphate	97.50%	> 4	Synthetic phosphate solutions	(Abo Markeb et al. 2016)
3	Hydrated ferric oxide nanoparticles	Phosphate	80%	23.55	Municipal wastewater	(Martin et al. 2009)
4	Biochar	Phosphorus	1.77% (Max)	326.63	Agricultural wastewater	(Fang et al. 2015)
5	Polymeric anion exchanger with iron oxide nanoparticles	Phosphorus	> 95%			(Dong et al. 2020)
6	HAIX, DOW-HFO, & DOW-HFO-Cu.	Phosphorus		23		(Sengupta and Pandit 2011)
7	Magnetic Fe_3O_4 nanoparticles	Erythromycin	97%	44.1	Industrial wastewater	(Zhang et al. 2016)
8	$Fe_xCo_{3-x}O_3$ nanoparticles	Congo Red Dye		128.6	Industrial wastewater	(Liu et al. 2019)
9	Hybrid nanostructures	Lanthanides	80%	16.2 (La), 17.9 (Eu), 23.0 (Tb)		(Afonso et al. 2019)
10	Magnetic Fe_3O_4 nanocrystals	Oil		23.01		(Ahamad et al. 2019)
11	Janus nanoparticles (M-Janus NPs)	Oil	91.50%			(He et al. 2020)
12	Alumina nanoparticles	Insecticides	80%	25.54	Tap and lake water	(Derbalah et al. 2015)

desorption process. 97.5 percent of phosphate was recovered after three washing cycles. It can be said that magnetite NPs immobilised on the cationic polymer have immense potential in phosphate recovery (Abo Markeb et al. 2016).

Martin et al. used a polymeric anion exchanger combined with hydrated ferric oxide nanoparticles for the removal and recovery of phosphate from municipal wastewater. The conventional ion-exchange method for phosphate removal using a hybrid anion exchanger (HAIX) is not quite efficient, but with the use of hydrated ferric oxide (HFO) nanoparticles, the adsorption capacity increases significantly. The material is the HAIX into which the HFO nanoparticles are dispersed. This combination was trialled with real effluents in fixed bed mini columns. The results showed that HAIX resin removed a significant amount of phosphate and reached its capacity after treating 100 bed volumes. The experiments demonstrated that phosphate was preferred in adsorption, and none of the other ions were eluted in the experiment. The regeneration study indicated that from a stream with phosphate concentration of 1250 mg/L, 80 percent of phosphorus was eluted in the first bed volume. The performance of HAIX resin did not deteriorate even after 10 partial regenerations, and the phosphate elution levels were consistently high. This regeneration capacity means that this process has a rather small operational footprint. The findings of this study suggest the used media is rich in potential for phosphate removal and recovery from wastewater streams (Martin et al. 2009).

Calcium and magnesium-loaded biochar (Ca–Mg/biochar) was created for P recycling in one such study. The characterization, adsorption efficiency, and selectivity of the Ca–Mg/biochar were investigated at 300°C (Ca–Mg/B300), 450°C (Ca–Mg/B450), and 600°C (Ca–Mg/B600). The availability of P in Ca–Mg/biochar after post sorption was examined. The synthesized Ca–Mg/biochar was rich in organic functional groups of hydroxyls, carboxyl, carbonyl, and methoxyl, which were useful for adsorption, according to FT-IR analysis in this work. The number of mesoporous structures grew as the synthesis temperature of Ca–Mg/biochar rose, and the distribution of mesoporous structures changed from ordered to disordered; according to TEM examination for Ca–Mg/biochar adsorption, the quantity and distribution of mesopores are critical. The mesoporous structures of Ca–Mg/biochar contained many nanoparticles. The adsorption of P by Ca–Mg/B300 and Ca–Mg/B450 was investigated. Nano-CaO and nano-MgO particles enhanced the P adsorption of Ca–Mg/biochar, which achieved adsorption equilibrium after 360 minutes. It was endothermic, spontaneous, and exhibited a rise in solid–liquid interface disorder. The highest levels of P adsorption were 294.22, 315.33, and 326.63 mg/g. The average pH level of biogas fermentation liquid had no significant effect on the P adsorption selectivity of Ca–Mg/biochar. The nano-CaO and nano-MgO particles in the biogas fermentation liquid might lessen the detrimental interaction effects of coexisting ions (Fang et al. 2015).

Phosphate removal and recovery from wastewater can help to reduce negative environmental effects while also supplementing fertilizer supplies. Phosphate can be removed from complicated wastewaters and concentrated phosphate solutions recovered using hybrid anion exchangers (HAIX) with doped ferric oxide nanoparticles (FeO NPs). Study combined HAIX with a weak acid cation exchanger (WAC) to enhance phosphate and calcium in moderate regenerants and precipitate both elements for recovery. This work catalyses the design of fit-for-purpose adsorbents favourable to resource-efficient electrochemical regeneration by enhancing understanding and demonstration of a hybrid ion exchange-electrochemical process. The following are some of the key findings of the study. The pH-dependent capacity of both hybrid anion exchange (HAIX) and weak acid cation exchange (WAC) resins may be regenerated using mild regenerants. Electrochemical water splitting was used to create mild regenerants with minimum chemical and energy inputs. In batch experiments, the majority of HAIX and WAC adsorption capacity (50 percent HAIX, 80 percent WAC) was maintained for numerous cycles. Using pH 11 catholyte, FeO np dramatically improved HAIX regeneration efficiency. When utilizing moderate regenerants, the length of the intraparticle diffusion channel influenced column regeneration effectiveness (Dong et al. 2020).

Another study describes a method for selectively removing P from wastewater and recovering it as a solid-phase fertilizer using a polymeric anion exchanger coated with iron oxide nanoparticles. HAIX, DOW-HFO, and DOW-HFO-Cu were three examples of hybrid materials that have been researched. Batch equilibrium tests were conducted in batch reactors by adding a known mass of the exchanger (varying from 0.5 g to 1.0 g) to a known volume of a synthetic solution. As a P source, Na_2HPO_4 was employed. The content of P in the sample ranged from 5.0 to 75.0 mg/L. The assays were carried out at pH 5.0 and 8.0 to see how pH affected the uptake of orthophosphate phosphorus. The three resins were tested at room temperature to determine their equilibrium P removal capability. To achieve appropriate mixing and balance, the reactors were stirred with a magnetic stirrer at 400 rpm for 48 h. The study's main findings are that all the three materials investigated are susceptible to effective regeneration. Within ten bed volumes, a single step regeneration using 2.5 percent sodium chloride and 2.0 percent sodium hydroxide reliably recovered over 95.0 percent of sorbed orthophosphate phosphorus. After ten cycles of exhaustion-regeneration, just a modest capacity reduction (about 1.5 percent) was found. After compensating for the hydroxide lost during regeneration, the spent regenerant can be reused. The phosphate free spent regenerant could be regenerated with the performance closer to that of fresh regenerant (Fig. 3). After adding calcium nitrate or a mix of ammonium chloride and magnesium sulphate to the wasted regenerant, orthophosphate phosphorus can be recovered as calcium phosphate or magnesium ammonium phosphate (Sengupta and Pandit 2011).

In a recent study, separation, and the recovery of erythromycin from wastewater by using imprinted magnetic nanoparticles was attempted. ERY-surface imprinted magnetic nanoparticles (Fe_3O_4) were synthesised by using Fe_3O_4 nanoparticles as the matrix and acryloyl modified β-CD (β-cyclodextrin) as a functional monomer, and ethyleneglycol dimethacrylate (EDGMA) as the cross-linking agent. Multi-β-CD molecules were made to react with a single template molecule for the preparation of imprinted β-CD polymers. Initiator modified Fe_3O_4 nanoparticles were prepared followed by the synthesis of Fe_3O_4@MIP. The Fe_3O_4@MIPs were mixed with varying concentrations of mother liquor for 12 h, and then magnetically separated. Ethanol/acetic acid was used for elution. The results of NMR spectroscopy indicated that a single ERY molecule could interact with three cyclodextrin (CD) molecules. With the β-CD to ERY ratio at 3:1, the Fe_3O_4@MIPs showed high adsorption capacity

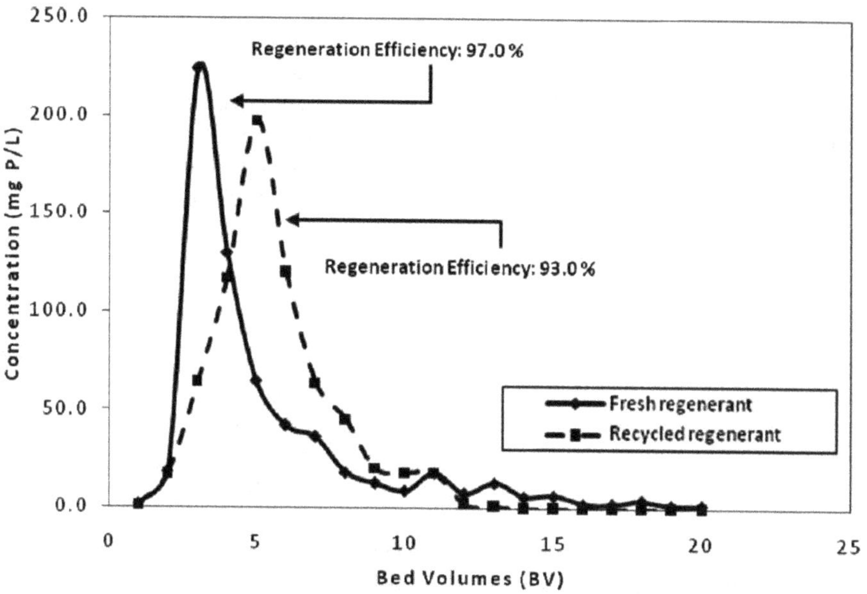

Fig. 3. Comparison of the regeneration efficiency of the fresh and recycled regenerant for HAIX column. Reproduced with permission from Sengupta, S. and Pandit, A. 2011. Selective removal of phosphorus from wastewater combined with its recovery as a solid-phase fertilizer. Water Research. 45(11): 3318–3330. https://doi.org/10.1016/j.watres.2011.03.044.

and maximum imprinting efficiency for ERY. The Fe_3O_4@MIPs showed the highest recovery of 97.34 percent. The ERY samples showed a purity of 89.23 percent as determined by HPLC. This study indicates Fe_3O_4@MIPs to be effective for ERY recovery from wastewater (Zhang et al. 2016).

Metal oxide nanoparticles were used for the adsorption of dyes present in textile and industrial wastewater. In this study, $Fe_xCo_{3-x}O_4$ nanoparticles are used for the adsorption of Congo Red. The $Fe_xCo_{3-x}O_4$ nanoparticles are prepared as per the techniques of (Hu et al. 2013) and (Wei et al. 2015). The synthesised nanoparticles were cubical and of 40–80 nm in size. The 5 percent Fe $Fe_xCo_{3-x}O_4$ nanoparticles were characterised with high pore size, which is suitable for adsorption of Congo Red, and it was experimentally found that 5 percent of nanoparticles exhibited the best Congo Red adsorption rate. The adsorption capacity of 0.05 g of 5 percent Fe $Fe_xCo_{3-x}O_4$ nanoparticles was determined to be 128.6 mg/g. It was observed that the maximum rate was achieved when a small amount of adsorbent was used. The adsorption data fit well with both Langmuir and Freundlich isotherms. It was found that the $Fe_xCo_{3-x}O_4$ (5 percent Fe) nanoparticles had good cyclic adsorption capacity and reusability. This study presented $Fe_xCo_{3-x}O_4$ (5 percent Fe) nanoparticles to be effective for the removal and recovery of Congo Red dye from wastewaters (Liu et al. 2019).

The synthesis, characterization, and application of hybrid nanostructures to extract and recover lanthanides from water were described in one such study, thereby enabling the recycling of these high-value elements. The nanocomposite combines the intriguing features of graphite nanoplatelets (GNPs) with the magnetic capabilities of magnetite, and it has good sorption properties for La (III), Eu (III), and Tb (III).

Characterization of the nanocomposite using TEM depicts the existence of spherical shape of $FeSO_4$ NPs, average size below 100 nm, and attached on the surface of GNPs (Fig. 4). All sorption experiments were conducted in unary or ternary solutions of La (III), Eu (III), and Tb (III), with an initial element concentration of 100 g/L, which was chosen based on lanthanide concentrations in rivers and waste streams. By adding the appropriate volume of certified reference standard solutions to ultra-pure water (UPW), the unary and ternary solutions were created. Batch tests were carried out in Schott glass flasks at 22°C for a maximum of 24 h, with mechanical stirring (250 rpm). The ability of the composite to remove TCE from water was tested by exposing 5 mg of Fe_3O_4/GNPs to 100 mL of ternary solutions of La (III), Eu (III), and Tb (III) for varied pH levels. The results reveal that electrostatic interactions dominate the sorption mechanism, making this technique effective for pH values greater than 5.2, the composite's experimental isoelectric point. At pH 7.5 and 2 mg/L composite, the greatest solid loading attained was 23 mg/g. The Fe_3O_4/GNPs removed Eu (III) and Tb (III) species preferentially in ternary TCE solutions relative to La (III), but this type of sorption affinity was not as noticeable in unary systems (Afonso et al. 2019).

In another study, oil recovery was done using a nanocomposite based on lignin prepared from date palm pits by adsorption. For oil removal and recovery from an aqueous medium, the adsorbent must be oleophilic and hydrophobic. It is known that the date palm pit is a source of carbon with a high surface area, suitable pore structure, and high adsorption capacity (Ahmed 2016). A novel adsorbent was fabricated with super-hydrophobic and oleophilic properties, microporous structure along with magnetic behaviour, and a magnetic macro porous carbon (Fe_3O_4@MPC). The SEM image of Fe_3O_4@MPC showed that the Fe_3O_4 nanoparticles were embedded in the carbon (graphite) matrix. The average size of the nanocomposite (Fe_3O_4@MPC) was determined to be approximately 22 nm using a TEM image. The pore size was found to be 8–10 nm. The porous morphology of the nanocomposite was favourable for oil recovery. The results of adsorption experiments showed that more than 85 percent of the total oils were adsorbed onto the Fe_3O_4@MPC surface in the first 30 minutes.

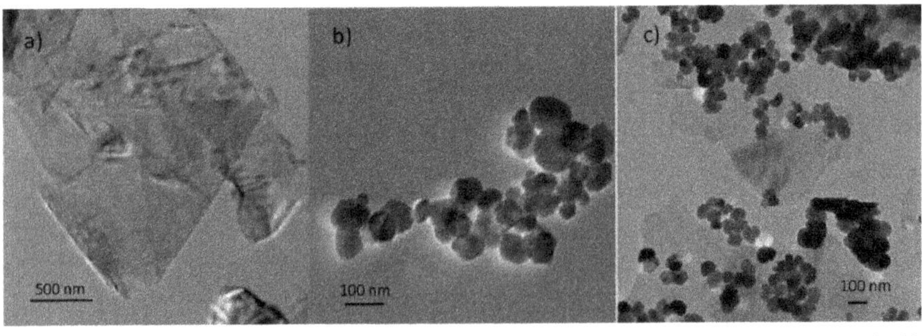

Fig. 4: TEM images of graphite nanoplatelets (a), Fe3O4 (b) and Fe3O4/GNPs (c). Reproduced with permission from Afonso, E.L., Carvalho, L., Fateixa, S., Amorim, C.O., Amaral, V.S., Vale, C., Pereira, E., Silva, C.M., Trindade, T. and Lopes, C.B. 2019. Can contaminated waters or wastewater be alternative sources for technology-critical elements? The case of removal and recovery of lanthanides. Journal of Hazardous Materials. 380 (June): 120845. https://doi.org/10.1016/j.jhazmat.2019.120845.

The maximum adsorption capacity of 22.37 g/g was recorded in 20/50, diesel with deionised water. After adsorption, the oil filled Fe_3O_4@MPCs were treated with ethanol to release the adsorbed oils, and the Fe_3O_4@MPCs were dried and used again. The nanocomposite exhibited excellent reusability without reduction in adsorption capacity. The findings in this study put forward the adsorption by Fe_3O_4@MPC nanocomposites as an efficient and economical technique for oil recovery from wastewaters (Ahamad et al. 2019).

In a similar study, Janus particles have been considered for oil removal and recovery from wastewaters. Janus particles possess asymmetric surface wettability and are more interfacially active than the particles with uniform wettability and, hence, are better for separating emulsified oil droplets. Based on this, magnetic-Janus nanoparticles (M-Janus NPs) were fabricated by controlled adsorption of two natural cellulosic materials having opposite nature of solubilities-hydrophilic and hydrophobic on magnetite NP surface (He et al. 2019). FE-SEM images revealed that M-Janus NPs formed "micelle-like" structures with an average diameter of 183 nm. To test the oil removal and recovery properties of M-Janus NPs, model oily wastewaters (surfactant stabilised crude oil-in-water and cooking oil-in-water emulsions) were used. It was found that M-Janus NPs deposited at the oil-water interface and hence lead to efficient oil removal and recovery from the wastewater. It was found that M-Janus NPs have firmer anchoring at the oil-water interface and superior interfacial activities. When M-Janus NPs were added to oily wastewaters, they anchor onto the oil droplet surfaces. The oil droplets tagged by M-Janus NPs were separated using a permanent magnet, leaving a clear aqueous phase. The removal/recovery of M-Janus was found to be > 91.5 percent. The M-Janus NPs were proven to be reusable and exhibited high removal/recovery efficiency for at least 5 cycles of regeneration. This study indicates M-Janus NPs to be a promising candidate for oil removal and recovery from oily wastewaters, which is sustainable, low cost and efficient (He et al. 2020).

For the removal and recovery of insecticide like Lindane from water samples, mesoporous alumina nanoparticles were used in a study. Mesoporous alumina nanoparticles (MA-NPs) were preferred because of turnable NPs, large surface area-volume ratio and open cylinder pores to adsorb organic molecules (Lindane). The formation of uniformly sized (100–250 nm) spherical NPs was observed by field-emission scanning electron microscopy (FESEM). NH_3-TPD (temperature-programmed desorption) and ^{27}Al NMR analyses showed that active acid on MA-NPs attracted Lindane molecules maximally at pH 4.5. It was found that the adsorption capacity of MA-NPs decreases with an increase in temperature. The high initial concentration of Lindane results in high adsorption. Increased dosage of MA-NPs leads to increased active sites which mean increased adsorption. Stripping agent like ethanol was used for collection or removal of Lindane from MA-NPs. MA-NPs can be reused for several cycles without incurring mesostructural change or damage. The findings in this study prove MA-NPs to be useful for the removal and recovery of organic pollutants (here Lindane) from real water sources (Derbalah et al. 2015).

5.4 Conclusion

In this chapter an attempt has been made to highlight the efficacy of nanomaterials like CNTs, MNPs, GNPs, nZVI, nano silica, nanocrystal, carbon nanosphere, and various nanocomposites in resource recovery from wastewater. Recovery of resources like Cesium (I) ions, Strontium (II) ions, lead, thallium, copper, Cadmium (II), molybdenum, Chromium (VI) and (III), Pb (II) and Mg (II), Au (I) and (II), Pd (II), Ag (II), Zn (II), Ni (II), Ni (II), neodymium, Hg (II), uranium, etc., has been achieved significantly. Studies also signify the recovery of resources like bisphenol F, phosphate, phosphorus, erythromycin, Congo Red Dye, lanthanides, oil, insecticides, etc., by employing nanomaterials like MWCNTs, nanocomposite of immobilized magnetite nanoparticles on cationic polymer, hydrated ferric oxide nanoparticles, biochar, polymeric anion exchanger with iron oxide nanoparticles, HAIX, DOW-HFO, and DOW-HFO-Cu, magnetic Fe_3O_4 nanoparticles, FexCo3-xO3 nanoparticles, hybrid nanostructures, M-Janus NPs, alumina nanoparticles, etc. With the progress in the field, successful reports are available on improving the performance of resource recovery by regenerant reuse. However, vast array of resources which are still being wasted in wastewater should be recovered using novel nano materials or composites. Optimization of resource recovery using these novel nanomaterials with highest performance level is necessary.

Abbreviations

3DOM	-	Three-dimensionally ordered macroporous nanomaterial
AA	-	Acrylic acid
AAS	-	Atomic absorption spectroscopy
AM	-	Acrylamide
AMD	-	Acid mine drainage
BM	-	Bulk mixing
BPF	-	Bisphenol F
CNFM	-	Chitin nanofibrous membrane
CNS	-	Carbon nanosphere
CNT	-	Carbon nanotube
CS	-	Chitosan
CTO	-	3-chloropropyltriethoxysilane
DE	-	Diatomaceous earth
DLS	-	Dynamic Light Scattering
DT	-	Dithiocarbamate
DT-MCM-41	-	Magnetic di-thio functionalized mesoporous silica nanoparticle
EDS/XEDS	-	Energy dispersive X-ray spectrometry
EGDMA	-	Ethylene glycol dimethacrylate
FESEM	-	Field Emission Scanning Electron Microscope
FT-IR	-	Fourier Transform Infrared Spectroscopy
GA	-	Gum Arabic
GNPS	-	Graphite Nanoplatelets

GO	-	Graphite oxide
HAIX	-	Hybrid anion exchanger
HF	-	Hydrofluoric acid
HFO	-	Hydrated ferric oxide
HRTEM	-	High-resolution transmission electron microscopy
HR-XPS	-	High-resolution X-ray photoelectron spectroscopy
ICPMS	-	Inductively coupled plasma mass spectrometry
ICP-OES	-	Inductively coupled plasma - optical emission spectrometry
ICPOES	-	Inductively coupled plasma optical emission spectrometry
MA-NP	-	Mesoporous alumina nanoparticle
MF	-	Microfluidic
MNA	-	Magnetic nanoadsorbent
MNHA	-	Magnetically modified hydroxyapatite nanoparticle
MNP-OPP	-	Magnetic nanoparticle surface modified with orange peel powder
MPC	-	Magnetic macroporous carbon
MPiX	-	Magnetic potential-responsive ion-exchange nanomaterial
MWCNT	-	Multi-walled carbon nanotube
MWCNT-TD2	-	Triazole dendrimer-functionalized Multi-walled carbon nanotube
MWCNT-TD2P	-	α-aminophosphonate derivative of MWCNT-TD2
NiHCF	-	Nickel hexacyanoferrate
NMR	-	Nuclear magnetic resonance
NP	-	Nanoparticle
nZVI	-	nanoscale zero-valent iron particles
PDMC	-	Poly methacryloxyethyltrimethyl ammonium chloride
QMNP	-	Quercetin-coated nanoparticle
RAFT	-	Reversible addition-fragmentation chain transfer
RMNP	-	Rhein-coated magnetic Fe_3O_4 nanoparticle
SEM	-	Scanning electron microscope
Si-NP	-	Silica nanoparticle
SNP	-	Silica nanoparticle
SWCNT	-	Single-walled carbon nanotube
TEM	-	Transmissions electron microscope
TiNT	-	Titanate nanotube
TPD	-	Temperature-programmed desorption
UA-MPSE	-	Ultrasonic-assisted magnetic solid phase extraction
UPW	-	Ultra Pure Water
VSM	-	Vibrating sampling magnetometry
WAC	-	Weak acid cation exchanger
XANES	-	X-ray absorption near edge structure
xGnP	-	Graphene Nanoplatelets
XPS	-	X-ray photoelectron spectroscopy
XRD	-	X-ray diffraction
β-CD	-	β-cyclodextrin

References

Abdel Ghafar, H.H., Ali, G.A.M., Fouad, O.A. and Makhlouf, S.A. 2015. Enhancement of adsorption efficiency of methylene blue on Co3O4/SiO2 nanocomposite. Desalination and Water Treatment. 53(11): 2980–2989. https://doi.org/10.1080/19443994.2013.871343.

Abo Markeb, A., Alonso, A., Dorado, A.D., Sánchez, A. and Font, X. 2016. Phosphate removal and recovery from water using nanocomposite of immobilized magnetite nanoparticles on cationic polymer. Environmental Technology (United Kingdom). 37(16): 2099–2112. https://doi.org/10.1080/09593330.2016.1141999.

Afonso, E.L., Carvalho, L., Fateixa, S., Amorim, C.O., Amaral, V.S., Vale, C. et al. 2019. Can contaminated waters or wastewater be alternative sources for technology-critical elements? The case of removal and recovery of lanthanides. Journal of Hazardous Materials. 380(June): 120845. https://doi.org/10.1016/j.jhazmat.2019.120845.

Agarwal, S., Sadegh, H., Monajjemi, M., Hamdy, A.S., Ali, G.A.M., Memar, A.O.H. et al. 2016. Efficient removal of toxic bromothymol blue and methylene blue from wastewater by polyvinyl alcohol. Journal of Molecular Liquids. 218: 191–197. https://doi.org/10.1016/j.molliq.2016.02.060.

Ahamad, T., Naushad, M., Ruksana and Alshehri, S.M. 2019. Ultra-fast spill oil recovery using a mesoporous lignin based nanocomposite prepared from date palm pits (*Phoenix dactylifera* L.). International Journal of Biological Macromolecules. 130: 139–147. https://doi.org/10.1016/j.ijbiomac.2019.02.038.

Ahmed, M.J. 2016. Preparation of activated carbons from date (*Phoenix dactylifera* L.) palm stones and application for wastewater treatments: Review. Process Safety and Environmental Protection. 102: 168–182. https://doi.org/10.1016/j.psep.2016.03.010.

Ajayan, P.M. and Ebbesen, T.W. 1997. Nanometre-size tubes of carbon. Reports on Progress in Physics. 60(10): 1025–1062. https://doi.org/10.1088/0034-4885/60/10/001.

Ali, E., Hadis, D., Hamzeh, K., Mohammad, K., Nosratollah, Z., Abolfazl, A. et al. 2014. Carbon nanotubes: properties, synthesis, purification, and medical applications. Nanoscale Research Letters. 9(1): 393.

Apul, O.G., Wang, Q., Zhou, Y. and Karanfil, T. 2013. Adsorption of aromatic organic contaminants by graphene nanosheets: Comparison with carbon nanotubes and activated carbon. Water Research. 47(4): 1648–1654. https://doi.org/10.1016/j.watres.2012.12.031.

Arias, M., Barral, M.T. and Mejuto, J.C. 2002. Enhancement of copper and cadmium adsorption on kaolin by the presence of humic acids. Chemosphere. 48(10): 1081–1088. https://doi.org/10.1016/S0045-6535(02)00169-8.

Baby, J., Raj, J., Biby, E., Sankarganesh, P., Jeevitha, M., Ajisha, S. et al. 2011. Toxic effect of heavy metals on aquatic environment. International Journal of Biological and Chemical Sciences. 4(4): 939–952. https://doi.org/10.4314/ijbcs.v4i4.62976.

Balandin, A.A., Ghosh, S., Bao, W., Calizo, I., Teweldebrhan, D., Miao, F. et al. 2008. Superior thermal conductivity of single-layer graphene. Nano Letters. 8(3): 902–907. https://doi.org/10.1021/nl0731872.

Banerjee, S.S. and Chen, D.H. 2007. Fast removal of copper ions by gum arabic modified magnetic nano-adsorbent. Journal of Hazardous Materials. 147(3): 792–799. https://doi.org/10.1016/j.jhazmat.2007.01.079.

Biškup, B. and Subotić, B. 2005. Removal of heavy metal ions from solutions using Zeolites. III. Influence of Sodium Ion Concentration in the Liquid Phase on the Kinetics of Exchange Processes between Cadmium Ions from Solution and Sodium Ions from Zeolite A. Separation Science and Technology. 39(4): 925–940. https://doi.org/10.1081/ss-120028454.

Chu, W. 1999. Lead metal removal by recycled alum sludge. Water Research. 33(13): 3019–3025. https://doi.org/10.1016/S0043-1354(99)00010-X.

Crane, R.A. and Sapsford, D.J. 2018. Towards "Precision Mining" of wastewater: Selective recovery of Cu from acid mine drainage onto diatomite supported nanoscale zerovalent iron particles. Chemosphere. 202: 339–348. https://doi.org/10.1016/j.chemosphere.2018.03.042.

Das, N. 2010. Recovery of precious metals through biosorption - A review. Hydrometallurgy. 103(1–4): 180–189. https://doi.org/10.1016/j.hydromet.2010.03.016.

Derbalah, A., El-Safty, S.A., Shenashen, M.A. and Abdel Ghany, N.A. 2015. Mesoporous alumina nanoparticles as host tunnel-like pores for removal and recovery of insecticides from environmental samples. ChemPlusChem. 80(7): 1119–1126. https://doi.org/10.1002/cplu.201500098.

Dong, H., Wei, L. and Tarpeh, W.A. 2020. Electro-assisted regeneration of pH-sensitive ion exchangers for sustainable phosphate removal and recovery. Water Research. 184: 116167. https://doi.org/10.1016/j.watres.2020.116167.

Ekmekyapar, F., Aslan, A., Bayhan, Y.K. and Cakici, A. 2006. Biosorption of copper(II) by nonliving lichen biomass of *Cladonia rangiformis hoffm*. Journal of Hazardous Materials. 137(1): 293–298. https://doi.org/10.1016/j.jhazmat.2006.02.003.

El-Maghrabi, H.H., Younes, A.A., Salem, A.R., Rabie, K. and El-shereafy, E. sayed. 2019. Magnetically modified hydroxyapatite nanoparticles for the removal of uranium (VI): Preparation, characterization and adsorption optimization. Journal of Hazardous Materials. 378(October 2018): 120703. https://doi.org/10.1016/j.jhazmat.2019.05.096.

Ernhart, C.B. 1992. A critical review of low-level prenatal lead exposure in the human: 1. Effects on the fetus and newborn. Reproductive Toxicology. 6(1): 9–19. https://doi.org/10.1016/0890-6238(92)90017-N.

Fang, C., Zhang, T., Li, P., Jiang, R., Wu, S., Nie, H. et al. 2015. Phosphorus recovery from biogas fermentation liquid by Ca-Mg loaded biochar. Journal of Environmental Sciences (China). 29: 106–114. https://doi.org/10.1016/j.jes.2014.08.019.

Feng, N., Guo, X. and Liang, S. 2009. Adsorption study of copper (II) by chemically modified orange peel. Journal of Hazardous Materials. 164(2-3): 1286–1292. https://doi.org/10.1016/j.jhazmat.2008.09.096.

Fu, L., Zhang, L., Wang, S., Zhang, B. and Peng, J. 2017. Selective recovery of Au(III) from aqueous solutions by nanosilica grafted with cationic polymer: Kinetics and isotherm. Journal of the Taiwan Institute of Chemical Engineers. 80: 342–348. https://doi.org/10.1016/j.jtice.2017.07.020.

Gautam, R.K. and Chattopadhyaya, M.C. 2016. Desorption, regeneration, and reuse of nanomaterials. Nanomaterials for Wastewater Remediation. 297–309. https://doi.org/10.1016/b978-0-12-804609-8.00012-1.

Geng, Y., Wang, S.J. and Kim, J.K. 2009. Preparation of graphite nanoplatelets and graphene sheets. Journal of Colloid and Interface Science. 336(2): 592–598. https://doi.org/10.1016/j.jcis.2009.04.005.

Grancharov, S.G., Zeng, H., Sun, S., Wang, S.X., O'Brien, S., Murray, C.B. et al. 2005. Bio-functionalization of monodisperse magnetic nanoparticles and their use as biomolecular labels in a magnetic tunnel junction based sensor. Journal of Physical Chemistry B. 109(26): 13030–13035. https://doi.org/10.1021/jp051098c.

Gugushe, A.S., Mpupa, A. and Nomngongo, P.N. 2019. Ultrasound-assisted magnetic solid phase extraction of lead and thallium in complex environmental samples using magnetic multi-walled carbon nanotubes/zeolite nanocomposite. Microchemical Journal, 149(May). 103960. https://doi.org/10.1016/j.microc.2019.05.060.

Guler, U.A. and Ersan, M. 2016. *S. cerevisiae* cells modified with nZVI: a novel magnetic biosorbent for nickel removal from aqueous solutions. Desalination and Water Treatment. 57(16): 7196–7208. https://doi.org/10.1080/19443994.2015.1013992.

Gupta, V.K. and Nayak, A. 2012. Cadmium removal and recovery from aqueous solutions by novel adsorbents prepared from orange peel and Fe 2O 3 nanoparticles. Chemical Engineering Journal. 180: 81–90. https://doi.org/10.1016/j.cej.2011.11.006.

Gupta, Vinod Kumar, Tyagi, I., Agarwal, S., Sadegh, H., Shahryari-Ghoshekandi, R., Yari, M. et al. 2015. Experimental study of surfaces of hydrogel polymers HEMA, HEMA-EEMA-MA, and PVA as adsorbent for removal of azo dyes from liquid phase. Journal of Molecular Liquids. 206: 129–136. https://doi.org/10.1016/j.molliq.2015.02.015.

He, X., Liang, C., Liu, Q. and Xu, Z. 2019. Magnetically responsive Janus nanoparticles synthesized using cellulosic materials for enhanced phase separation in oily wastewaters and water-in-crude oil emulsions. Chemical Engineering Journal, 378(June). 122045. https://doi.org/10.1016/j.cej.2019.122045.

He, X., Liu, Q. and Xu, Z. 2020. Treatment of oily wastewaters using magnetic Janus nanoparticles of asymmetric surface wettability. Journal of Colloid and Interface Science. 568: 207–220. https://doi.org/10.1016/j.jcis.2020.02.019.

Ho, Y.S. and McKay, G. 1998. Sorption of dye from aqueous solution by peat. Chemical Engineering Journal. 70(2): 115–124. https://doi.org/10.1016/S1385-8947(98)00076-X.

Hosseinzadeh, H., Hosseinzadeh, S. and Pashaei, S. 2019. Fabrication of novel magnetic graphene oxide nanocomposites for selective adsorption of mercury from aqueous solutions. Environmental Science and Pollution Research. 26(26): 26807–26821. https://doi.org/10.1007/s11356-019-05918-0.

Hu, J., Chen, G. and Lo, I.M.C. 2005a. Removal and recovery of Cr(VI) from wastewater by maghemite nanoparticles. Water Research. 39(18): 4528–4536. https://doi.org/10.1016/j.watres.2005.05.051.

Hu, J., Lo, I.M.C. and Chen, G. 2005b. Fast removal and recovery of Cr(VI) using surface-modified jacobsite (MnFe2O4) nanoparticles. Langmuir. 21(24): 11173–11179. https://doi.org/10.1021/la051076h.

Hu, L., Huang, Y. and Chen, Q. 2013. FexCo3-xO4 nanoporous particles stemmed from metal-organic frameworks Fe3[Co(CN)6] 2: A highly efficient material for removal of organic dyes from water. Journal of Alloys and Compounds. 559: 57–63. https://doi.org/10.1016/j.jallcom.2013.01.095.

Iannazzo, D., Pistone, A., Ziccarelli, I., Espro, C., Galvagno, S., Giofré, S.V. et al. 2017. Removal of heavy metal ions from wastewaters using dendrimer-functionalized multi-walled carbon nanotubes. Environmental Science and Pollution Research. 24(17): 14735–14747. https://doi.org/10.1007/s11356-017-9086-2.

Ige, P.P., Baria, R.K. and Gattani, S.G. 2013. Fabrication of fenofibrate nanocrystals by probe sonication method for enhancement of dissolution rate and oral bioavailability. Colloids and Surfaces B: Biointerfaces. 108: 366–373. https://doi.org/10.1016/j.colsurfb.2013.02.043.

Jahangir, M.A., Imam, S.S., Muheem, A., Chettupalli, A., Al-Abbasi, F.A., Nadeem, M.S. et al. 2020. Nanocrystals: characterization overview, applications in drug delivery, and their toxicity concerns. Journal of Pharmaceutical Innovation. https://doi.org/10.1007/s12247-020-09499-1.

Jang, B.Z. and Zhamu, A. 2008. Processing of nanographene platelets (NGPs) and NGP nanocomposites: A review. Journal of Materials Science. 43(15): 5092–5101. https://doi.org/10.1007/s10853-008-2755-2.

Jing, J. and Shi, C. 2020. Reusable magnetic nanoparticle immobilized nitrogen-containing ligand for classified and easy recovery of heavy metal ions. Molecules. 25(14): 1–11. https://doi.org/10.3390/molecules25143204.

Kalaitzidou, K., Fukushima, H. and Drzal, L.T. 2007. Multifunctional polypropylene composites produced by incorporation of exfoliated graphite nanoplatelets. Carbon. 45(7): 1446–1452. https://doi.org/10.1016/j.carbon.2007.03.029.

Khlifi, R. and Hamza-Chaffai, A. 2010. Head and neck cancer due to heavy metal exposure via tobacco smoking and professional exposure: A review. Toxicology and Applied Pharmacology. 248(2): 71–88. https://doi.org/10.1016/j.taap.2010.08.003.

Kristianto, H., Putra, C.D., Arie, A.A., Halim, M. and Lee, J.K. 2015. Synthesis and characterization of carbon nanospheres using cooking palm oil as natural precursors onto activated carbon support. Procedia Chemistry. 16: 328–333. https://doi.org/10.1016/j.proche.2015.12.060.

Kurniawan, T.A. and Lo, W. hung. 2009. Removal of refractory compounds from stabilized landfill leachate using an integrated H2O2 oxidation and granular activated carbon (GAC) adsorption treatment. Water Research. 43(16): 4079–4091. https://doi.org/10.1016/j.watres.2009.06.060.

Li, X.Q. and Zhang, W.X. 2007. Sequestration of metal cations with zerovalent iron nanoparticles—A study with high resolution x-ray photoelectron spectroscopy (HR-XPS). Journal of Physical Chemistry C. 111(19): 6939–6946. https://doi.org/10.1021/jp0702189.

Li, Y.J., Yang, Z.M., Chen, Y.C., Huang, L. and Tang, H.Y. 2019. Adsorption, reclaim, and regeneration of Cd by magnetic calcium dihydrogen phosphate nanoparticles. Huanjing Kexue/Environmental Science. 40(4): 1849–1856. https://doi.org/10.13227/j.hjkx.201808020.

Ling, L. and Zhang, W.X. 2015. Enrichment and encapsulation of uranium with iron nanoparticle. Journal of the American Chemical Society. 137(8): 2788–2791. https://doi.org/10.1021/ja510488r.

Ling, L., Huang, X.Y. and Zhang, W.X. 2018. Enrichment of precious metals from wastewater with core–shell nanoparticles of iron. Advanced Materials. 30(17): 1–6. https://doi.org/10.1002/adma.201705703.

Liu, J., Wang, N., Zhang, H. and Baeyens, J. 2019. Adsorption of Congo red dye on Fe x Co 3-x O 4 nanoparticles. Journal of Environmental Management, 238(January). 473–483. https://doi.org/10.1016/j.jenvman.2019.03.009.

Madhava Rao, M., Ramesh, A., Purna Chandra Rao, G. and Seshaiah, K. 2006. Removal of copper and cadmium from the aqueous solutions by activated carbon derived from *Ceiba pentandra hulls*. Journal of Hazardous Materials. 129(1–3): 123–129. https://doi.org/10.1016/j.jhazmat.2005.08.01.

Marathe, K. and Doshi, P. 2015. Magnetic nanoparticles: preparation, physical properties, and applications in biomedicine. IEEE International Conference on Intelligent Robots and Systems, 2015-Decem, 2550–2555. https://doi.org/10.1109/IROS.2015.7353724.

Martin, B.D., Parsons, S.A. and Jefferson, B. 2009. Removal and recovery of phosphate from municipal wastewaters using a polymeric anion exchanger bound with hydrated ferric oxide nanoparticles. Water Science and Technology. 60(10): 2637–2645. https://doi.org/10.2166/wst.2009.686.

Mehdinia, A., Akbari, M., Baradaran Kayyal, T. and Azad, M. 2015. High-efficient mercury removal from environmental water samples using di-thio grafted on magnetic mesoporous silica nanoparticles. Environmental Science and Pollution Research. 22(3): 2155–2165. https://doi.org/10.1007/s11356-014-3430-6.

Moad, G., Rizzardo, E. and Thang, S.H. 2013. RAFT polymerization and some of its applications. Chemistry - An Asian Journal. 8(8): 1634–1644. https://doi.org/10.1002/asia.201300262.

Moradi, O., Shahryari-Ghoshekandi, R., Sadegh, H. and Norouzi, M. 2016. Application of carbon nanotubes in nanomedicine: New medical approach for tomorrow. Medical Imaging: Concepts, Methodologies, Tools, and Applications, 2021–2062. https://doi.org/10.4018/978-1-5225-0571-6.ch082.

Müller, R.H., Gohla, S. and Keck, C.M. 2011. State of the art of nanocrystals—Special features, production, nanotoxicology aspects and intracellular delivery. European Journal of Pharmaceutics and Biopharmaceutics. 78(1): 1–9. https://doi.org/10.1016/j.ejpb.2011.01.007.

Nassar, N.N. 2010. Rapid removal and recovery of Pb(II) from wastewater by magnetic nanoadsorbents. Journal of Hazardous Materials. 184(1–3): 538–546. https://doi.org/10.1016/j.jhazmat.2010.08.069.

Nemati, Y., Zahedi, P., Baghdadi, M. and Ramezani, S. 2019. Microfluidics combined with ionic gelation method for production of nanoparticles based on thiol-functionalized chitosan to adsorb Hg (II) from aqueous solutions. Journal of Environmental Management. 238(October 2018): 166–177. https://doi.org/10.1016/j.jenvman.2019.02.124.

Raza, M.A., Westwood, A., Brown, A., Hondow, N. and Stirling, C. 2011. Characterisation of graphite nanoplatelets and the physical properties of graphite nanoplatelet/silicone composites for thermal interface applications. Carbon. 49(13): 4269–4279. https://doi.org/10.1016/j.carbon.2011.06.002.

Ryu, J., Kim, S., Hong, H.J., Hong, J., Kim, M., Ryu, T. et al. 2016. Strontium ion (Sr2+) separation from seawater by hydrothermally structured titanate nanotubes: Removal vs. recovery. Chemical Engineering Journal. 304: 503–510. https://doi.org/10.1016/j.cej.2016.06.131.

Sadegh, H., Shahryari-Ghoshekandi, R., Tyagi, I., Agarwal, S. and Gupta, V.K. 2015. Kinetic and thermodynamic studies for alizarin removal from liquid phase using poly-2-hydroxyethyl methacrylate (PHEMA). Journal of Molecular Liquids. 207: 21–27. https://doi.org/10.1016/j.molliq.2015.03.014.

Sadegh, H., Zare, K., Maazinejad, B., Shahryari-Ghoshekandi, R., Tyagi, I., Agarwal, S. and Gupta, V.K. 2016. Synthesis of MWCNT-COOH-Cysteamine composite and its application for dye removal. Journal of Molecular Liquids. 215: 221–228. https://doi.org/10.1016/j.molliq.2015.12.042.

Sadegh, H., Ali, G.A.M., Gupta, V.K., Makhlouf, A.S.H., Shahryari-ghoshekandi, R., Nadagouda, M.N. et al. 2017. The role of nanomaterials as effective adsorbents and their applications in wastewater treatment. Journal of Nanostructure in Chemistry. 7(1): 1–14. https://doi.org/10.1007/s40097-017-0219-4.

Selvam, P., Bhatia, S.K. and Sonwane, C.G. 2001. Recent advances in processing and characterization of periodic mesoporous MCM-41 silicate molecular sieves. Industrial and Engineering Chemistry Research. 40(15): 3237–3261. https://doi.org/10.1021/ie0010666.

Sengupta, S. and Pandit, A. 2011. Selective removal of phosphorus from wastewater combined with its recovery as a solid-phase fertilizer. Water Research. 45(11): 3318–3330. https://doi.org/10.1016/j.watres.2011.03.044.

Sha, L., Xueyi, G., Ningchuan, F. and Qinghua, T. 2009. Adsorption of Cu2+ and Cd2+ from aqueous solution by mercapto-acetic acid modified orange peel. Colloids and Surfaces B: Biointerfaces. 73(1): 10–14. https://doi.org/10.1016/j.colsurfb.2009.04.021.

Shahbazi, A., Younesi, H. and Badiei, A. 2011. Functionalized SBA-15 mesoporous silica by melamine-based dendrimer amines for adsorptive characteristics of Pb(II), Cu(II) and Cd(II) heavy metal ions in batch and fixed bed column. Chemical Engineering Journal. 168(2): 505–518. https://doi.org/10.1016/j.cej.2010.11.053.

Shen, Y.F., Tang, J., Nie, Z.H., Wang, Y.D., Ren, Y. and Zuo, L. 2009. Preparation and application of magnetic Fe3O4 nanoparticles for wastewater purification. Separation and Purification Technology. 68(3): 312–319. https://doi.org/10.1016/j.seppur.2009.05.020.

Singh, P., Srivastava, S. and Singh, S.K. 2019. Nanosilica: recent progress in synthesis, functionalization, biocompatibility, and biomedical applications [Review-article]. ACS Biomaterials Science and Engineering. 5(10): 4882–4898. https://doi.org/10.1021/acsbiomaterials.9b00464.

Singh, S., Barick, K.C. and Bahadur, D. 2011. Novel and efficient three dimensional mesoporous Zno nanoassemblies for envirnomental remediation. International Journal of Nanoscience. 10(4-5): 1001–1005. https://doi.org/10.1142/S0219581X11008654.

Song, X., Gunawan, P., Jiang, R., Leong, S.S.J., Wang, K. and Xu, R. 2011. Surface activated carbon nanospheres for fast adsorption of silver ions from aqueous solutions. Journal of Hazardous Materials. 194: 162–168. https://doi.org/10.1016/j.jhazmat.2011.07.076.

Stefaniuk, M., Oleszczuk, P. and Ok, Y.S. 2016. Review on nano zerovalent iron (nZVI): From synthesis to environmental applications. Chemical Engineering Journal. 287: 618–632. https://doi.org/10.1016/j.cej.2015.11.046.

Su, Y., Adeleye, A.S., Huang, Y., Sun, X., Dai, C., Zhou, X. et al. 2014. Simultaneous removal of cadmium and nitrate in aqueous media by nanoscale zerovalent iron (nZVI) and Au doped nZVI particles. Water Research. 63: 102–111. https://doi.org/10.1016/j.watres.2014.06.008.

Sublet, R., Simonnot, M.O., Boireau, A. and Sardin, M. 2003. Selection of an adsorbent for lead removal from drinking water by a point-of-use treatment device. Water Research. 37(20): 4904–4912. https://doi.org/10.1016/j.watres.2003.08.010.

Sun, X., Xu, L., Jiang, W., Xuan, Y., Lu, W., Li, Z. et al. 2021. Adsorption mechanism of rhein-coated Fe3O4 as magnetic adsorbent based on low-field NMR. Environmental Science and Pollution Research. 28(1): 1052–1060. https://doi.org/10.1007/s11356-020-10541-5.

Tabatabaei, S., Forouzesh Rad, B. and Baghdadi, M. 2020. Semicontinuous enhanced electroreduction of Cr(VI) in wastewater by cathode constructed of copper rods coated with palladium nanoparticles followed by adsorption. Chemosphere. 251: 126309. https://doi.org/10.1016/j.chemosphere.2020.126309.

Tu, Y.J., Chan, T.S., Tu, H.W., Wang, S.L., You, C.F. and Chang, C.K. 2016. Rapid and efficient removal/recovery of molybdenum onto ZnFe2O4 nanoparticles. Chemosphere. 148: 452–458. https://doi.org/10.1016/j.chemosphere.2016.01.054.

Tu, Y.J., Lo, S.C. and You, C.F. 2015. Selective and fast recovery of neodymium from seawater by magnetic iron oxide Fe3O4. Chemical Engineering Journal. 262(2015): 966–972. https://doi.org/10.1016/j.cej.2014.10.025.

Tuzen, M. and Soylak, M. 2007. Multiwalled carbon nanotubes for speciation of chromium in environmental samples. Journal of Hazardous Materials. 147(1-2): 219–225. https://doi.org/10.1016/j.jhazmat.2006.12.069.

Wang, L., Li, J., Jiang, Q. and Zhao, L. 2012. Water-soluble Fe 3O 4 nanoparticles with high solubility for removal of heavy-metal ions from waste water. Dalton Transactions. 41(15): 4544–4551. https://doi.org/10.1039/c2dt11827k.

Wang, Zhiguo, Li, P., Fang, Y., Yan, L., Zhou, W., Fan, X. and Liu, H. 2019. One-step recovery of noble metal ions from oil/water emulsions by chitin nanofibrous membrane for further recycling utilization. Carbohydrate Polymers, 223(May): 115064. https://doi.org/10.1016/j.carbpol.2019.115064.

Wang, Zhongde, Guo, S., Wu, Z., Fan, H., Guan, G. and Hao, X. 2017. A smart potential-responsive ion exchange nanomaterial with superparamagnetism for cesium ion separation and recovery. Separation and Purification Technology. 187: 199–206. https://doi.org/10.1016/j.seppur.2017.06.033.

Wei, J., Feng, Y., Liu, Y. and Ding, Y. 2015. MxCo3-xO4 (M = Co, Mn, Fe) porous nanocages derived from metal-organic frameworks as efficient water oxidation catalysts. Journal of Materials Chemistry A. 3(44): 22300–22310. https://doi.org/10.1039/c5ta06411b.

Wu, W., He, Q.G., Hu, R., Huang, J.K. and and Chen, H. 2007. Preparation and characterization of magnetite Fe3O4 nanopowders. Rare Met. Mater. Eng. 36: 238.

Yan, W., Lien, H.L., Koel, B.E. and Zhang, W.X. 2013. Iron nanoparticles for environmental clean-up: Recent developments and future outlook. Environmental Sciences: Processes and Impacts. 15(1): 63–77. https://doi.org/10.1039/c2em30691c.

Zare, K., Gupta, V.K., Moradi, O., Makhlouf, A.S.H., Sillanpää, M., Nadagouda, M.N. et al. 2015. A comparative study on the basis of adsorption capacity between CNTs and activated carbon as adsorbents for removal of noxious synthetic dyes: a review. Journal of Nanostructure in Chemistry. 5(2): 227–236. https://doi.org/10.1007/s40097-015-0158-x.

Zelmanov, G. and Semiat, R. 2011. Iron (Fe+3) oxide/hydroxide nanoparticles-based agglomerates suspension as adsorbent for chromium (Cr+6) removal from water and recovery. Separation and Purification Technology. 80(2): 330–337. https://doi.org/10.1016/j.seppur.2011.05.016.

Zhang, B., Fu, L., Wang, S. and Zhang, L. 2018. Adsorption of palladium(II) from aqueous solution using nanosilica modified with imidazoline groups. Materials Chemistry and Physics. 214(Ii): 533–539. https://doi.org/10.1016/j.matchemphys.2018.04.120.

Zhang, L., Pan, F., Liu, X., Yang, L., Jiang, X., Yang, J. et al. 2013. Multi-walled carbon nanotubes as sorbent for recovery of endocrine disrupting compound-bisphenol F from wastewater. Chemical Engineering Journal, 218: 238–246. https://doi.org/10.1016/j.cej.2012.12.046.

Zhang, Y., Li, J., Wang, F., Wu, G., Qv, X., Hong, H. et al. 2016. Recovery and separation of erythromycin from industrial wastewater by imprinted magnetic nanoparticles that exploit β-cyclodextrin as the functional monomer. Journal of Separation Science. 39(2): 450–459. https://doi.org/10.1002/jssc.201500927.

Zhao, Y.G., Shen, H.Y., Pan, S.D., Hu, M.Q. and Xia, Q.H. 2010. Preparation and characterization of amino-functionalized nano-Fe 3O4 magnetic polymer adsorbents for removal of chromium(VI) ions. Journal of Materials Science. 45(19): 5291–5301. https://doi.org/10.1007/s10853-010-4574-5.

6

A Short Note on Naturally-derived Materials for Oil-water Separation

S. Sutha[1] and *N.R. Dhineshbabu*[2,*]

6.1 Introduction

Marine ecosystems are very important for the health of both marine and terrestrial environments. However, discarded plastics, residential wastes, industrial wastes, and shipping accidents involving oil spills eventually cause devastating consequences for marine life and the habitats. Among the various causes, one of the challenging causes comes from oil spills. It destroys the basic adaptation abilities of marine species, thus exposing these organisms to lethal risks. Besides affecting the marine ecology, it imparts serious health risks to the people in nearby coastal zones (Zhang et al. 2019). Several oil spillage disasters took place in the Indian nation and have raised serious environmental concerns. For example, oil spillage in Ennore, Chennai (Fig. 1).

Spilled oil in the water bodies can take months or even years to remove, thus known to be one of the most expensive processes in disaster management. Thus, oil/water separation is an important field in scientific research to address environmental and socio-economic issues. The current methods used to address such oil spillage problems are employing microorganisms, skimmers, separators, sorbents, and chemical dispersants. Still, the conventional techniques are not satisfactory in terms of performance.

In recent years, membrane separation technology, one of the nanotechnology-based approaches, has been proved to be one of the most effective methods in terms of oil separation efficiency, stable effluent quality, and cost-effective technique. Regarding this, specially wettable materials are one of the most primitive techniques

[1] Department of Electronics and Communication Engineering, Sri Venkateswara College of Engineering, Sriperumbudur, Tamil Nadu – 602117. Email: suthasenthil@gmail.com

[2] Department of Electronics and Communication Engineering, Aditya Engineering College, Surampalem, Andhra Pradesh – 533437.

* Corresponding author: babudhinesh2009@gmail.com

Fig. 1: Oil spill at Ennore, Chennai, Retrieved from https://www.thehindu.com.

for oil-water separation. These materials are specially wettable whose water contact angle exceeds 150° and have oil contact angle near to 0°. These materials can be classified into two categories: filter type, and absorption type. The filter type includes 2D materials such as textile fabrics, filtration membranes, etc. (Ge et al. 2016). The adsorption type of materials comprises 3D materials such as porous sponge, foam, etc. In addition, smart materials are currently being used, which mainly utilize magnetic separation technology, that can regulate the separation process (Singh et al. 2020).

Among the materials and methods listed, sorbents are highly appealing since the oil can be removed completely without harming the atmosphere. One of the foremost characteristics of a sorbent material is that it should be superhydrophobic and superoleophilic, have a high oil absorption/adsorption capability, and be cost-effective (Ge et al. 2016). Commercially available synthetic sorbents suffer from severe disadvantages, such as complexity, expensive and toxic chemicals used, and generation of secondary pollution. Thus, seeking environment-friendly and biodegradable sorbent materials to replace synthetic sorbents remains imperative to address the environmental problems due to oil spillage. In the present work, a critical review is presented, focused mainly on bio-based materials such as silk, cuttlebone, seashell, milkweed, sugarcane bagasse wastes, Balsa wood, and sand for oil spill clean- up applications (Fig. 2). This chapter focuses on research completed in the last two years.

Fig. 2: (a) Cuttlebone (retrieved from www.thesprucepets.com), (b) Sugarcane bagasse (retrieved from www.czarnikow.com), (c) Milkweed floss (retrieved from www.britannica.com), (d) Balsa wood, (e) Silk cocoon (retrieved from www.123rf.com), (f) sSeashell (retrieved from www.nbcnews.com).

6.2 Naturally derived sorbents/filter

This section covers the most important and attractive naturally derived materials for oil–spill cleanup application. The numerical results of the research are summarized in Table 1 for reference.

6.2.1 Seashell

A seashell is a hard, protective outer layer (composed of calcium carbonate) of soft-skinned animals living in the sea (Currie et al. 2007). Empty seashells can be found in the seashores, as the animal has been eaten by another animal or has decomposed. The seashell contains abundant amount of protein, namely 3,4-dihydroxyl-phenylalanine, which can easily adhere to different surfaces via covalent and noncovalent bonds (Jamsaz and Goharshadi 2020).

Inspired from the structural characteristics of seashells, the authors developed a specially wettable sponge made of naturally derived seashell, epoxy resin, and polyurethane (EP/SPU) for oil-water separation (Jamsaz and Goharshadi 2020). In the work, seashells were cleaned, grounded, sieved (760 μmaverage particle size), and calcined (CS). Then the cleaned PU sponge was dipped in CS—water suspension and subsequently in epoxy resin. The interconnected 3D porous structure increased the WCA to superhydrophobic without any surface functionalization (Fig. 3(a)). Further, on functionalization with EP, the WCA increased greater than 170°. It was observed from the results that SPU behaved indifferently concerning oil (same performance as PU) and organic solvents (poor performance compared to PU) and the reason behind the observation was missing in the publication. In addition, EP/SPU tolerated 3.63 kPa without any volume change and performed well on an average of 10 cycles.

In a separate study, the authors used seashells to manufacture 3D foams of graphene using chemical vapor deposition technique (Fig. 3(b)), which have applications in

Table 1: Summary of the naturally-derived materials and its performance in oil-water separation.

S. No.	Material	Morphological features	WCA	Oil – water separation	Reusability	Mechanical durability	References
1	Seashell/epoxy resin/ polyurethane (EP/ SPU) sponge	Rough interconnected 3D porous structure, well-defined and fine porous structure	94.01 ± 1.23° to 174.10 ± 0.47°	Cooking, crude oils, and organic solvents. PU—2.79–39.41 g/g. EP/SPU—28.03 –42.17 g/g	10	Tolerate 3.63 kPa pressure	(Jamsaz and Goharshadi 2020)
2	3D graphene foams	3D foams, highly porous and interconnected structures	-	Organic solvents, motor oils	-	Stable after 200 bending cycles	(Shi et al. 2016)
3	Milkweed floss fibers	Outer and inner surface possess smooth and nearly uniform surface	140°	Absorption capacity - 90%	Low viscosity oils – 40 times	-	(Panahi et al. 2020)
4	Sugarcane bagasse	3D porous network structure	Around 150.5° and 148.3°	Oil absorption capacity - 25 g/g	-	Young's modulus of about 88 kPa	(Thai et al. 2020)
5	Balsa wood sponge	Porous structure and hierarchical rough	140°	Separation efficiency – 99%	-		(Huang et al. 2020)
6	Delignified wood	cellular structure with tubule-like channels	153.5°	Oil sorption capacity 37 g g-1	10	90% compressive strain, and 97% height retention after 100 cycles	(Yang et al. 2020)
7	EC	Porous structure	150.1°	Oil absorption capacity (49.94–140.90 g/g)			(Li et al. 2016)
8	Magnetic sawdust	Layered structure	156.3°	Oil sorption capacity (3.59 to 7.83 g/g)	20	-	(Yin et al. 2020)
9	Superhydrophobic pellets	Rough	152°	Separation efficiency higher than 90%	-	-	(Latthe et al. 2020)
10	Silk fibers	Fibers covered with flakes of varying thicknesses	150 ± 3°	Oil sorption capacity - 46.83 g/g (crude oil) and 84.14 g/g (motor oil)	5	-	(Patowaryet al. 2016)
11	Cuttlebone	Internal lamellar structure	154°	90% separation efficiency	10	-	(Xu et al. 2020)
12	Sand@soot	Rough spherical nanoparticles along with irregularly shaped microparticles	134 ± 5°	Separation efficiency - 99%.	10	-	(Mosayebi et al. 2020)

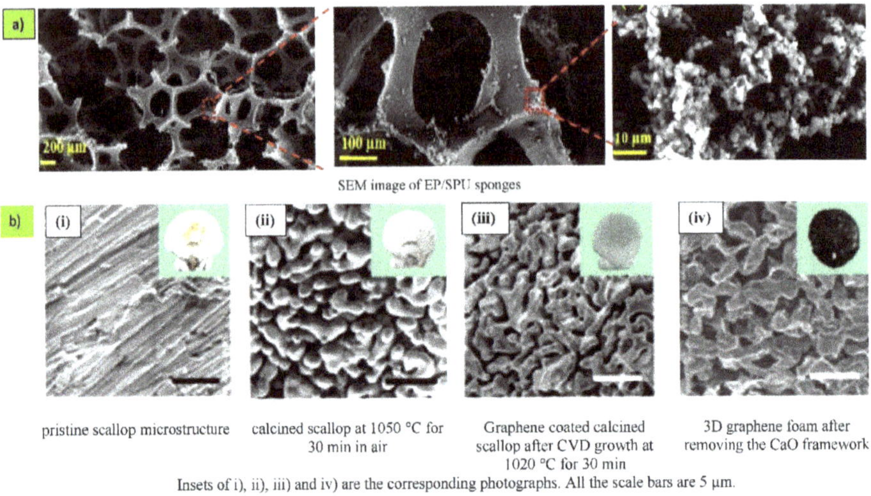

Fig. 3: SEM image of (a) Seashell/EP/PU (Jamsaz et al. 2020), reproduced with permission, Copyright 2021, Elsevier. (b) Seashell/graphene foams (Shi et al. 2016), reproduced with permission, Copyright 2021, Elsevier.

oil—water separation (Shi et al. 2016). The graphene foams were created using nickel foam models, which have exhibited favorable properties. Shi et al., on the other hand, used calcium carbonate in the shells of oceanic mollusks, which had specially wettable properties to water and organic liquids. Furthermore, the graphene foams exhibited absorption of 200 to 250 times of their weight, and the adsorbates were quickly removed with acetone, suggesting the graphene foams' recyclability.

6.2.2 *Milkweed floss*

The milkweed floss is a seed fibre of milkweed (Asclepias syriaca) containing 80% of cellulose and hemicelluloses (Richard et al. 2019). Milkweed floss is used in life jackets and belts and supports 30 times its weight (Crews et al. 1991). Inspired by the special properties of milkweed floss, the authors developed a sorbent based on milkweed floss for oil spill cleanup (Fig. 4(a)) (Panahi et al. 2020). The outcomes indicated that the sorbent based on milkweed floss had an absorption capacity greater than 100 times of its own weight. It was also mentioned that capillary diameter and porosity played a significant role in functional properties of the sorbent suitable for oil spillage clean up. It was reported that when oil contact angle is low compared to water contact angle, it results in high capillary forces for oil, thus showing the differential behavior. It is also mentioned that in a hydrophobic hollow fiber of milkweed floss, initially oil was adsorbed on the surface via Van der Waals forces and then penetrated into the fibrous channel.

6.2.3 *Sugarcane bagasse*

They are cellulose waste materials, which have promising properties of high porosity and low density, and are highly sought for oil spill clean-up applications. Stimulated

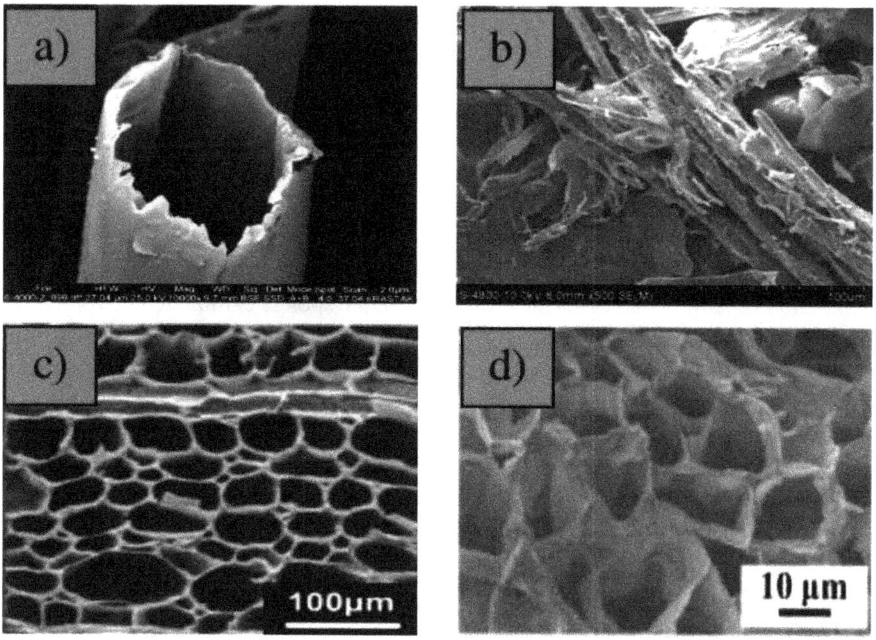

Fig. 4: SEM image of (a) Milkweed floss (Panahi et al. 2020), reproduced with permission, Copyright 2021, Elsevier. (b) Sugarcane bagasse (Thai et al. 2020), reproduced with permission, Copyright 2021, Elsevier. (c) DES-delignified wood (Yang et al. 2020), reproduced with permission, Copyright 2021, Elsevier. (d) F-rGO@WS (Huang et al. 2020), reproduced with permission, Copyright 2021, Elsevier.

by their properties, Thai et al. utilized sugarcane bagasse-based aerogel for oil-water separation as shown in Fig. 4(b) (Thai et al. 2020). To fabricate the aerogel, bagasse was washed, grounded into fibers, and dried. The dried bagasse was immersed in PVA, followed by freeze dried to produce aerogel with 3D porous network structure, self-assembled through hydrogen bonding. Since the 3D porous structure of bagasse aerogel contains hydroxyl groups on its surface, surface modification with silane coating improved its hydrophobicity without modifying the porous nature. It was observed that on increasing the fiber concentration, there was an increase in the mass/density of the aerogel, and corresponding decrease in the porosity. The results showed that the porous structure had an absorption capacity of 25 times its own weight and withstood 88 kPaYoung's modulus, proposing a promising candidate for oil spill clean-up applications.

6.2.4 Delignified natural wood

Recently, delignified natural wood with cellular architecture and high porosity have become great candidates for oil absorption. In general, cell walls of the wood consist of cellulose, hemicelluloses, and lignin in their structure. However, the delignified wood has oriented porous cellulose structure, deprived of hemicelluloses and lignin. In the work done by (Yang et al. 2020), balsa wood samples were used for this application as given in Fig. 4(c). The wood samples were dried, lyophilized, and surface modified with silane to obtain super hydrophobic porous honeycomb-like cellular structure. The sample showed initial linear elastic deformation at low strain and allowed 90% of compressive strain after 100 cycles, which emphasized the robust structure of the delignified wood. The delignified wood demonstrated better absorption capacity, implying a stable absorption performance.

Huang et al. reported the performance of wood sponge based on balsa wood covered with reduced graphene oxide (F-rGO@WS) for vicious crude oil removal (Fig. 4(d)) (Huang et al. 2020). The unobstructed longitudinal channels of treated balsa wood exhibited hydrophobic characteristics. The F-rGO@WS exhibited 99% separation capability and efficiency. In addition, the sponge exhibited electrothermal capability and improved the fluidity of the oil. In line with the work, Huang et al. fabricated balsa wood sponge by employing delignification and freeze-drying process, which exhibited superhydrophobic characteristics. The longitudinal and transverse surface of F-rGO@WS exhibited differential behavior towards silicone oil, exemplifying the significance of the channel surface structure over the functional properties. It was also demonstrated that when the sponge was dipped in hexane/water mixture, the hexane was selectively absorbed with a separation efficiency of 99.0%.

6.2.5 *Eichhornia crassipes (E.C.)*

It is a major freshwater weed, generally regarded as the most bothersome aquatic plant, which adversely affects human activities such as fishing, and water transport. However, towards the positive side, recent studies revealed that the processed E.C. adsorbed cationic dyes efficiently, owing to the porous structure (Li et al. 2016). Admired by the properties, Sun et al., utilized carbonized E.C/Fe3O4 for oil—water separation. To prepare carbonized E.C., E.C's stem was cleaned, freeze-dried, and carbonized under vacuum (CEC) in nitrogen atmosphere. Then, Fe3O4 magnetic particles were allowed to diffuse into the porous structure of CEC and finally modified with silane surface modification. The adsorption capacity was analyzed concerning Fe_3O_4 material and observed that the presence of Fe_3O_4 increased the adsorption capacity. The results showed that the absorption volume remained stable even after three cycles and disrupted beyond that due to low mechanical strength.

6.2.6 *Sawdust*

Sawdust, composed of small chippings of wood, is getting attention in waste-water treatment due to its porous structure. Yin et al. believed that the sawdust could be used for treating oil spillage clean-up by fabricating superhydrophobic magnetic sawdust. Instead of employing synthetic hydrophobic material, the authors used candelilla wax (CW) (Yin et al. 2020). The absorbent exhibited better absorption capacity and magnetic behavior, owing to the porous structure of sawdust as depicted in Fig. 5(a). The absorbent recovered the oil in the form of cake as well as filter. The raw sawdust had a layered structure with numerous vessels between the layers to store oil. The magnetic hysteresis loops of CW-coated Fe_3O_4@PS exhibited superparamagnetic behaviors with no obvious change in remanence and coercivity.

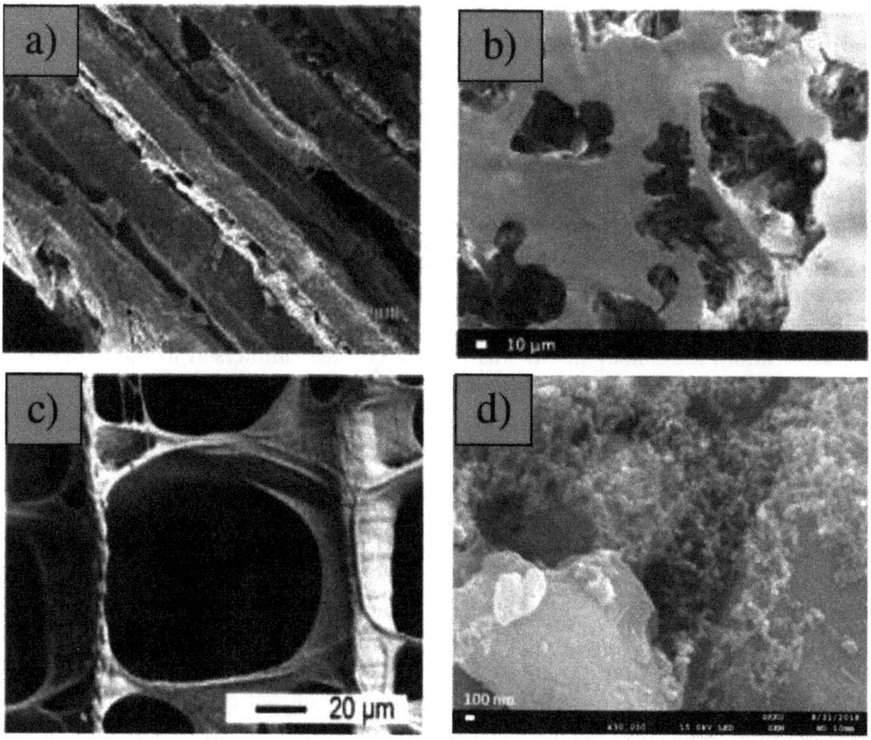

Fig. 5: SEM image of (a) CW-coated Fe_3O_4@PS (Yin et al. 2020), reproduced with permission, Copyright 2021, Elsevier. (b) Sawdust composite pellets with different concentrations of sawdust (Latthe et al. 2020), reproduced with permission, Copyright 2021, Elsevier. (c) Cuttlebone (Xu et al. 2020), reproduced with permission, Copyright 2021, Elsevier. (d) Sand @ soot (Mosayebi et al. 2021), reproduced with permission, Copyright 2021, Elsevier.

Similarly, Latthe et al. employed sawdust of different woods such as neem, babul, and sal from the local sawdust mill for oil—spill clean-up (Fig. 5(b)) (Latthe et al. 2020). At first, the sawdust was washed, sieved (average particles size—125 μm), dried, and grinded to achieve fine powder of sawdust. Pellets of sawdust were prepared using polystyrene/chloroform. The pellets performed better separation efficiency for hexane even after 36 repeated cycles. The results implied that the pellets showed better performance towards low viscosity oil than high viscous oils, due to ease of infiltration through the voids. The stability of the pellets was studied using an industrial oil-water mixture and warm water. The results demonstrated that separation efficiency was good under gravity and in different temperature conditions. The pellets maintained excellent superhydrophobic stability even after 100 peel-off cycles and showed only few scratches upon sand abrasion tests and in different pH conditions.

6.2.7 *Silk Degumming*

Silk is a natural fiber, mainly composed of proteins such as fibroin and sericin, and is usually obtained from the cocoons of the larvae of mulberry silkworms.

Patowary et al. developed a superhydrophobic natural sorbent prepared from silk fibroin fibers. The sorbent fibers had better selectivity and sorption capacity (Patowary et al. 2016). The results demonstrated better reusability and oil sorption ability than wool-based sorbents.

6.2.8 *Cuttlebone*

It is a bone in the cuttlefish, that comprises of dorsal shield and lamellar matrix. The lamellar matrix exhibits the characteristics such as low dense porous structure and good mechanical stability, which are desirable for oil spill removal. Xu et al. utilized the cuttlebone for effective oil spill cleanup (Fig. 5(c)) (Xu et al. 2020). For the fabrication, the dorsal shield was peeled, cut, and immersed in polydopamine and fluorinated silica nanoparticles solutions subsequently. The structure was sub-structured into several chambers separated by lamellae, which revealed a quasi-periodic porous structure. The surface demonstrated a good superhydrophobic property, better thermal stability, and abrasion resistance. The surface displayed excellent separation efficiency (> 99%) and for continuous oil separation, a "cephalopod" electric device was designed.

6.2.9 *Sand*

Sand grains form a filtration membrane, preventing larger particles from passing through the gaps between grains. Sand particles strain out the floc and expel the majority of the solids as flocculated water is passed under gravity (Yong et al. 2016). The sand was used by Mosayebi et al. as a desirable substrate for oil–water separation. The sand was exposed to flame to deposit hydrophobic nanosoot particles with a core–shell structure (Fig. 5(d)) in this study (Mosayebi et al. 2020). Sand grains were sieved, washed, dried, and flame treated before being used in the fabrication. The micro and nanoscale roughness of the sand@soot surface makes it ideal for water repellency and oil-water separation.

6.3 Summary

While several attempts have been made to fabricate superhydrophobic-superoleophilic sorbents, fundamental aspects of developing the optimal structure for the removal of particular oil molecules have yet to be clarified. Since the fibers were naturally derived, the surface alteration had an effect on their porosity. Pore volume and packing factor can be used to calculate sorption capacity. The packing factor rises as the ratio of fiber volume to total sorbent volume rises, resulting in a reduction in pore volume and sorption capacity. Internal capillary storage and capillary bridging between fibers were found in naturally derived fibers, which increased the packing factor and damaged the capillary bridging, resulting in degradation in oil sorption performance. In terms of oil viscosity, currently available sorbents have been documented to perform well in low viscous oil spills. They can't be used for highly viscous oils because they enter the pores at a snail's pace. Furthermore, the pH of the water has a significant impact on the sorbents' efficiency. Some researchers have looked into the thermo- and magnetic-response behavior of sorbents, which are more useful in retrieving sorbed oil from the sorbents. In a nutshell, the researchers are tasked with developing highly efficient oil-sorbent materials, oil-sorbent robots, and retrieving various types of sorbed oils.

References

Crews, P.C, Sievert, S.A. and Woeppel, L.T. 1991. Evaluation of milkweed floss as an insulative fill material. Textile Research. 61(4): 203–210.

Currie, J., Harrison, N., Wang, L., Jones, M. and Brooks, M. 2007. A preliminary study of processing seafood shells for eutrophication control. J. Dispersion Sci. Technol. 2: 460–467.

Ge, J., Zhao, H.-Y., Zhu, H.-W., Huang, J., Huang, J., Shi, L.-A. and Yu, S.-H. 2016. Advanced sorbents for oil-spill cleanup: Recent advances and future perspectives. Adv. Mater. 28: 10459–10490.

Huang, W., Zhang, L., Lai, X., Li, H. and Zeng, X. 2020. Highly hydrophobic F-rGO@wood sponge for efficient clean-up of viscous crude oil. Chemical Engineering Journal. 386: 123994.

Jamsaz, A. and Goharshadi, E.K. 2020. An environmentally friendly superhydrophobic modified polyurethane sponge by seashell for the efficient oil/water separation, Process Safety and Environmental Protection. 139: 297–304.

Latthe, S., Vishnu, K., Rajaram, S., Appasahe, B., Saravanan, N., Chang-Sik, H. et al. 2020. Sawdust-based superhydrophobic pellets for efficient oil-water separation. Materials Chemistry and Physics, 243: 122634.

Li, Q., Zhan, J., Chen, B., Meng, X. and Pan, X. 2016. Removal of Pb, Zn, Cu, and Cd by Two Types of *Eichhornia crassipes*. Environ. Eng. Sci., 33: 88–97.

Mosayebi, E., Azizian, S., Cha, B.J., Woo, T.G. and Kim, Y.D. 2021. Fabrication of highly hydrophobic sand@soot with core–shell structure and largescale production possibility for oil/water separation, Journal of Physics and Chemistry of Solids. 150: 109815.

Panahi, S., Moghaddam, M.K. and Moezzi, M. 2020. Assessment of milkweed floss as a natural hollow oleophilic fibrous sorbent for oil spill cleanup. J. Environ. Manage. 15(268): 110688. doi: 10.1016/j.jenvman.2020.110688.

Patowary, M., Pathak, K. and Ananthakrishnan, R. 2016. Robust superhydrophobic and oleophilic silk fibers for selective removal of oil from water surface. RSC Adv. 6: 73660–73667.

Richard, C., Cousin, P., Foruzanmehr, M., Elkoun, S. and Robert, M. 2019. Characterization of components of milkweed floss fiber. Separ. Sci. Technol. 54: 3091–3099.

Shi, L., Chen, K., Du, R., Bachmatiuk, A., Rümmeli, M.H., Xie, K. et al. 2016. Scalable seashell-based chemical vapor deposition growth of three-dimensional graphene foams for oil–water separation. J. Am. Chem. Soc. 138(20): 6360–6363.

Singh, B., Surender Kumar, Kishore, B. and Narayanan, T.N. 2020. Magnetic scaffolds in oil spill applications. Environ. Sci.: Water Res. Technol. 6: 436–463.

Thai., Q.B., Nguyen, S.T., Ho, D.K., Tran, T.D., Huynh, D.M., Do, N.H.N. et al. 2020. Cellulose-based aerogels from sugarcane bagasse for oil spill-cleaning and heat insulation applications. CarbohydrPolym. 228: 115365.

Xu, J., Cao, R., Li, M., Chen, G. and Tian, J. 2020. Superhydrophobic and superoleophilic cuttlebone with an inherent lamellar structure for continuous and effective oil spill cleanup. Chemical Engineering Journal. 127596.

Yang, R., Cao, Q., Liang, Y., Hong, S., Xia, C., Wu, Y. et al. 2020. High capacity oil absorbent wood prepared through eco-friendly deep eutectic solvent delignification. Chemical Engineering Journal. 401: 126150.

Yin, Z., Li, Y., Song, T., Bao, M., Li, Y., Lu, J. et al. 2020. Preparation of superhydrophobic magnetic sawdust for effective oil/water separation. Journal of Cleaner Production. 253: 120058.

Yong, J., Chen, F., Yang, Q., Bian, H., Du, G., Shan, C. et al. 2016. Oil-water separation: a gift from the desert. Communication. 3(7): 1500650.

Zhang, B., Matchinski, E.J., Chen, B., Ye, X., Jing, L. and Lee, K. 2019. Marine oil spills—oil pollution, sources and effects. World Seas: an Environmental Evaluation (Second Edition), Volume III: Ecological Issues and Environmental Impacts, 391–406.

7

Recovery of Various Commercial Energy Forms from Wastewater

K. Krishna Koundinya,[1] *G. Karthik Pavan Kumar,*[2] *Surajit Mondal*[3,*] and *Paulami Ghosh*[4]

7.1 Wastewater generation

Water is a major component on Earth, which is an abundant, naturally available resource that occupies 75% of the globe. Being an essential element, the quality of water plays a crucial role in many biological activities such as reproduction, excretion, respiration, especially for marine lives, also for blood circulation and its availability made it have several applications in both domestic and commercial applications like electricity and power production which required water for thermoelectric cooling, hydro power, fuel cells, mining, and so on. These tend to include various biological, physical, and chemical impurities such as fertilisers, organic and inorganic components, microbial contaminants, toxic heavy metals, radio nuclei, and so on.

The chemicals removed from the industrial wastewater will be in different forms, such as salts, sludges, metal, and other polymers.This water is altered considering the acidity, which must be neutralised at the effluent treatment plants to reuse the water if suitable and to make sure that the pH of the water is around 7 before dumping the water into nearby lakes or rivers or any other water bodies. Purified water can be used for various energy generation purposes like electrical energy, thermal energy in the form of steam, super-heated steam and also used as a coolant in process

[1] Dept. of Electrical and Electronics, School of Engineering, UPES.
[2] Dept. of Microbiology, School of Health Sciences, UPES.
[3] Dept. of Electrical and Electronics, School of Engineering, UPES.
[4] Department of Allied Health Sciences, School of Health Science and Technology, UPES, India.
* Corresponding author: surajitmondalee@gmail.com

equipment like furnaces, reactors, heat exchangers, hydrogen generation and other manufacturing units. Every country is now focusing on wastewater treatment as portable or drinking water has also been under extinction due to the climatic change, population increment and industrial usage of water in large quantities. So global scenario of water demand and treatment also plays a crucial role in developing waste management methodologies that can prevent the loss of water to the oceans and seas. Developed countries such as the UAE, the United Kingdom, and the United States have made it their priority to extract energy from water in various forms, and the global statistics of wastewater generation and energy recovery are also described in this chapter, which explains the future scope, advancements, and innovations in energy recovery schemes.

7.2 Global statistics of wastewater generation and energy recovery

Global wastewater generation is nearly equal to global water consumption, and it is clear that the majority of global wastewater is neither collected nor treated; it is discharged directly into the environment without treatment to remove contaminants, and agricultural wastewater is almost never collected or treated, making data collection for these types of wastewater flows practically impossible. The statistics of water consumption and wastewater generation by various sectors has been explained in Fig. 1 and the change in water demand and wastewater generation with respect to time has been plotted in Fig. 2.

Larger quantities of wastewater has been produced as per the report given by UN. The global demand of water has been increasing year by year. By 2050, the total water supplied to the irrigation will be increased to an appreciable level as per a study in 2010 (Assocham 2019).

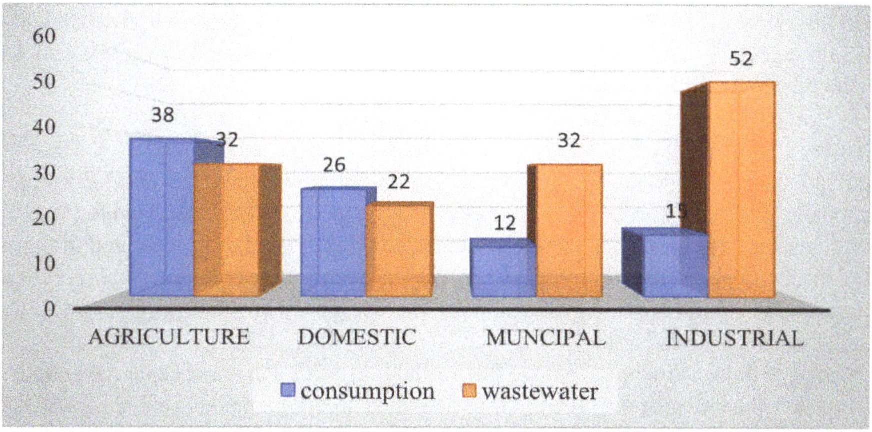

Fig. 1: Global statistics of water consumption and wastewater generation by various sectors (Assocham 2019).

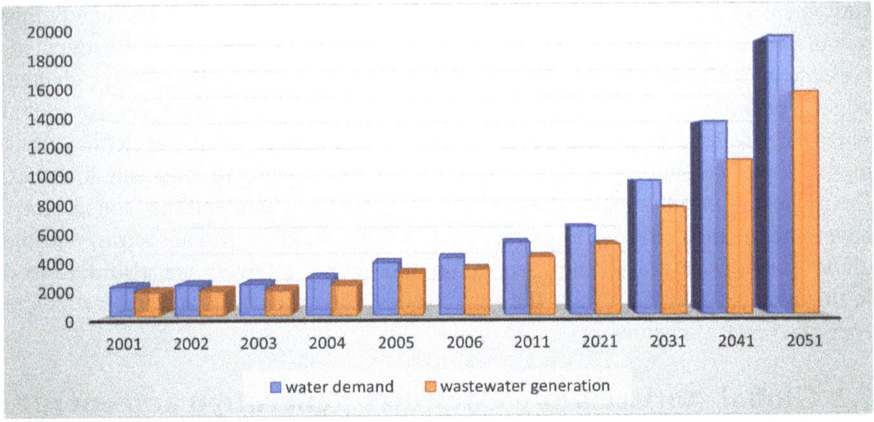

Fig. 2: The change in water demand and wastewater generation with respect to time (Assocham 2019).

7.3 Requirement of energy recovery from wastewater

The COD of the wastewater will be around 0.5 kg/m^3 which can be expressed as the amount of oxygen required for reducing substances in water that can be oxidized and the energy density of wastewater is nearly 74000 KJ/m^3 (McCarty et al. 2011). The chemical internal energy of the mixture of municipal wastewater that includes both domestic and industrial wastewater is experimentally determined to be 16800 KJ/m^3 where only domestic waste has the internal chemical energy of 76000 KJ/m^3 (Owen 1982, Logan 2011). The energy density is theoretically predicted to be 10^7 J/m^3 which is much larger than the energy consumed for wastewater treatments (Heidrich et al. 2011). Municipal wastewater in general consists of both organic and inorganic matter that can be converted into various energy forms; specifically, the latent heat present in the organic matter is the source of energy, which can be utilized for many operations and various waste treatment mechanisms. As time goes on, there will be a gradual rise in the water demand in various sectors with respect to time which is plotted in Fig. 3 till 2050.

Various operations have been implemented to recover the energy from wastewater in many energy forms such as organic sludges, electrical energy, thermal energy, etc. The sludge formed after the treatment of wastewater has a very good calorific value and can be classified as primary and waste activated sludge (WAS), which have an appreciable energy content. The primary sludge is generated through the mechanical wastewater treatment process. It occurs after the wastewater is flown through the screen and the grid chamber. The energy content in the primary sludge is about 23.2–29 MJ/Kg and WAS is the secondary sludge that is generated after the treatment of the effluent to remove the residual organics and some remaining solids. The energy content in the WAS (waste activated sludge) is about 18.6–23.2 MJ/Kg (Weemaes and Vestraete 1998).

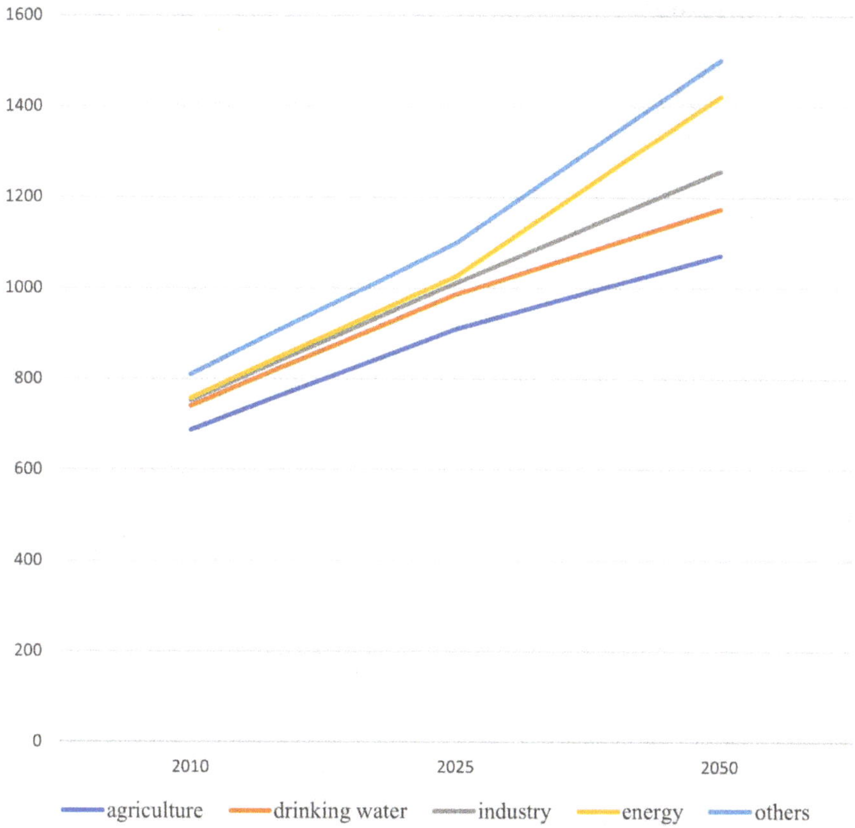

Fig. 3: The water demand variation by different sectors with respect to time (McCarty 2011).

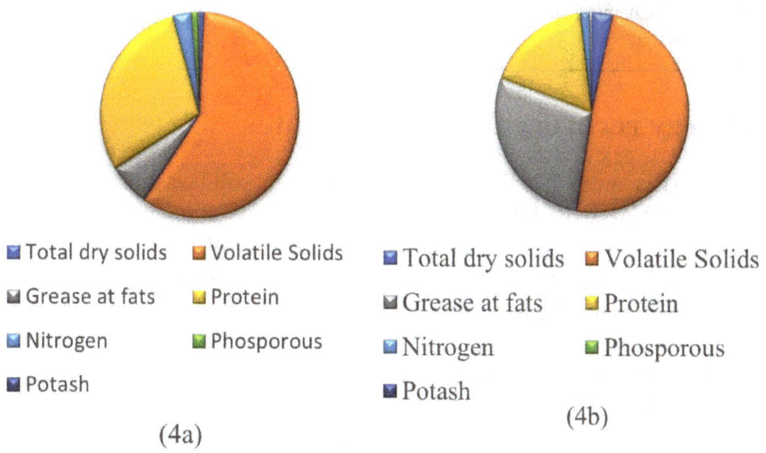

Fig. 4: (a) Composition of WAS and (b) Composition of Primary Sludge (Weemaes 1998).

7.4 Technologies for energy recovery from wastewater

The water should be operated through some unit operations depending on the contaminants and impurities in it. The materials like plastics, surgical disposals, paper, heavy metals and other substances can be separated through mechanical operations like filtration and are represented as per their concentration percentage in Fig. 4; these can be treated as municipal solid wastes from which the energy can be extracted. Various treatment stages and equipment used for wastewater are described in Table 1.

Waste water may be produced from households, commercial purposes and process industries consisting of biological contaminants like suspended solids, dissolved solids/liquids, harmful antibiotic-resistant microbes that cannot be removed by traditional purification techniques and numerous nonbiological contaminants like paper, plastics, inorganic matter such as ceramics, toxic heavy metals, dissolved salts, chlorinated hydrocarbons, metal complex hydrocarbons and other polymers that can be removed by various mechanical unit operations. The energy can be recovered from water but the energy, in turn, is required for wastewater treatment in the form of chemical energy, thermal energy, biological form from various microbial, electrochemical energy in the form of Micro Fuel Cells (MFCs). The potential of energy recovered from the wastewater is 5 times the energy that is required for wastewater treatment. Many techniques in application to extract the numerous energy forms from wastewater are represented in Table 2.

Table 1: Treatment stages involved in energy recovery from wastewater [Weemaes and Vestraete 1998].

S. No	Treatment stages	Equipment used
1	Primary (Mechanical)	Primary Clarifiers, Primary Settling Tanks
2	Secondary (Biological)	Bio Reactors, Aeration Tanks, Bio Contactors, Bio Towers, etc.
3	Tertiary (Chemical and thermal)	Micro filters with synthetic membrane, Expanded Granular Sludge Bed (EGSB) Reactor, Up-flow Anaerobic Reactors and Evaporators

7.5 Energy recovery from solid wastes

On a global basis, around 70:19 ratio of the municipal solid waste is being landfilled and recycled where only 11% is converted from waste to energy forms (Nowling 2018). The solids separated from the wastewater usually may be considered as municipal solid wastes which include garbage wastes, food scraps, waste woods, surgical disposals, etc. The general composition of municipal solid waste on a global basis is shown in the below Fig. 5.

The organic content in the wastewater is again classified as biodegradable which can be converted to energy forms using biological reactions with microbial activity and non-biodegradable compounds are converted to various energy forms like fuel oil and light gases that also can be used as fuels with the help of thermal and chemical operations as represented in the Fig. 6. The biochemical processes implemented are the most economic as only anaerobic conditions are to be maintained at lower

Table 2: Various techniques for energy recovery from the organic contaminants of wastewater.

S. No.	Methodology (Type of Conversion)	Technological Option	Mechanism	Output Energy Form	Applicable Feed Stocks
1	Thermal	Incineration	Thermal degradation is followed by the oxidation of waste at high temperature and optimum aeration	Steam, pure flue gas and ash	Industrial liquid feedstocks, Urban solid feedstocks
2	Biochemical	Anaerobic digestion	Enzymatic decomposition of organic waste having high moisture with biodegradable content in absence of oxygen using suitable micro-organisms at the required temperature	Bio-methane	Industrial liquid feedstocks, Urban solid and liquid feedstocks
3	Thermochemical	Pyrolysis and Gasification	Catalytic/thermal degradation of organic wastes with low moisture content and high non-biodegradable content	Fuel oil and gas	Industrial and urban solid wastes
4	Electrochemical	Microbial Fuel Cell	Electro decomposition of wastes using microbes as charge conducting medium and microbial cells as electrodes	Electricity and bio hydrogen	Industrial and urban liquid feedstocks

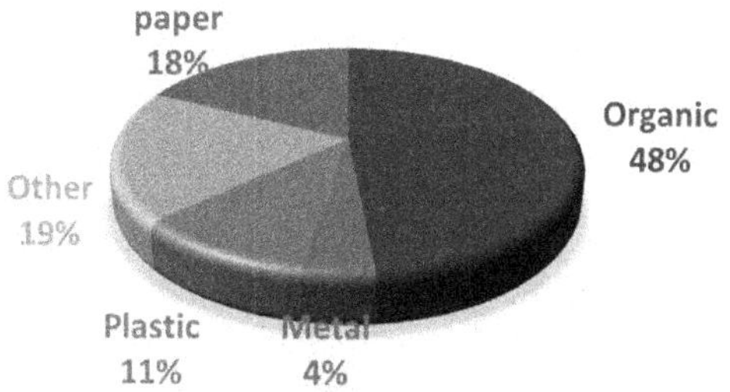

Fig. 5: General composition of global municipal solid wastes (Carmen and Bisanz 2009).

temperatures when compared to the thermal conversion processes like pyrolysis and gasification but the crucial complication of the biological processes is the requirement of preprocessing the solid waste as the biological or biochemical treatment can be done economically for biodegradable compounds.

7.6 Thermochemical techniques

In thermochemical operations, both biodegradable and non-biodegradable compounds can be converted to fuel oil and fuel gases, so urban municipal wastes which are dumped into the water can be separated and treated using thermochemical

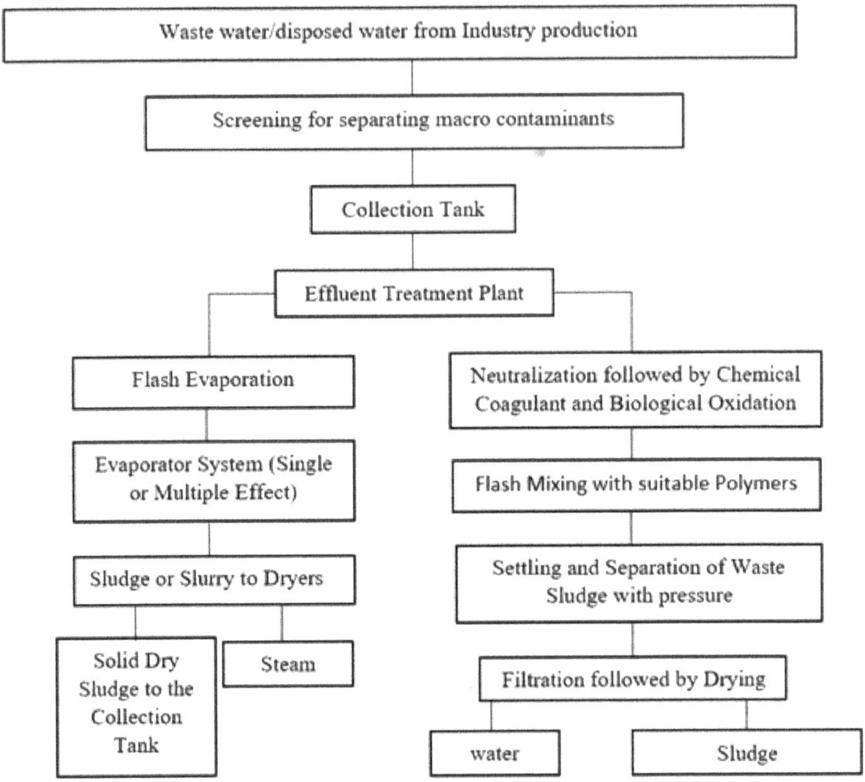

Fig. 6: Flow sheet representing the water-sludge separation methods.

mechanisms as they have the main advantage of generating the process waste at lower volumes and they would also require much lesser place than landfills. They are operated under anaerobic conditions which tend to the prevention of emitting greenhouse gases like carbon dioxide, carbon monoxide, NOx and other oxides of sulfur even if they decompose some organic pollutants like halogenated hydrocarbons which are highly toxic and acidic. The products obtained from these operations can be used as the raw materials for secondary fuels and these fuels have appreciable calorific values and thermochemical operations can be recommended for energy generation by making an economic design of the equipment as the cost analysis plays a major role in the economy of the operation as well as the system.

7.6.1 *Incineration*

- Reduces the volumes of waste with higher efficiencies
- Effective removal of contaminants
- Emits CO_2, CO, N_2O, SF and other greenhouse gases like CFS
- The pollutants affect the effectiveness of the steam cycle
- To get rid of dioxin content, there must be optimum residence time of raw materials in the incinerator

- The performance can be improved by making the equipment free from exhaust gases.

7.6.2 *Gasification*

- Converts organic wastes to the fuel gases and syn-gas
- Wooden waste, biomass, coal and other solid contaminants can be converted to gases
- The output products would be 15–40% syngas (Lower Heating Value is 4000–13000 KJ/Nm^3) and 4.5–9% by mass of CH_4 (Násner et al. 2017)
- Energy in the form of biofuel, electricity and heat can be obtained from syn-gas and many chemical intermediates with numerous industrial applications can be obtained
- An eco-friendly process with higher efficiencies
- The main disadvantage of gasification is that the variable size of raw materials varies the calorific value, and in order to maintain the optimum calorific value consistency, the area available for gasification must be kept as large as possible, which tends to load the raw materials with the smallest possible size.
- Techniques like plasma gasification help to melt the glass, metals and decompose the complex hydrocarbons at high temperatures from 2700 to 4500°C in the inert medium and electrical energy and the syngas is produced at higher temperatures which completely decomposes the dioxins and furans (Fabry et al. 2013, Sanlisoy and Carpinlioglu 2017). Table 3 explains the domination of gasification technology over incineration and pyrolysis.

7.6.3 *Pyrolysis*

- Thermal degradation of organic wastes is done at the temperature of 400-900 °C depending on the type of raw material taken
- An inert atmosphere is maintained to avoid the oxidation of hydrocarbons (CO_X, NO_X, SO_X)

Table 3: Summary of the performances of the MSW treatment technologies (Wilson et al. 2013).

	Incineration	**Pyrolysis**	**Plasma Gasification**	**Conventional Gasification**
Capacity (tons/day)	250	250	250	250
Yield Efficiency (KWh/kg)	0.5	0.3	0.4	0.8-0.9
Power Generation Capacity (KWh/day)	160000	180000	108000	224000
Unit Cost INR/KWh installed	31560	16106	72500	9068.75
Unit Cost (INR/ nominal ton/day)	36275	11600	69600	8120

- Classified as slow, fast and flash pyrolysis, depending on the operating conditions
- Mostly used reactors for pyrolysis are fixed bed reactors, fluidized bed reactors and rotary kilns
- A fluidized bed can be continuously operated but the gas produced would have a low calorific value and the fixed bed leads to uneven heating, which may corrupt the reaction and produce unwanted char sludge which requires more energy to be removed.

7.7 Biological techniques

7.7.1 Gas production from sludge digestion

Sludge digestion, as shown in Fig. 7, is the process of digestion of bacteria in the organic waste that produces gas which can be used as a fuel to produce energy. Anaerobic digestion of this sludge content produces primarily carbon dioxide and methane. The material that is only left after the digestion is called the digestate which is rich in nutrients and can be used as a fertilizer for crops. Biogas is produced to serve as a fuel for the combustion engines, which convert it to mechanical energy to generate power. This is the most common method used to produce electricity; apart from this, there are also different methods to produce power.

Economic advantages
Reduction of manure purchase, in the context of an increase in the cost of fossil energies.

Agronomic advantages
Organic waste at competitive prices, insect elimination at the storage pit and odours' suppression.

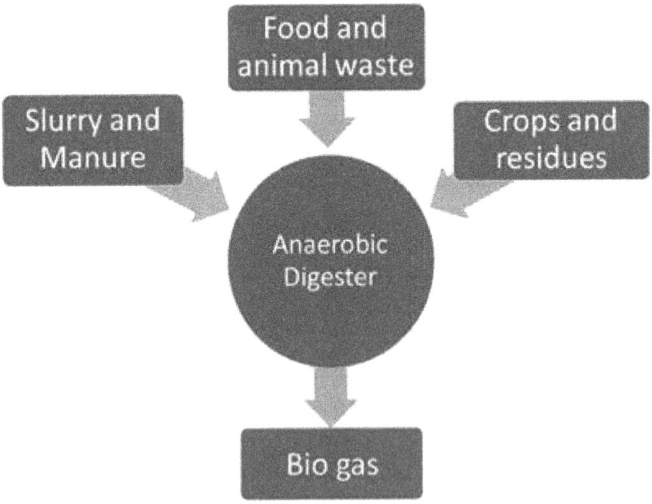

Fig. 7: Outline of the anaerobic digestion of sludge for producing the biogas.

Environmental advantages

Biogas produced by anaerobic digestion is a source of renewable energy. There is sustainable management of organic waste and reduction in the pollution by nitrogen fertilizers.

7.7.2 *Sludge Incineration*

Incineration of the sludge involves the complete or partial conversion of organic solids to oxidised end products as shown in Fig. 8. The sludge is heated in a furnace which allows complete combustion of the sludge and the surplus heat energy. From some research work, it has been proven that complete reduction has been taking place and about 5–7% of ash is left. The heat produced can be used to heat the water by converting the water into steam and thereby to the turbine and generating electricity.

Advantages of sludge incineration

- Destruction of pathogens and toxic compounds
- Energy recovery potential

Disadvantages of sludge incineration

- High capital and operating costs
- Proper O&M practices

Organic matter that is still present in the water, such as mother liquors from different pharmaceutical industries like aromatic compounds, pesticides, plasticizers, and other chlorinated compounds, can cause immune bacterial strains to develop that can have a variety of negative effects on people and marine life even at trace levels,

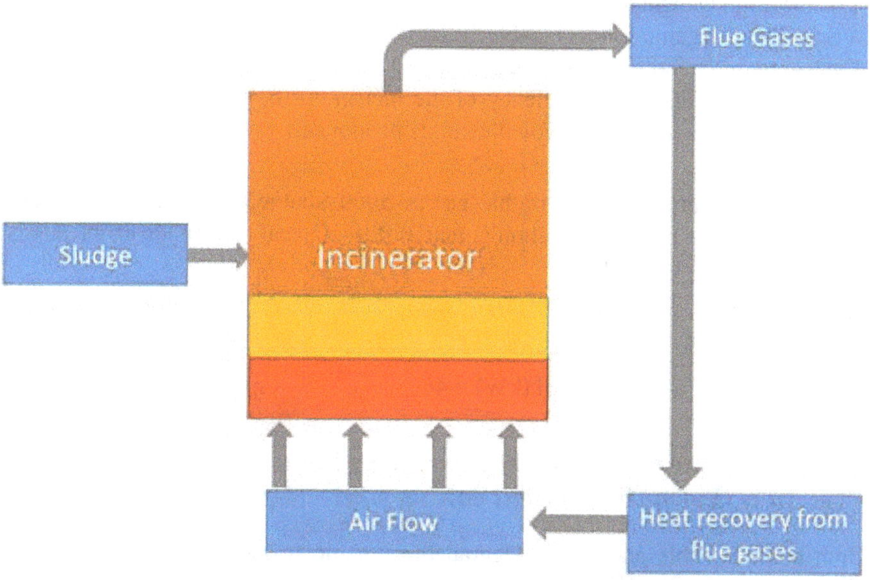

Fig. 8: Incineration of the sludge which results in the production of heat.

which can resist the antibiotic activity and the water might have higher levels of toxicity. These organic contaminants can be degraded or decomposed by various operations that are listed below

(a) Air Stripping
(b) Bio Degradation
(c) Chemical Degradation by Sorption
(d) Volatilization

Nucleophilic displacement reactions like hydrolysis help in the biodegradation of longer chain hydrocarbons and other unsaturated aliphatic compounds using adsorption of polar components which are soluble in water or hydrophilic adsorbents to separate them from the other compounds based on polarity. The other complex organic compounds undergo aerobic techniques using biogenic procedures with the help of macrophytes, hydrophytes and carps which have rapid growth tendency and short reproductive nature by the application of light and thermal energy.

7.7.3 *Assessment of energy recovery potential of a biodegradable organic compound*

The biodegradable organic waste present in the water can be used to get an output energy form, and the assessment of a biodegradable organic compound can be done by taking a basic case study of a biochemical operation.

The benefit of biodegradable organic waste present in municipal wastewater is that this could be converted into various commercial energy forms. This can be elaborated using a case study as follows:

Let the total weight of waste be W (tons) and Total Organic/Volatile Solids: VS is 50%, Organic biodegradable fraction be 66% of Volatile Solids, then the biodegradable fraction will be = $0.33 \times W$ Let the typical digestion efficiency of the incinerator be 60% where the typical bio-gas yeild: B (m3)/kg of Volatile solids evaporated = $B \times 0.60 \times 0.33 \times W \times 1000 = 198 \times B \times W$
The typical Calorific Value of bio-gas is 5000 kcal/m3, the energy recovery potential and the power generation potential if the typical conversion efficiency is η% will be determined as:
Energy recovery potential (kWh) = $B \times 5000/860 = 1151.25 \times B \times W$
Power generation potential (kW) = $1151.25 \times B \times W/24 = 46.97 \times B \times W$
Net power generation potential (kW) = $46.97 \times B \times W \times \eta$

7.7.4 *Assessment of energy recovery potential of organic waste in a thermal-chemical conversion*

Both biodegradable and non-biodegradable organic wastes can contribute to the energy output in case of a thermochemical operation for energy recovery; in such a scenario, the assessment of energy recovery potential of organic waste in a thermal/

chemical conversion is explained with a detailed case study of thermal-chemical conversion.

Let the total organic waste from a thermal-chemical operation be W tons, Net Calorific value be k-Cal/kg
Thus, Energy recovery potential (kWh) = LCV x M x 1000/860 = 1.16 x LCV x M
Potential for power generation (kW) = 1.16 x LCV x M/ 24 = 0.048 x LCV x M
Let the typical yield or efficiency for conversion be η %
Thus, Net power generation potential (kW) = 0.048 x η x LCV x W
If the NCV = 1200 k-Cal/kg., then Net power generation potential (kW) = 57.6 x η x LCV X M

7.8 Mechanical unit operations for energy recovery from waste water

Many Asian countries have been focusing on the mechanical unit operation for 100% energy recovery and water sustainability which mandated treatment operations like Zero Liquid Discharge (ZLD) and Minimal Liquid Discharge (MLD) as they eliminate or minimize the wastewater dumping by removing all the contaminants with higher efficiency. As the scope in implementing ZLD and MLD has been increasing, energy efficiency has also become one of the major focal points of the industrial sectors, especially in Asia. Many technologies are innovated and have come into implementation for energy recovery and also resource recovery.

7.8.1 Ultra-High Pressure Reverse Osmosis (UHPRO)

Ultra-High Pressure Reverse Osmosis (UHPRO), as shown in Fig. 9, can help in minimizing both the installation cost and operational cost as the traditional ZLD and MLD operation would be operated at high intense thermal treatments which can affect the cost as well as the working efficiency of the equipment to time. So, there will be good progress in adopting UHPRO for both ZLD and MLD to reduce the energy intensity to treat the wastewater; this system can be associated with the Ultra-High Pressure Pumping System for the enhancement of performance efficiency and reliability. There is a device known as Ultra PX that makes UHPRO even more cost-effective by reducing the waste of excess energy during the UHPRO process; thus, the size of the thermal plant will be reduced as the system returns pressure energy, which ultimately minimizes the total energy costs and the emissions can also be reduced.

7.8.2 Evaporator Systems-Single Effect (SEE) and Multiple Effect Evaporator (MEE) for ZLD

Evaporators usually play a crucial role in recovering energy as well as resources using ZLD to purify the water to 100% for which various unit operations like

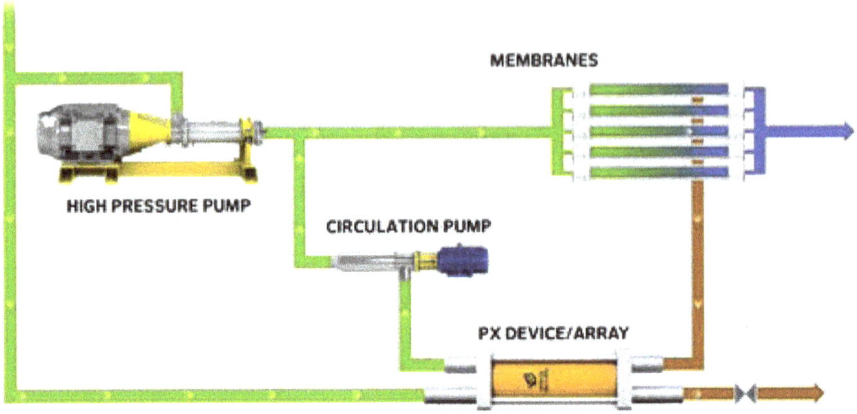

Fig. 9: Ultra-High Pressure Reverse Osmosis Outline (KYsearo 2015).

Flash evaporation and drying techniques can be used for enhancing the system performance with higher efficiencies. The wastewater produced from the production of pharmaceutical and other process industries has high quantities of toxic and acidic content which cannot be directly drained out to the other water bodies and is sent to the effluent treatment plants which has the maximum potential of extracting pure water from the wastewater mixture.

Effluent Treatment Plants (ETPs), represented in Fig. 10, have large tanks to which the wastewater is channelled and settled for many hours to separate the high dense contaminants from the wastewater by sedimentation. The water is then transferred to the effluent treatment plant which usually uses the unit operations like evaporation followed by drying. Based on the characteristics of wastewater, type of effect is selected. The evaporator systems are categorised into two forms that are distinguished in Table 4.

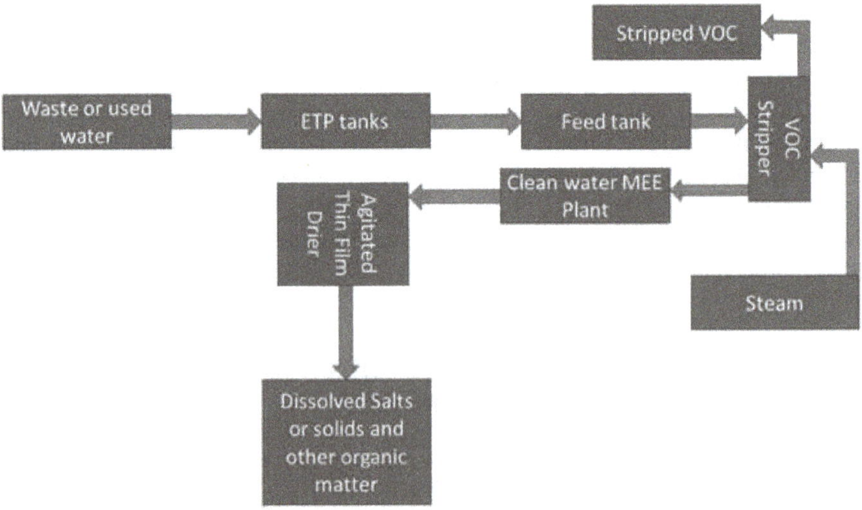

Fig. 10: Separation of dissolved salts and other organic matter from industrial waste water in ETP.

Table 4: Summarizing the differences between Single Effect Evaporator System and Multiple Effect Evaporator System.

Single Effect Evaporators	Multiple Effect Evaporators
• Used when cheap steam is supplied • Low economy as steam is not reused • Mass and energy balance contributes to the heat requirements for a single effect • Perfect mixing should be done to make sure that the concentrations of evaporator product and the solution inside evaporators are to be the same • Uses 1:1 steam to evaporate water • Used for large capacity operations • Runs at constant temperature and pressure conditions • More wastage of steam	• Used for heavy applications by producing and using the high-quality steam • More economy as steam is produced in one effect, which is used as heat input to the consecutive effect • Perfect mixing should be done to make sure that the concentrations of evaporator product and the solution inside each evaporator are to be same, respectively • Uses lesser than 1 kg of steam from external sources to produce 1 kg of water vapour • Uses large capacity operations just like single effect evaporator system • Each effect in a multiple-effect evaporator system runs with different pressures and temperatures • Steam wastage is the least possible thing that takes place as the steam generated in one effect is the heat input to the very next effect

The Multiple Effect Evaporator System is installed in maximum scenarios to recover the energy from wastewater with an appreciable economy and some additional energy regulation techniques like flash evaporation, vapour compression, vapour bleeding, etc., which enhance the waste sludge separation without using excess energy. Depending on the types of feedstocks from the collection tank of ETP, the evaporators are classified into numerous categories which are listed below:

(a) Horizontal Tube Evaporators
(b) Horizontal Spray Film Evaporators
(c) Short Tube Evaporators
(d) Basket Type Evaporators
(e) Long Tube Evaporators
(f) Climbing Film Evaporators
(g) Falling Film Evaporators
(h) Forced Circulation Evaporators
(i) Plate Type Evaporators

The main disadvantages of Multiple Effect Evaporators are that they must be calibrated regularly as the mother liquors are highly corrosive, so there might be a chance of choking which happens due to the sticky sludges deposited inside the shell and tubes of evaporators. The cleaning of such highly sticky deposits requires concentrated HNO_3 wash and high-pressure water wash, which might weaken the physical and mechanical properties of the evaporators.

Depending on the efficiency factor and feeding mechanism, the Multiple Effect Evaporator System is categorized as

(a) Forward Feed type
(b) Backward Feed type

(c) Mixed Feed type
(d) Parallel type

The outline of the machine following the above list of feed types of wastewater feeding to the evaporators and collecting the recovered sludge is explained in detail along with the steam flow through each effect in Fig. 11. The sludge is then sent to the drying system and the dry solids are collected and used for energy generation sources as they can be biomass if processed due to their appreciable calorific values.

The water is recovered in the form of steam which can be used as a heating medium and as an input material to produce many other energy forms such as electricity from steam turbines, hydrogen production, biogas production, etc. The water can also be condensed as the water-energy can be extracted through its motion in the form of water wheels, hydroelectricity, and water clocks. Water energy is classified into five categories and is described in Table 5.

(a) Hydro-Power
(b) Hydro-Electricity
(c) Ocean Thermal Energy Conversion
(d) Tidal Power and
(e) Ocean Wave Power

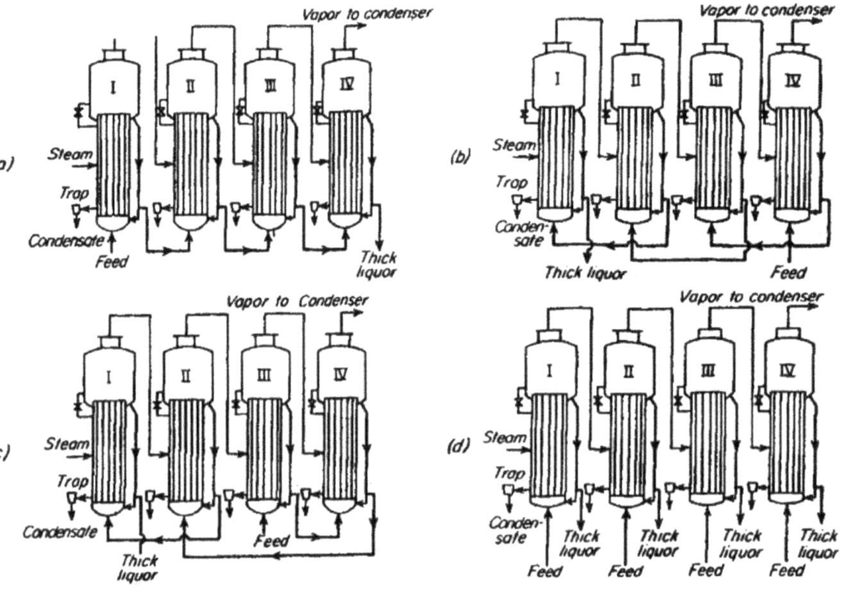

Fig. 11: Outline of Waste Water Feed Mechanism types delivered to the MEE systems (McCabe and Smith 1967).

Table 5: Overview of various characteristics of water energy forms.

	Description	Advantages	Disadvantages
Hydro-Power	Low-level electricity generation uses the kinetic energy of water to rotate turbine which converts water energy to mechanical energy and this is converted to electrical energy by generators connected to the turbines	Used for irrigation purposes, produce low-level electrical energy with higher efficiencies from 70% to 90%	• High maintenance cost • Cannot be installed in cold climatic zones as the water might be frozen • Efficiencies are water flow dependent
Hydro-Electric (Valenti 1992)	Constructing a hydro-electric dam to utilize the kinetic energy of water for producing electrical energy by the turbine rotation and generator	• Power generation without combustion of fossil fuels. • No particulate matter is emitted including other greenhouse gases. • No extinction and high-potential savings.	• More damming results in heavy floods to large areas • Water quality is affected • Catastrophes could take place if the dam fails • Nutrients and minerals are blocked due to the silts deposited • Possibilities of mercury and lead leakages may be accumulated in the tissues of fish which kill them and ultimately humans who consume them
Ocean Thermal Energy Conversion (Vega 2005)	Divided into two types of technologies: closed cycle and open cycle, where ammonia vapours produced from the heat received from the low-temperature waters are used as the working fluid in a closed cycle through which the turbines are rotated and electrical energy is generated where the ammonia is recovered and reused. In the case of the open cycle, water vapours are heated using heat from the atmosphere and they rotate the turbine and the electrical energy is generated	• A renewable form that is eco-friendly and also reduces the dependency on fossil fuel burning to generate energy. • No greenhouse gases generation. • Keeps the marine ecosystem healthy as it is always. • The water vapours produced are later recovered as portable and pure water as they are free of salts.	• Geographic: Plants should be installed at those places where the surface temperatures of water are 36–40°F • Economic drawbacks: Operating costs and initial costs are very high

Table 5 contd. ...

...Table 5 contd.

	Description	**Advantages**	**Disadvantages**
Tidal Power (Freeman 2004)	Use tides to generate electricity, assisted by the gravitational pull of the moon during nights and the sun during days. Tidal barrages can be used to extract the energy in a tidal power generation station. Barrages can be constructed at the mouth of any sea coast such that a minimum of 5 metres of the tide height from its foot should be taking place to make barrage work	• A renewable form that is eco-friendly and also reduces the dependency on fossil fuel burning for energy generation. • No greenhouse gases generation. • Keeps the marine ecosystem healthy as it is always. • An energy source that is free.	• Expensive for both installation, and maintenance. • Tides only take place in an optimum range, around 10 hours a day. • This is not consistent
Ocean Wave Power (Weiss and Peter 2001)	The alternate form of solar power as it uses sun pulls on a gravitational basis to cause a wave and the kinetic energy of the wave is used to move a turbine and using a generator the electrical energy is generated	• Wave power stations can be constructed at the mouth of any sea coast • A renewable form that is eco-friendly and also reduces the dependency on fossil fuel burning for energy generation. • No greenhouse gases generation. • Keeps the marine ecosystem healthy as it is always. • The energy source is free.	• Expensive for both the installation, and maintenance. • Disrupts the material of turbines and other equipment as they are constructed in the middle of sea or ocean

7.9 Metal recovery from waste water

The metal contaminants might be dissolved or segmented in the wastewater which might show some adverse impacts on the living beings who are consuming that water. Some of them are toxic enough to cause various health issues which are shown in Table 6 (Babel and Kurniawan 2003). These metals can be rcovered from the wastewater by treating the water by filtration, membrane separation, adsorption and other unit operations, and some of them are briefly elaborated in this chapter.

Table 6: Toxic metals found in the wastewater (Babel and Kurniawan 2003).

Metals	Toxic effects	Composition (mg/L)
Arsenic (Sb)	Skin manifestations, visceral cancer, vascular disease	0.050
Cadmium (Cd)	Dialysis issues, cancers	0.01
Chromium (Cr)	Headache, spewing, cancer	0.05
Copper (Cu)	Liver damage, insomnia	0.25
Nickel (Ni)	Dermatitis, nausea, asthma, coughing, carcinogen	0.20
Zinc (Zn)	Depression, lethargy, increased thirst	0.80
Lead (Pb)	Kidney disease, nervous system failure	0.006
Mercury (Hg)	Rheumatoid arthritis, kidney problems, mental disorder	0.00003

7.9.1 *Chemical precipitation*

Chemical precipitation is a process in which the metals dissolved in a liquid are separated from a solution by increasing the saturation of the liquid to super saturation stage so that the solubility of solids or metals will be minimized. Then these metals can be separated from the treated water as precipitates. This process is also used to de-ionize the water by reducing their solubility. In general, for commercial applications, lime and lime stone are used for purifying the water as they are available at low prices and the process of separation would be very simple with these agents. There are some limitations with these methods which include sludge formation, slow process and low efficiencies with respect to settling of the precipitates.

7.9.2 *Ion exchange method*

Another successful method is to remove metals from wastewater and is also used by many industries mainly to remove heavy metals. The disadvantages of the use of this process are that this process cannot handle concentrated metal solutions, is non-selective, and is also highly sensitive to the PH of the water.

7.9.3 *The electro-winning metal recovery process*

The metals dissolved in the wastewater are separated by electrolysis followed by a chemical reaction. The metals are deposited on the electrodes that are immersed in the water. The anode for this method will be insoluble in the water. A similar method is also implemented for separating the metal impurities by producing steam;

the metal impurities are charged as the plasma is passed through an electric field called corona. Electrostatic precipitators are used to attract the charged metals and they have electrodes to which the particulates get attracted. The main disadvantage for this kind of method is that they are operated at higher temperatures which require high installation and maintenance.

7.9.4 *Adsorption*

Nowadays, this is the new emerging technology to remove metals from wastewater. When a substance is transferred from the liquid phase to the surface, the solid particle with a mass transfer process is known as the adsorption process. The three main steps involved in the adsorption process are as follows:

- The transport of the metal from the wastewater to the surface of the same.
- Adsorption on the surface substance.
- Transport within the sorbent particle.

The key factors in the selection of a suitable adsorbent are its cost-effectiveness and applicability. Adsorption is divided into different parts according to the sources of wastewater which are explained below:

7.9.5 *Adsorption for industrial wastewater*

The metals are recovered from the wastewater produced from the industries by several methods and among them adsorption plays a crucial role in separating the metals. Processes like physisorption are subjected to remove the metal impurities by the means of physical separation without any chemical reaction. The chemisorption, on the other hand, uses a chemical reaction to separate the adsorbate from the water. The adsorbents like activated carbon, silica gel, and other commercial materials are used to clean the industrial water.

7.9.6 *Adsorption for biological waste and agriculture waste*

This type of adsorption is mostly known as bio-sorption. Resources such as hazelnut shells, rice husk, jackfruit, maize, etc., can be used as the metal recovery adsorbent after modifying chemically or converting to activated carbon by heating. Resources like cross-linked chitosan, starch gel, alumina, etc., are removed from wastewater by this process.

7.9.7 *Membrane Filtration*

Membrane filtration is also given a huge amount of attention for wastewater treatment as it is not only capable of removing suspended solids and organic compounds but is also used to remove metals from the wastewater. Ultrafiltration, which is a type of membrane filtration, uses a permeable membrane that is used to separate metals, macromolecules and suspended solids from the wastewater. Metal will be removed or not depending upon the pore size of the membrane, which varies from 2 mm to 20 mm. The molecular weight of the compound also has a range of 1000–1000000 Da.

7.9.8 Electrodialysis

It is a membrane separation process that might be a combination of both electrolysis and dialysis. In this process, ionized ions are passed through an ion-exchange membrane by applying an electric potential to remove the metals from the wastewater.

7.9.9 Photocatalysis

Environmental pollutants in wastewater are quickly and effectively reduced using this method. Charge carriers that travel to semiconductor surfaces that can undergo reduction and oxidation are used in this process. By using reducing and oxidising species in the wastewater that have the suitable redox potential, the metals separate from the wastewater and are subsequently extracted from it.

7.10 Recent advances in energy recovery from waste water

The increase in the demand for renewable energy from the past few years has led the biofuels to come into the picture, which can be broadly classified as solid, gaseous, and liquid fuels, which are in turn produced by bio-renewable resources, but, they have a lower energy density in terms of volume or mass. Their higher heating value is about 15–20 MJ/kg, which is less than coal's and oil's respective values of 30 and 40 MJ/kg. Thus, with a proper conversion method, an improved energy content from biomass can be obtained; depending on the technology, there are three generations classified to produce biofuels. Table 7 represents the types of generations and their respective raw materials used for energy recovery and various biomass conversions are elaborated in Fig. 12.

Biogas and bio-oils are produced from the biochemical and thermochemical conversion processes. Biogas is commonly produced by anaerobic digestion of chemicals and consists of methane as a major constituent with other gases which can be used as a fuel for a wide range of production applications including heat, steam, electricity generation and also as a vehicle fuel whereas bio-crude oils on the other side are produced from thermochemical conversion. There are two procedures, mainly pyrolysis and Hydro-thermal Liquefaction (HTL), in which HTL is the preferred one as it has an appreciable potential for converting high-water-content feedstocks

Table 7: Types of generations and their respective raw materials used for energy recovery.

Type of Generation	Raw materials
First Generation	Food crops (Sugar, Starch, Vegetable oils)
Second Generation	Non-food (Agroforestry Residues)
Third Generation	Algae

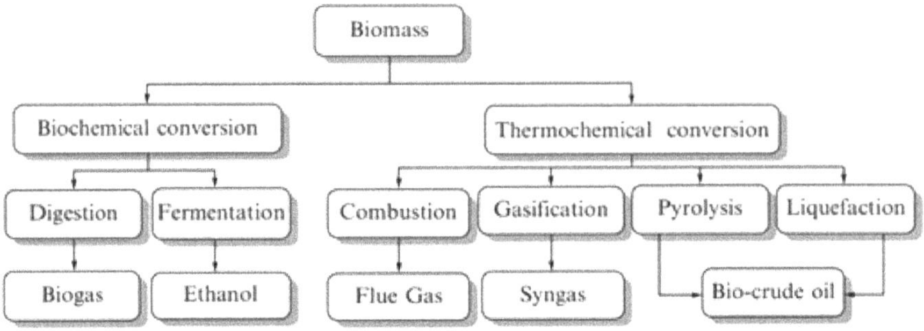

Fig. 12: Various biomass conversion methods (Wen et al. 2009, Devi 2005).

such as wastewater sludge and microalgae at their mild operating temperature of 200–400°C but at elevated pressure.

7.10.1 Hydro-thermal liquefaction

Hydrothermal Liquefaction is a technique applied to convert biomass to biofuels, operates at high pressure and relatively low temperature and mainly uses water at subcritical conditions. HTL is mostly performed under batch reactors. But when it comes to large applications, the continuous flow reactors are more desirable. The continuous flow reactors are used as the bio-oil can be produced in shorter reaction times. Table 8 shows the case studies on energy recovery from wastewater.

7.10.2 Comparison of batch and continuous for the respective feedstock

From Figs. 13 and 14, it is evident that the continuous reactors are far better in terms of giving the oil yield for the respective feedstocks given. They also are the basis for the energy-efficient utilisation reactors and make the process economically feasible.

Table 8: Case studies on the energy recovery from wastewater (Nazari et al. 2021).

Reported Years	Related Case Studies
1940	An early study suggested that cornstalks, corncobs, sugarcane, etc., could be converted to petroleum-like products (Theegala and Midgett 2012).
1984	HTL of wastewater sludge was reported by Kranich and Eralp. Sewage sludge was converted to oil at different reaction temperatures (Behrendt et al. 2008).
1986	Battelle Memorial Institute in the US has published a patent on the conversion of sludge to bio-oil titled as sludge-to-oil reactor system (Goudriaan and Peferoen 1990, Ruiz et al. 2013).
1990	The Shell Research Laboratory in Amsterdam published work on the hydrothermal upgrading process of waste biomass (Kalogo and Monteith 2008, Tyagi and Lo 2013).

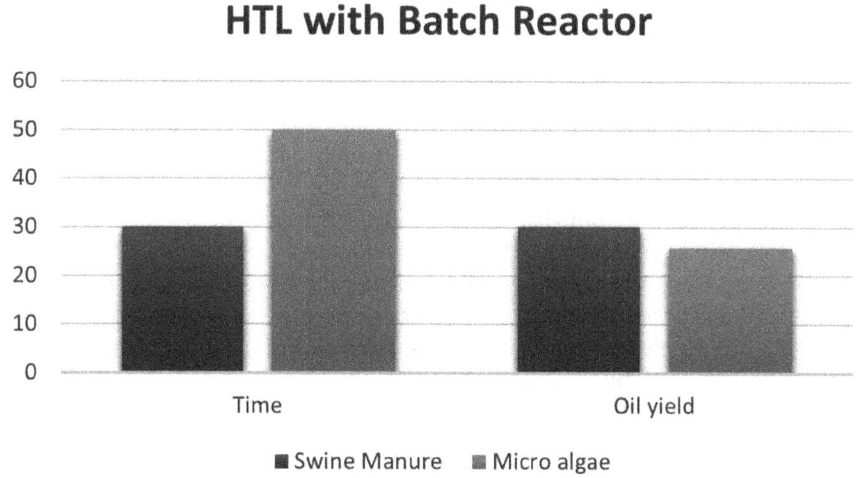

Fig. 13: Batch reactor operations for feedstocks of swine manure and microalgae (Shuping et al. 2010, Vardon et al. 2011).

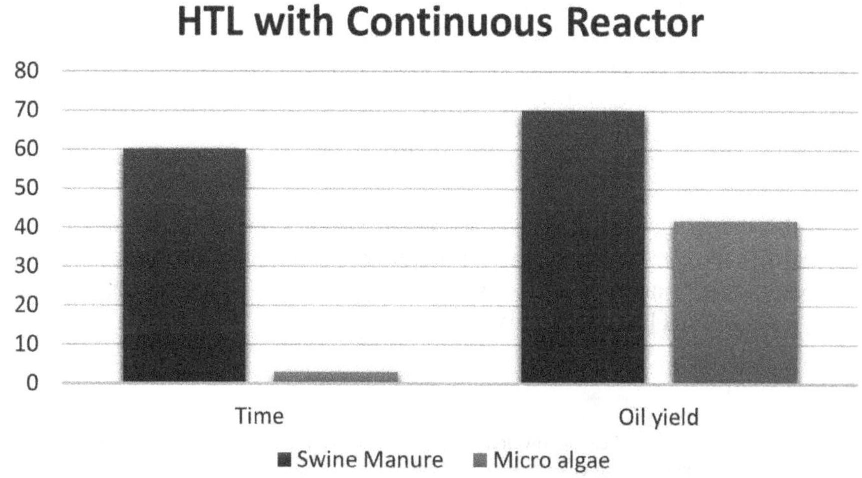

Fig. 14: Continuous reactor operations for feedstocks of swine manure and microalgae (Ocfemia et al. 2006, Jazrawi et al. 2013).

7.11 Conclusion

The extinction of fossil fuels has led every country in the search and exploration of alternative energy forms including conventional and non-conventional or renewable sources, among which water has a very crucial role in generating the energy. Water is used for many domestic and industrial applications, ultimately from which the wastes and contaminants are dissolved and come along with water. They will have certain calorific value which can be recovered in the form of biofuel, gas, electricity, etc. Even the water purified can be reused or can be converted into steam by using mechanical unit operations and can be used as an energy source for electricity and thermal energy. Processes like incineration, gasification and pyrolysis play a vital role in converting the wastewater into various energy forms because they are more efficient. But when it comes to cost analysis, thermochemical operations usually cost more than biological operations. Still, many research studies have been undergoing to extract energy with the help of renewable energy sources as energy inputs as they are reliable and by utilizing them, the dependency on fossil fuels and other conventional can be reduced.

The water from oceans has a great contribution towards energy recovery in the form of tides and waves to extract and convert the kinetic energy and thermal energy of ocean water to mechanical energy using turbines and then to electrical energy using generators. But still, OCET must have many developments as it is the lowest efficient energy conversion area with a large amount of investment which cannot be considered to be in the economy. Many studies and researches have been undergoing to improve the efficiency and cleanness of the ocean technology in the area of energy recovery because ultimately wastewater is drained into the seas from numerous ships and cargos after using the water for maritime purposes.

Metals and toxic heavy metal complexes dissolved in water from various operations like mining, and pharmaceutical mother liquors can be processed to recover the metals as they can be added to some other materials to improve their physical and mechanical properties and this can be the research area for innovating more economic and efficient procedures to recover the metals by considering the environmental impact due to the processes that are implemented for recovering the energy wastewater.

If the energy generation using wastewater becomes one of the primary energy sources, then the cost of energy can be minimized by modifying and innovating some methods as water energy is a kind of renewable energy which does not have any extinction problems like fossil fuels.

References

Assocham; http://www.spml.co.in/Download/Reports/Think-Blue-Effective-Water-Management-June-2019-EY-Assocham-Report.pdf.

Babel, S. and Kurniawan, T.A. 2003. Various treatment technologies to remove arsenic and mercury from contaminated groundwater: an overview. In: Proceedings of the First International Symposium on Southeast Asian Water Environment, Bangkok, Thailand, 24–25 October, pp. 433–440.

Behrendt, F., Neubauer, Y., Oevermann, M., Wilmes, B. and Zobel, N. 2008. Direct liquefaction of biomass. Chem. Eng. Technol. 31(5): 667–77.

Carmen and Bisanz 2009; WEC-World Energy Council. World Energy Resources: Waste to Energy. London: World Energy Council; 2013. pp. 1–14. DOI: 10.1080/09297040802385400; from: https://www.worldenergy.org/wp.content/uploads/2013/10/WER_2013_7b_Waste_to_Energy.pdf

Devi, L. 2005. Catalytic removal of biomass tars; olivine as prospective in-bed catalyst for fluidized-bed biomass gasifiers (Thesis). Technische Universiteit Eindhoven.

Fabry, F., Rehmet, C., Rohani, V. and Fulcheri, L. 2013. Waste gasification by thermal plasma: A review. Waste and Biomass Valorization. 4: 421–439. DOI: 10.1007/s12649-013-9201-7.

Freeman, Kris. Tidal Turbines: Wave of the Future? Environmental Health Sciences (January 1, 2004): 26.

Gautam, R. Rajneesh-Gautam/publication/335228238 Sewage_Generation and Treatment Status for the City of Delhi its Past Present and Future Scenario. A Statistical Analysis /links /5d59131e299bf151badce21f /Sewage-Generation-and-Treatment-Status-for-the-City-of-Delhi-its-Past-Present-and-Future-Scenario-A-Statistical Analysis.pdf? origin=publication detail.

Goudriaan, F. and Peferoen, D.G.R. 1990. Liquid fuels from biomass via a hydrothermal process. Chem. Eng. Sci. 45(8): 2729–34.

Heidrich, E.S., Curtis, T.P. and Dolfing, J. 2011. Determination of the internal chemical energy of wastewater. Environ. Sci. Technol. 45: 827.

Jazrawi, C., Biller, P., Ross, A.B., Montoya, A., Maschmeyer, T. and Haynes, B.S. 2013. Pilot plant testing of continuous hydrothermal liquefaction of microalgae. Algal Res. 2(3): 268–77.

Kalogo, Y. and Monteith, H. 2008. State of science report: energy and resource recovery from sludge.

KYsearo, https://kysearo.com/how-the-px-pressure-ex-changer-works/; 2015.

Logan, B.E. 2008. Microbial Fuel Cell (Wiley, 2008).

McCabe, W.L. and Smith, J.C. 1967. 2nd ed., pg. 464., 1967 McGraw-Hill.

McCarty, P.L., Bae, J. and Kim, J. 2011. Domestic wastewater treatment as a net energy producer - can this be achieved? Environ. Sci. Technol. 45: 7100.

Modified from Weemaes, M.P.J. and Verstraete, W.H. 1998. Evaluation of current wet sludge disintegration techniques. J. Chem. Technol. Bio-technol. 73(2): 83–92.

Násner, A., Lora, E., Palacio, J., Rocha, M., Restrepo, J., Venturini, O. et al. 2017. Refuse derived fuel (RDF) production and gasification in a pilot plant integrated with an Otto cycle ICE through aspen plusTM modelling: Thermodynamic and economic viability. Waste Management. 69: 187–201. DOI: 10.1016/j.wasman.2017.08.006.

Nazari, L., Sarathy, S., Santoro, D., Ho, D., Ray M.B. and (Charles) Xu, C. 2021. Western University, London, ON, Canada, Trojan Technologies, London, ON, Canada. Recent advances in energy recovery from wastewater.

Nowling U. 2016. Waste to Energy: An Opportunity Too Good to Waste, or a Waste of Time? [Internet]. Available from: http://www.powermag.com/waste-energy-opportunity-good-waste-waste time/?pagenum=1 [Accessed: 20-04-2018].

Ocfemia, K.S., Zhang, Y. and Funk, T. 2006. Hydrothermal processing of swine manure to oil using a continuous reactor system: effects of operating parameters on oil yield and quality. Am. Soc. Agric. Biol. Eng. 49(6): 1897–904.

Owen, W.F. 1982. Energy in Wastewater Treatment (Prentice-Hall, 1982).

Ruiz, Ha, Rodríguez-Jasso, R.M., Fernandes, B.D., Vicente Aa and Teixeira Ja. 2013. Hydrothermal processing, as an alternative for upgrading agriculture residues and marine biomass according to the biorefinery concept: a review. Renew. Sust. Energ. Rev. 21: 35–51.

Sanlisoy, A. and Carpinlioglu, M. 2017. A review on plasma gasification for solid waste disposal. International Journal of Hydrogen Energy. 42: 1361–1365. DOI: 10.1016/j.ijhydene.2016.06.008.

Shuping, Z., Yulong, W., Mingde, Y., Kaleem, I., Chun, L. and Tong, J. 2010. Production and characterization of bio-oil from hydrothermal liquefaction of microalgae Dunaliella tertiolecta cake. Energy. 35(12): 5406–11.

Theegala, C.S. and J.S. Midgett. 2012. Hydrothermal liquefaction of separated dairy manure for production of bio-oils with simultaneous waste treatment. Bioresour. Technol. 107: 456–63.

Treatment of Organic Recalcitrant Contaminants in Wastewater, by Asmita Gupta and Indu Shekhar Thakur Additional information, is available at the end of the chapter http://dx.doi.org/10.5772/66346; 2017.

Tyagi, V.K. and Lo, S.-L. 2013. Sludge: a waste or renewable source for energy and resources recovery. Renew. Sust. Energ. Rev. 25(71): 708–28.

Valenti, Michael. 1992. Storing Hydroelectricity to Meet Peak-Hour Demand. Mechanical Engineering (April 1, 1992): 46.

Vardon, D.R., Sharma, B.K., Scott, J., Yu, G., Wang, Z., Schideman, L. et al. 2011. Chemical properties of biocrude oil from the hydrothermal liquefaction of Spirulina algae, swine manure, and digested anaerobic sludge. Bioresour. Technol. 102(17): 8295–303.

Vega, L.A. 2005. Ocean Thermal Energy Conversion. http://www.hawaii.gov/dbedt/ert/otec/index.html (accessed on September 13, 2005).

Weiss, Peter. Oceans of Electricity. Science News Online (April 14, 2001). http://www.sciencenews.org/articles/20010414/bob12.asp (accessed on September 13, 2005).
Wen, D., Jiang, H. and Zhang, K. 2009. Supercritical fluids technology for clean biofuel production. Prog. Nat. Sci. 19(3): 273–84.
Wilson, B., Williams, N., Liss, B. et al. 2013. A Comparative Assessment of Commercial Technologies for Conversion of Solid Waste to Energy. Boca Raton, Florida: Enviro Power Renewable, Inc; 2013. Available from: https://pdfs.semanticscholar.org/92ba/d2a1a1d4870a57b6fc2263e2e9a9fd882647.pdf.

8

Biofuel Recovery from Wastewater Using Nanomaterial

Muhammad Rizwan,[1,*] *Nayab Raza,*[1] *Ghulam Mujtaba,*[2] *Naveed Ahmed*[1] and *Sheraz Ahmed Memon*[3]

8.1 Introduction

Depletion of natural resources had resulted in an energy crisis which triggered research on finding alternate energy resources (Ren et al. 2021). The researchers are focusing on finding an eco-friendly and sustainable method for the treatment of wastewater to produce bioenergy and clean water. Renewable energy sources are alternative energy sources in comparison to fossil fuels (Sikkema et al. 2021). To come out from such situation, first, second, third, and fourth generation biofuels were explored (Cuong et al. 2021). Figure 1 shows the biofuel types and their sources. The potential of these biofuels needs to be enhanced to make them feasible at the commercial level. Several techniques are used for increasing the potential of biofuels; among them is the use of nanomaterials. Nanocatalysts are used for the enhancement of the biofuel potential of biomass.

8.2 Nanomaterial's characterization

Nanoparticles are characterized based on their physical properties, including size, shape, surface properties, crystallinity, and dispersion state (Kaliva and Vamvakaki 2020). The techniques used for the characterization of nanoparticles include Scanning

[1] U.S.-Pakistan Centers for Advanced Studies in Water (USPCASW), Mehran University of Engineering and Technology, Jamshoro 76062, Pakistan.

[2] Department of Energy and Environment Engineering, Dawood University of Engineering and Technology, Karachi, 74800, Pakistan.

[3] Institute of Environmental Engineering and Management, Mehran University of Engineering & Technology, Jamshoro, 76062, Sindh, Pakistan.

* Corresponding author: drmrizwan.uspcasw@faculty.muet.edu.pk

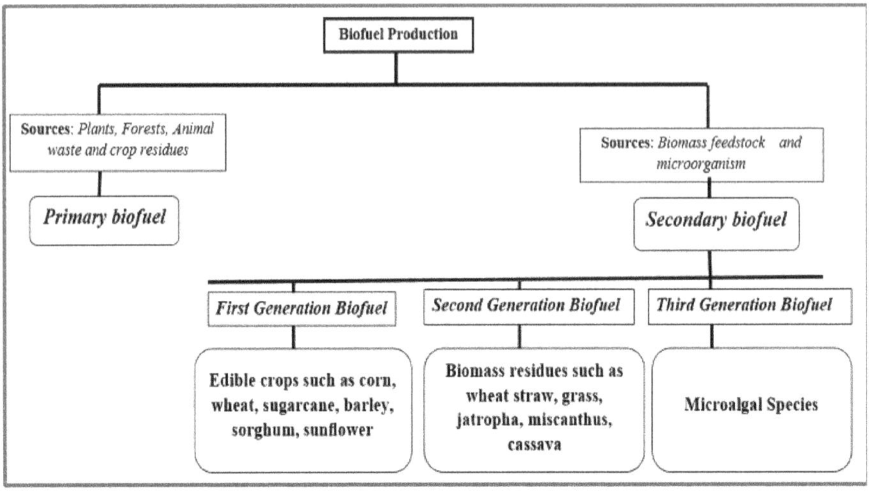

Fig. 1: Sources of biofuel.

Electron Microscope (SEM), Energy Dispersive X-Ray Analysis (EDX), Transmission electron microscopy (TEM), Fourier Transform Infrared Spectroscopy (FTIR), X-ray diffraction (XRD), particle size analyzer (PSA), UV-Visible Spectroscopy, and Thermo Gravimetric Analysis/Differential Thermal Analyzer (TG/DTA) (Akhtar et al. 2018). Figure 2 shows the characterization of nanoparticles based on structure and composition. Table 1 shows wastewater treatment using nanoparticles.

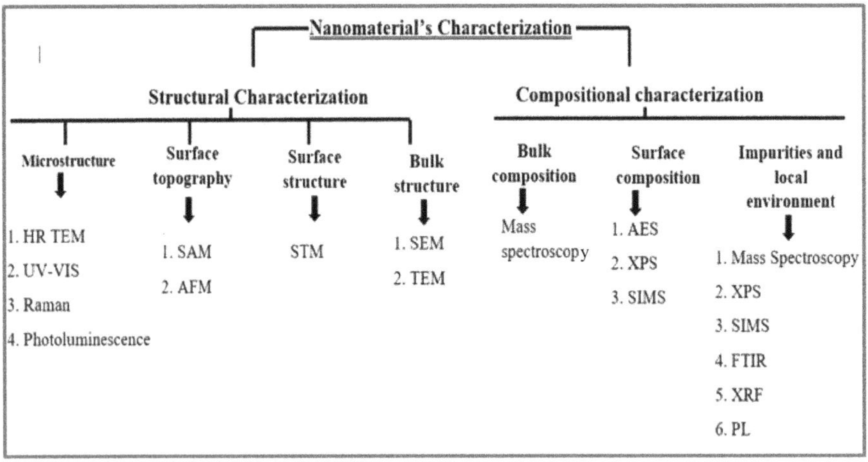

Fig. 2: Nanomaterials characterization: a. (Structural characterization) and b. (Compositional characterization).

Table 1: Wastewater treatment using different nanoparticles.

S. no	Wastewater type	Nanoparticles type	Pollutant Removal	Characterization	References
1	Secondary Industrial Effluent (SIE)	Iron oxide magnetic nano particles	Heavy metals	XRD /TEM	(Almomani et al. 2020)
2	Hazardous wastewater	Carbon nanotubes	Chloramine-T degradation	SEM, EDX/XRD	(Abd Elkodous et al. 2021)
3	Industrial wastewater	Polyethylenimine-functionalized pyroxene nanoparticles	Commercial Red dye	XRD, HRTEM, IR, TGA)	(Hethnawi et al. 2017)
4	Wastewater Effluents	Tungsten oxide nano particles	Degradation of the dye Rhodamine-B	FESEM	(Rahman et al. 2021)
5	Textile wastewater	Nitrogen-doped graphene quantum dot (NGQD)	Chlorothiazid, Carbamazepine, Azo dye, Methylene blue dye	FTIR, XRD, SEM, TEM, and Raman spectroscopy	(Kaur et al. 2018)
6	Tannery Wastewater treatment	Maghemite nanoparticles	Chemical oxygen demand total dissolved solids, and total suspended solids	FTIR, XRD, TEM Dynamic light scattering, and X-ray photoelectron spectroscopy	(Fouda et al. 2021)
7	Metal surface processing wastewater	Zero valent iron nanoparticle	Cr (VI) removal	XRD, Infrared spectroscopy, Scanning electron microscopy	(Bounab et al. 2021)
8	Wastewater	Active chitosan-coated magnetic iron oxide nanoparticles	Heavy metal	TEM, XRD, and XPS	(Radwan et al. 2019)
9	Secondary run-off wastewater (SERWW)	Functionalized Cellulose Nanomaterials	Pb(II) ions	XRD, FTIR, SEM, TGA-DTG	(Tsade et al. 2021)
10	Thermal power plant wastewater	Iron oxide nanoparticles (IONPs	Heavy Metals	DLS, FTIR, Raman spectroscopy, XRD, FESEM, EDS	(Yadav et al. 2020)
11	Pharmaceutical Wastewater	Zinc oxide, Graphene oxide nanoparticles	Levofloxacin	TEM, SEM, FTIR, XRD, absorption spectra and Brunauer Emmett-Teller	(El-Maraghy et al. 2020)
12	Textile Manufacturing Industry	Zero valent iron nanoparticles	dye color	XRD, SEM	(Nigam Ahuja et al. 2020)
13	Distillery wastewater	Ni-graphene nanocomposite (Ni-Gr NC)	MEG-containing wastewater	TEM, XRD	(Elreedy et al. 2017)
14	Wastewater treatment plant (WWTP)	Metal oxide nanoparticles (ZnO and TiO_2)	Dissolved organic matters (DOMs)	TEM	(Zhou et al. 2015)

8.3 Properties of nanoparticles

(a) Optical property

The most important characteristic of nanomaterials is their optical property which can be measured using spectrophotometer (Kim et al. 2021). Optical properties show adsorption and emission of electrons and their transition between two states during nanoparticles' synthesis. Optical activity and property can be seen through the color changes during the nanoparticle's synthesis. When light falls on nanoparticles, photons get excited, and particles jiggle at a specific wavelength and absorb light (Tsegay et al. 2021).

(b) Magnetic property

The nanomaterial's magnetic properties are due to electromagnetic elements such as nickel, iron, chromium, cobalt, and manganese (Chakrabarty et al. 2021). These compounds act superparamagnetic because of their nano size and have great potential to interact with other chemicals (Lu et al. 2021). These nanoparticles are used for specific purposes, having a variety of bare and coated surface coating with its functional group. The magnetic nanoparticles selectively attach to the functional molecules to facilitate movement to desired locations (Chethana et al. 2021).

(c) Size-dependent material

Cohesion between molecules and unit cell volume are two issues governed by the size and density of nanoparticles or nanostructured materials (Zou et al. 2010). It is observed that the cohesion energy per atom decreases with the size of nanoparticles. In nanostructure material, the constant lattice increases with the decreasing size of nanoparticles (Chatterjee et al. 2003). The surface and volume ratio increases with the size of nanoparticles, so the energy increases with a decrease in contact angle (Barisik et al. 2014).

(d) Mechanical property

The mechanical properties of nanoparticles include hardness, elasticity, brittleness, ductility, toughness, and stress (Liu et al. 2019). Nanomaterials have great potential in mechanical properties due to quantum effects and their volume (Wei et al. 2020).

(e) Electrical property

Mishra et al. 2013 studied the electric properties of materials at nanoscale level. Large or bulky material conductivity does not depend upon the dimensions or area of cross-section (Gao et al. 2011). A change in conductivity was observed when shear force was applied to nanotubes (Pandey et al. 2017).

(f) Catalytic property

Recently, researchers have been focusing on nanostructured catalysts with growing physiochemical properties of nanoparticles (Tripathi et al. 2015). The nanoscale catalysts have high-specific surface area and energy that bring about high catalytic

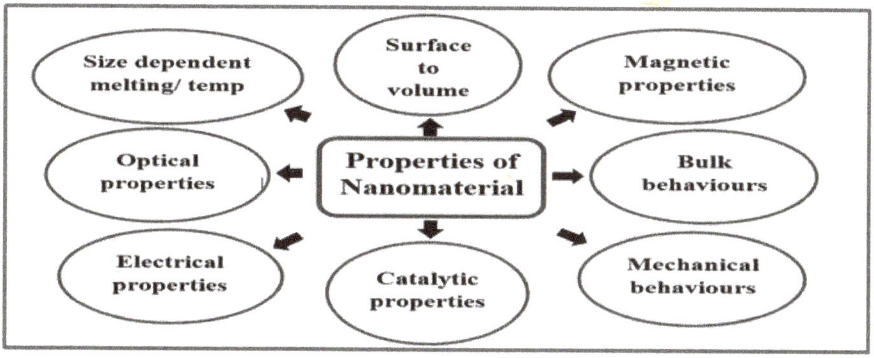

Fig. 3: Nanoparticles' physiochemical properties.

activity of nanoparticles. In nano-crystal selective reactions occur at very low temperature and contain side reactions that help in higher recycling rate and retrieve energy consumption (Wang et al. 2017). Figure 3 shows the physicochemical properties of nanoparticles.

8.4 Synthesis of Nanomaterials

8.4.1 How does the synthesis of nanoparticle take place?

There are two approaches typically used for synthesizing nanoparticles, i.e., top-down, and bottom-up approaches. The bulk material is broken down into the smallest fragments using spark and laser ablation in the top-down approach (Rane et al. 2018), whereas in the bottom-up approach, solvents and chemicals are involved in the synthesis of nanoparticles (Krishnadasan et al. 2007). The top-down and bottom-up approaches are shown in Figure 4.

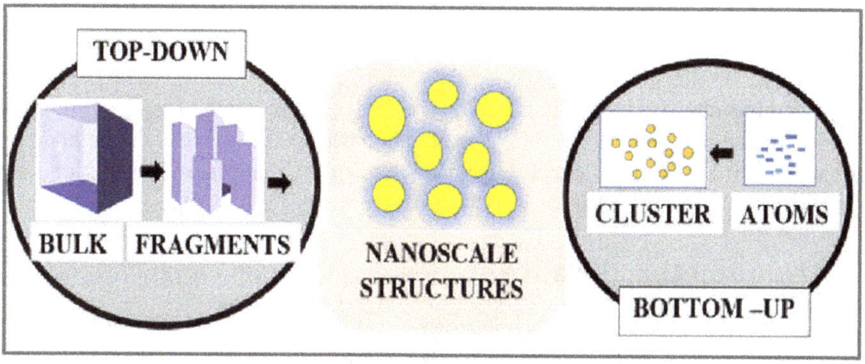

Fig. 4: Approaches for nanoparticles' synthesis: (1) Top-down (Bulk into Fragments), (2) Bottom-up (Cluster into smaller atoms).

8.4.2 Methods for nanomaterial synthesis

Three methods can synthesize nanomaterials.

I. Chemical synthesis
II. Biological synthesis
III. Physical synthesis

I. Chemical synthesis

In chemical synthesis, two types of solvent, i.e., organic and inorganic, act as reducing agents. The most common solvents used to synthesize nanoparticles are sodium citrate, ascorbate, sodium borohydride, polyol, tollen reagent, and N,N-dimethylformamide (DMF) (Hyeon 2003). In the chemical reduction of nanoparticles, these reducing agents can be used in the presence of stabilizers. Table 2 shows the chemical synthesis of nanoparticles. Chemical synthesis of nanoparticles using nanoparticles precursor and stabilizer is shown in Fig. 5.

II. Biological synthesis

The biological synthesis of nanoparticles is sustainable, economical, and simple (Srivastava et al. 2021). The biological synthesis of nanoparticles can be done by using plant, fungi, bacteria, algal extraction, and biomass after wastewater treatment. Several stabilizing agents are also used to synthesize nanoparticles, i.e., sodium citrate, ascorbate, sodium borohydride, polyol, tollen reagent, and N,N-dimethylformamide (DMF) (Singh et al. 2021). Different biological processes used for the synthesis of nanoparticles are shown in Fig. 6. Literature on biological synthesis of nanoparticles is shown in Table 3.

a. Bacterial mediated
b. Fungi mediated
c. Algae mediated
d. Plant mediated

III. Physical synthesis

The nanoparticles can be synthesized by using physical methods. The most common physical methods used for the synthesis of nanoparticles are mechanical method, ion beam deposition, sputter deposition, vapor deposition, electric arc deposition, and molecular beam epitaxy. Figure 7 shows different physical methods.

8.5 Nanomaterial's advantages

Nanotechnology provides considerable improvement in the industrial sector, helping revolutionize many technologies in terms of consuming less time and enhancing working efficiency (Holzinger et al. 2014). The advantages and application of nanotechnology, particularly nanoparticles, promise benefits globally in both expected and unexpected ways. Nanoparticle based technologies are eco-friendly and economical. The nanoparticles are highly efficient in the removal of pollutants (organic and inorganic pollutants) and recovery of ions and biomass from different

Table 2: Chemical synthesis of nanoparticles.

S.No	Chemical method	Wastewater	Nano Particles' type	Pollutant Removal	References
			Reduction Method		
1	**Polyol**	Wastewater	$Ag/Al(OH)_3$ nanocomposite	Bacterial pathogen	(Seo et al. 2012)
2	**Organic acid**	Aqueous solution	Iron oxide nanocomposites	Tetracycline (TC)	(Guo et al. 2016)
3	**Sodium borohydride**	Cow and human urine	Al NPs	Total carbon, total organic carbon, total nitrogen and total phosphorus	(Malek et al. 2021)
4	**Sugar**	Sugar industry effluent	Nickel moly date ($NiMoO_4$) coated NF	54.38% COD removal	(Jayabalan et al. 2021)
5	**Photo reduction**	Pharmacy	$ZnFe_2O_4$ nanoparticles	Sulfanilamide (SAM)	(Liu et al. 2021)
6	**Gamma radiolysis**	Textile effluent	Silver nanoparticle- (AgNP-)	Organic dye	(Vu and Nguyen 2021)
7	**Ultrasonic waves**	Dye solution	Green synthesized magnetite nanoparticles	Degradation of Acid Blue 15	(Prakash et al. 2021)
8	**Microwave**	Dye effluent	Ceria nanoparticles on carbon nanotubes	Dye removal	(López-Tinoco et al. 2021)
			Sedimentation method		
9	**Sol Gel**	Dyes in effluent	$Mg_{0.5}Zn_{0.5}$ $FeMnO_4$ magnetic nanoparticles (MNPs)	96% of RB21 dye was photo 48% removal of TOC	(Moradnia et al. 2021)
10	**Co-precipitation**	Textile waste water	Lead tungstate ($PbWO_4$) nanoparticles (NPs)	Thiazine dye	(Asha et al. 2021)
11	**Alkaline Precipitation**	-	ZnO nanoparticles	Rhodamine B and reactive orange	(Alshammari et al. 2020)
12	**Hydrothermal**	Dye effluent	Iron(II) nitrate u	Congo Red Dye	(Ikram et al. 2020)

wastewater effluents (Vandenabeele and Lucas 2020). This is because nanoparticles have a large surface area and have electromagnetic properties, which provide a large surface area for pollutants removal (Yaqoob et al. 2020).

The use of nanoparticles in materials and processes make these materials strong, durable, elastic and best electric conductor (Bannov et al. 2020). In electronics and information technology, nanoparticles are used in transistors, the basic switches and single tiny chip (Yu et al. 2021). Nanoparticles' application in the health sector is on the move. Gold nanoparticles are clinically investigated as potential treatments

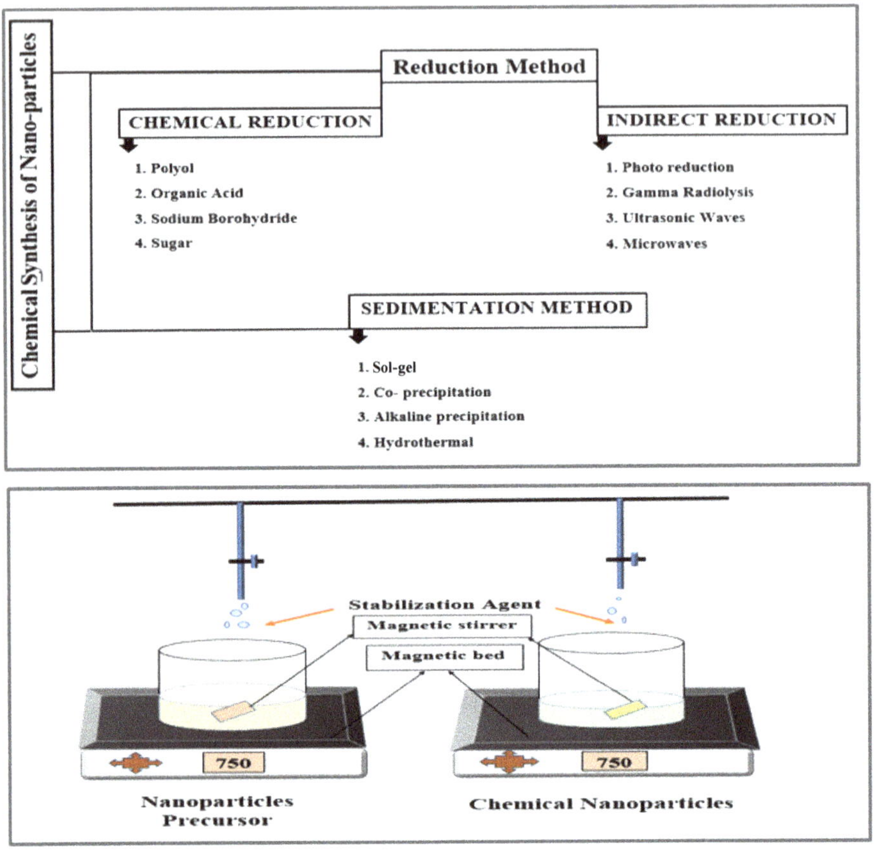

Fig. 5: Chemical synthesis of nanoparticles' using nanoparticles' precursor and stabilizer.

for cancer and other diseases (Gowri et al. 2021). Nanoparticles can overcome the energy crisis by providing clean, affordable, and cost-effective renewable energy (Wang et al. 2020).

8.6 Application of nanomaterials in biofuel production

Nanoparticles are widely used in the production of different biofuels such as bioethanol, biodiesel and biohydrogen. The use of nanoparticles in biofuel production enhances the biofuel productivity of different biomasses.

8.6.1 Bioethanol

Bioethanol is the most commonly used (90%) biofuel in the world. It can be produced from corn, barley, sugar beet, sugar cane, and roots of the cassava plant. Fermentation is the process that is used to produce bioethanol (Hulwan and Joshi 2011). During bioethanol production, starch is converted into sugar by fermentation process in the presence of enzymes (Hashmi et al. 2016).

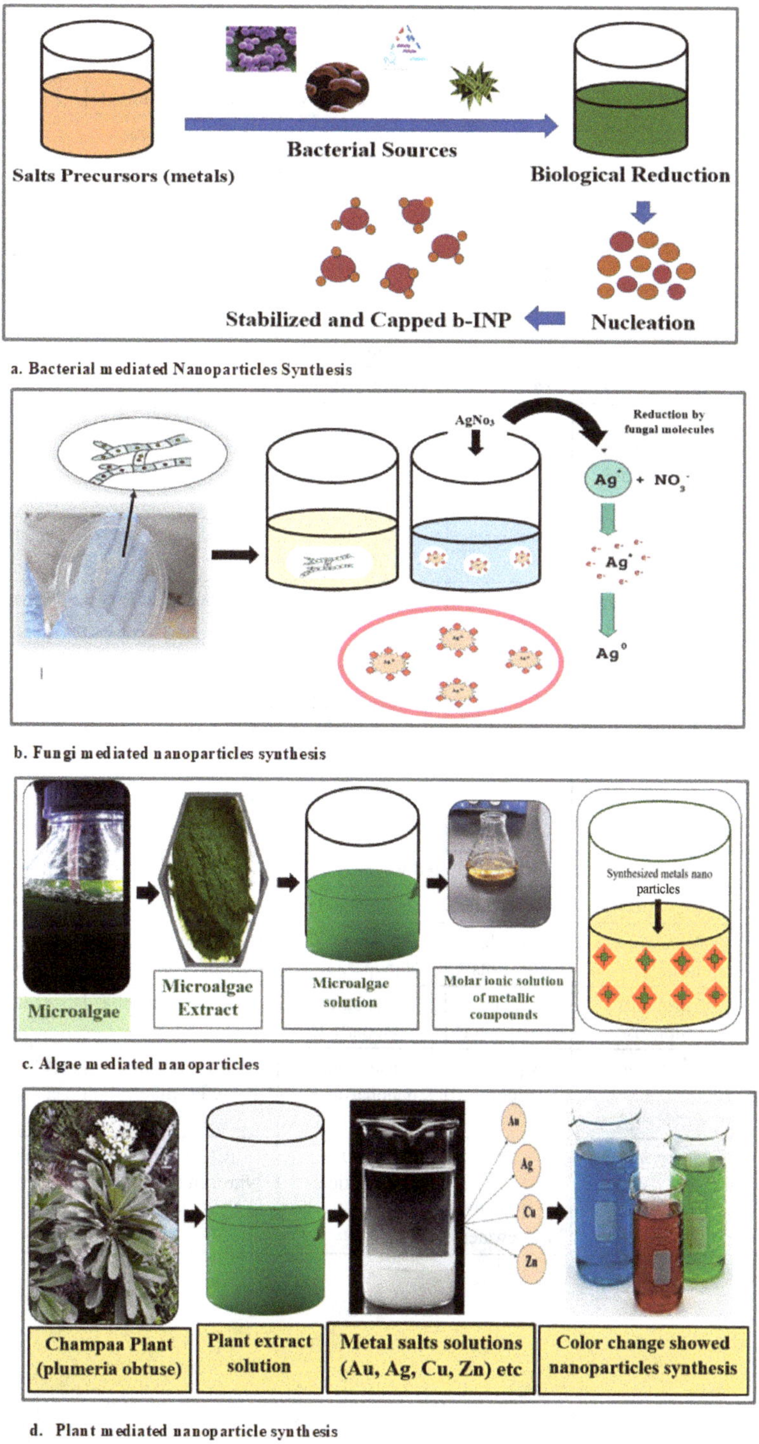

Fig. 6: Biological synthesis of nanoparticles using (a) bacteria (b) fungi (c) algae (d) plant.

Table 3: Biological synthesis of nanoparticles.

Bacterial Mediated Synthesis					
S #	**Bacterial Species**	**Wastewater**	**Nanoparticles**	**Pollutant Removal**	**References**
1	*Bacillus subtillis*	Fresh textile effluents	ZnO nanostructure	Methylene blue, Methyl orange, Rhodamine-B	(Dhandapani et al. 2020)
2	*Bacillus marisflavi*		AgNPs	Degradation of azo dyes	(Ahmed et al. 2020)
3	*Clostridium butyricum*	Anaerobic reactor	Silver nanoparticles	-	
Algae mediated nanoparticle synthesis					
S #	**Algae Species**	**Wastewater**	**Nanoparticles**	**Pollutant Removal**	**References**
6	*Cystoseira myrica, Sargassum latifolium* and *Padina australis*	Wastewater	Cu_2O nanoparticles	Antibacterial and anticancer activity	(Shukla et al. 2021)
7	*Helichrysum graveolens*	Wastewater	AgNPs	(Methylene Orange)	(Yazdi et al. 2020)
8	*Streptomyces* sp.	-	AgNPs	Bactericidal activity	(Manivasagan et al. 2015)
9	*Scenedesmus* sp.	-	Nanoparticles (Fe_3O_4 NPs)		(Markeb et al. 2019)
10	*Chlorosarcinopsis* sp.	-	Zinc oxide nanoparticles (ZnO- NPs)	Chemical Oxygen Demand (COD), Total Nitrogen (TN), Total Phosphorous (TP)	(Vasistha et al. 2021)
Fungi mediated nanoparticles synthesis					
S #	**Fungi Species**	**Wastewater**	**Nanoparticles**	**Pollutant Removal**	**References**
11	*Penicillium Oxalicum*	Textile Effluents	Iron Nanoparticles	Methylene Blue Dye	(Mathur et al. 2021)
12	*Cordyceps militaris*	Textile wastewater	Zinc oxide nanoparticle	Dye Methylene blue	(Li et al. 2019)
13	*Aspergillus niger*	Aqueous solution	Iron oxide nanoparticles (IONPs) (Fe_3O_4)	Cr(VI) removal efficiency	(Chatterjee et al. 2020)
14	*Mycorrhizal Fungi* (AMF)	Metallic-NPs-polluted wastewater	CuO nanoparticles (CuO-NPs)-	(COD), Total Nitrogen (TN), and CuO-NPs	(Ban et al. 2021)

Table 3 contd. ...

...Table 3 contd.

Plant mediated nanoparticles synthesis					
S #	**Plant species**	**Wastewater**	**Nanoparticles**	**Pollutant removal**	**References**
16	*Madhuca longifolia*	Waste Water Treatment	Cupric oxide nanoparticles	Degradation Of Methylene Blue (MB)	(Das et al. 2018)
17	**Minidium leavigatum**		$Cu_{0.5}Zn_{0.5}Fe_2O_4$ nanoparticles	Reactive Blue 222 Dye	(Yeganeh et al. 2020)
18	**Azadirachta indica**	Domestic Waste Water	Iron nano particles	Phosphate, Ammonia, Nitrogen Chemical Oxygen Demand	(Devatha et al. 2016)
19	*Musa acuminata*	-	Zinc oxide	Basic Blue 9 Crystal Violet And Cresol Red Dyes	(Abdullah et al. 2021)
20	*Moringa oleifera*	Industrial Wastewater	Silver nanoparticle	Dye and removal of bacteria	(Mehwish et al. 2021)

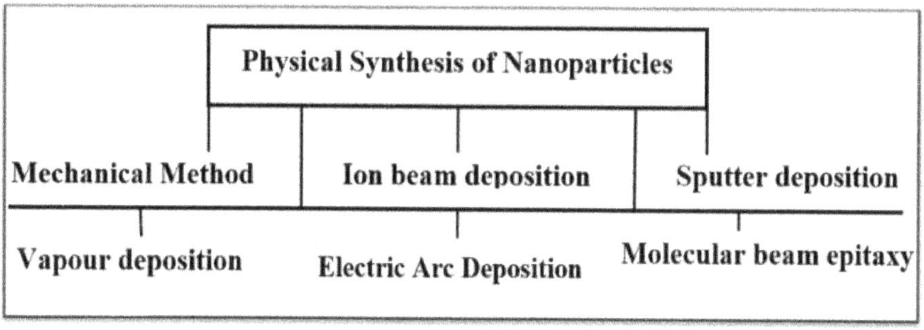

Fig. 7: Physical synthesis of nanoparticles.

8.6.1.1 Use of nanoparticles for bioethanol production

Recently, nanoparticles have been used in the production of biofuels. The conventional processes used for bioethanol production have several issues, such as the high cost of biomass production, lower reaction rate, and low productivity. These challenges can be overcome by using nanoparticles for bioethanol production, which will enhance bioethanol productivity. Nanoparticles also improve the pretreatment efficiency, enzymatic hydrolysis and improve the fermentation reaction rate. The size of nanoparticles, their morphology surface area and nature affect the reaction rate and productivity (Chaturvedi et al. 2012). Catalytic reusability is another major issue in the production of bioethanol (Miletić et al. 2012). Nanoparticles can be used in the complex bioethanol production process, and they can also be used for the recovery of cellulose catalysts (Rodríguez-Couto 2019). In bioethanol production, nanocatalysts

Table 4: Nanoparticles used for bioethanol production.

Type of nanoparticle	Biomass source	Bioethanol enhancement	References
Silver	*Chlorella vulgaris*	15.26%	(Razack et al. 2016)
Magnetic	Wheat straw	66.3%	(Pena et al. 2012)
Silica	*Clostridium ljungdahlii*	166.1%	(Kim et al. 2014)
Cobalt-ferrite-silica	*Clostridium ljungdahlii*	227.6%	(Kim and Lee 2016)
Iron oxide	*Saccharomyces cerevisiae* BY4743	0.26 g/g	(Sanusi et al. 2019)
MnO_2	*Aspergillus fumigatus* JCF	-	(Cherian et al. 2015)
Zinc oxide	*Fusarium oxysporum*	0.0359 g/g	(Gupta and Chundawat 2020)
NiO	Potato peels	32 g/L	(Sanusi et al. 2020)
Magnetic	-	91%	(Ivanova et al. 2011)

are recently used due to their reusability (Kim and Lee 2016). However, there is a need for further research on the role of nanocatalysts for bioethanol production. Table 4 shows the use of different nanoparticles for bioethanol production.

8.6.1.2 *Effect of nanoparticles on bioethanol production*

The use of nanoparticles during pretreatment can improve the chemistry at the molecular level. It also enables targeting of biocatalyst and removal of pollutants caused by chemical pretreatment (Razack et al. 2016). Metal-based nanoparticles are incredibly effective due to their small structure. They can easily interact with the biomolecules present in the biomass and release carbohydrates used for bioethanol production. Nanoparticles also affect the biochemical conversion activity by improving the enzymatic activity required to produce bioethanol. The most common nanoparticles used in renewable energy are oxides of iron, copper, cobalt, manganese, etc.

8.6.1.3 *Nanoparticles and bioethanol enhancement*

Gupta et al. (2020) developed zinc oxide nanoparticles by using *Fusarium oxysporum* to enhance bioethanol production in rice straw. The maximum bioethanol production in the presence of zinc oxide nanoparticles was 0.0359 g/g of dry weight-based plant biomass (Gupta and Chundawat 2020). In a study performed by Sanusi et al. (2020), nickel oxide was used as a biocatalyst *to enhance* cell growth and bioethanol production in *Saccharomyces cerevisiae* BY4743. The maximum bioethanol and biomass production in the presence of nickel oxide nanoparticles was 0.26 g/g and 2.04 g/L, respectively (Sanusi et al. 2020). In a recent paper published in Biotechnology Reports, Sanusi et al. (2021) used NiO nanoparticles as nanobiocatalysts to enhance bioethanol production in waste potato peels. The addition of NiO nanoparticles in the process resulted in the highest bioethanol concentration of 32 g/L (Sanusi et al. 2020). In a study conducted by Razack et al. (2016), silver nanoparticles were used to produce bioethanol from microalgae *Chlorella vulgaris*. The use of nanoparticles resulted in a 15.26% increase in total carbohydrate production (Razack et al. 2016).

In another study conducted by Pena et al. (2012), acid-functionalized magnetic nanoparticles produced bioethanol from wheat straw. Conversion of hemicellulose to oligosaccharides was increased by 66.3% in the presence of nanoparticles as compare to the control (Pena et al. 2012). Iron nanoparticles were used for improving the enzymatic activity and stability by providing support for enzyme immobilization. These nanoparticles were highly influential during the production of bioethanol (Srivastava et al. 2015). In another study, Jordan et al. (2011) immobilized cellulase enzyme complex onto magnetic particles by carbodiimide activation process for improving bioethanol production (Jordan et al. 2011). Graphene-based nanoparticles are highly durable and provide a large surface area for the immobilization of catalytic sites and enzymes responsible for producing bioethanol (Gokhale et al. 2013, Kakaei et al. 2016). In a study conducted by Kim et al. (2014), metal functionalized silica nanoparticles were effectively used to enhance bioethanol production up to 166.1% (Kim et al. 2014). The use of methyl-functionalized cobalt-ferrite-silica nanoparticles enhanced bioethanol production by nearly 213.5% (Kim and Lee 2016). Kim et al. (2014) used six types of nanoparticles to enhance bioethanol production in *Clostridium ljungdahlii*. Out of these nanoparticles, silica nanoparticles were highly efficient in the enhancement of bioethanol production. The surface area of silicon and particles were modified by using methyl and isopropyl. The modified silica nanoparticles enhanced ethanol, biomass, and acetic acid production in *C. ljungdahlii* by 166.1%, 34.5%, and 29.1%, respectively (Kim et al. 2014). Magnetic nanoparticles were used for the enhancement of bioethanol production in *Clostridium ljungdahlii*. The two types of nanoparticles which were used in this study were metal functionalized silica, and methyl functionalized cobalt ferrite–silica. Acetic acid, ethanol, and biomass production were enhanced by 59.6%, 213.5%, and 227.6%, respectively, in the presence of cobalt ferrite silica (Kim and Lee 2016). Cherian et al. (2015) immobilized cellulase from *Aspergillus fumigatus* JCF on to MnO_2 nanoparticles for the enhancement of bioethanol production. The immobilization of MnO_2 nanoparticles resulted in higher bioethanol production (Cherian et al. 2015). Ivanova et al. (2011) used magnetic NPs for the enhancement of bioethanol production. The theoretical bioethanol yield was increased by 91% (Ivanova et al. 2011).

8.6.2 Biodiesel production

Biodiesel carries a wide range of characteristics which makes it an efficient energy resource (Shameer et al. 2017). Biodiesel can be of different generations such as 1st generation, 2nd generation, 3rd generation, and 4th generation, depending upon the feedstock used (Singh et al. 2020). The steps involved in producing biodiesel from feedstock are extraction of oils from the biomass, refining of oil to remove any impurities, and conversion of oil to fuel. These processes can be optimized with the use of nanomaterials. Various nanomaterials are studied for the transesterification reaction. Nanomaterials can be used to enhance the biological reactions during a bioprocess. Nanomaterials can be employed as a catalyst (nanocatalysts) to boost microbes' activity, reduce the inhibitors generated during the bioprocess, and aid the reactions to produce specific yields (Khoo et al. 2020). Table 5 shows use of nanoparticles for biodiesel production.

Table 5: Nanoparticles used for biodiesel production (Sekoai et al. 2019).

Type of nanoparticle	Biomass source	Biodiesel enhancement	References
Ni doped ZnO	Castor	95.2%	(Baskar et al. 2018)
Ni0.5Zn0.5Fe_2O_4 doped with Cu	Soybean	85%	(Dantas et al. 2017)
Calcite/Au	Sunflower	97.58%	(Bet-Moushoul et al. 2016)
Carbon nanohorn dispersed with $Ca_2Fe_2O_5$	Tricaprylin	100%	(Sano et al. 2017)
Hydrotalcite particles with Mg/Al	Jatropha	95.2%	(Deng et al. 2011)
Soybean oil ZrO_2 loaded with $C_4H_4O_6HK$	Soybean	98.03%	(Qiu et al. 2011)
KF/CaO	Chinese tallow seed	96.8%	(Wen et al. 2010)
TiO_2-ZnO	Palm	98%	(Madhuvilakku and Piraman 2013)
CaO	Rice bran	93.5%	(Mazaheri et al. 2018)
KOH/Calcium Aluminate	Canola	91%	(Nayebzadeh et al. 2017)

8.6.2.1 Nanoparticles and biodiesel enhancement

Siva et al. (2015) calcirated the eggshell powder to obtain the nano-CaO catalyst using a sonochemical reactor. This nanocatalyst was then utilized in the transesterification process of algal oil. They concluded that the weight transesterification process improved when the catalyst loading was increased from 0.5 to 1.25%, resulting in a low triglyceride content biodiesel. The study reported that nano-CaO facilitated the conversion of algal oil to biodiesel with 96.3% efficiency (Siva and Muthu 2015). In another such study, the biodiesel from the algae was enhanced by using calcium-based nanocatalyst, and Teo et al. (2016) used nano-$Ca(OCH_3)_2$ as a catalyst. With a 30:1 methanol to oil ratio and 3% by weight of catalyst loading, they achieved 99% biodiesel yield (Teo et al. 2016).

Thangaraj and Piraman prepared biodiesel from *Madhuca indica* oil using a ZnO nanocatalyst coated with heteropoly acid. The 95% biodiesel yield is the proof that heteropoly acid coated ZnO based nanocatalyst has the crystallinity and morphology required for the process and can be used in transesterification process to obtain desired results (Thangaraj and Piraman 2016). Baskar et al. (2017) were fascinated by the idea that transition metal oxides used in the transesterification of the fatty acids yield high rates of conversion to biodiesel. They studied the effects of manganese-doped ZnO nanocatalyst on biodiesel production from *Mahua* oil. With 7:1 (w/v) methanol to oil ratio and 8% (w/v) catalyst loading at 50°C, the biodiesel yield was found to be 97% (Baskar et al. 2017). In another study, copper doped ZnO nanocatalyst was used to prepare biodiesel from waste cooking oil. The experiment was conducted under varying conditions such as 2–8% (w/w) catalyst loading,

30–60 minutes reaction time, 40–70°C temperature, etc. The results achieved exhibit that 97.7% of biodiesel was obtained from waste cooking oil (Sandhya et al. 2020).

Gardy et al. (2016) used $Ti(SO_4)O$ as a nanocatalyst for the production of biodiesel from waste cooking oil. Under 9:1 methanol to oil ratio, 1.5% by weight catalyst loading, and 75°C reaction temperature, titanium-based nanocatalyst achieved 97% of biodiesel (Gardy et al. 2016). In a study conducted by Ambat et al. (2018), potassium ion modified with TiO_2 was used for biodiesel enhancement. The maximum output of 98.5% biodiesel was obtained under the 6:1 methanol to oil ratio, 6% by weight catalyst loading, at 60°C reaction temperature (Ambat et al. 2018).

Alaei et al. (2018) conducted a study on the transesterification of vegetable oil using $MgO/MgFe_2O_4$ catalysts. The experimental conditions employed in their research included a reaction temperature of 110°C, a methanol to oil ratio of 12 (indicating the use of 12 parts methanol per part of vegetable oil), and a catalyst concentration of 4% by weight (based on the total weight of the reactants). The maximum biodiesel yield was 91.2% (Alaei et al. 2018). Rahmani Vahid and his group prepared $MgO/MgAl_2O_4$ using the combustion process. They used this nanocatalyst for the transesterification of sunflower oil to produce biodiesel. The maximum biodiesel production was 95% (Rahmani Vahid et al. 2017). Feyzi et al. (2017) studied the effects of $MgO\text{-}La_2O_3$ on the biodiesel production of sunflowers. They concluded that nanocatalyst exhibited high catalytic activity, which resulted in 97.7% biodiesel yield (Feyzi et al. 2017).

Na_2O impregnated carbon nanotube nanocatalysts were used for biodiesel production from waste cooking oil. These nanocatalysts contain lewis acids, which promotes catalytic activity and assists the transesterification reaction. The nanocatalyst then experimented with waste cooking oil to observe the biodiesel yield. At 65°C reaction temperature, 20:1 methanol to oil ratio, 3% by weight catalyst loading, the maximum biodiesel yield was 97% (Ibrahim et al. 2020). Asri et al. (2021) recently used multi-walled carbon nanotubes (MWCNT) as nanocatalysts to produce biodiesel from *Schleichera oleosa* L. 41% of biodiesel yield was achieved during the process at 4% by weight catalyst loading. Although a low amount of biodiesel was obtained, this study contributed much to the field of nanocatalysts broadening the horizons of using carbon-based nanoparticles (Asri et al. 2021).

8.6.3 Biohydrogen production

Hydrogen (H_2) has always been considered an alternative fuel because of its higher energy potential, no post-combustion GHG emissions, and its carbon-free nature (Kothari et al. 2012, Kumar et al. 2017b, Sivagurunathan et al. 2017). Photosynthesis and dark fermentation are the two ways by which H_2 can be produced. The photosynthetic bacterial pathway is more suitable theoretically, but it is not practical due to problems in the suitable reactor design and low efficiency. On the other side, dark fermentation by hydrogenogens is sustainable with a good carbon balance (Xia et al. 2015) and less energy requirement than the conventional H_2 production processes (Khan et al. 2016). The lower efficiency in substrate conversion is the disadvantage of this method, which should be minimized by improving process design to enhance final yields (Bundhoo and Mohee 2016).

Generally, the use of glucose or simple sugars as a substrate is favored due to biodegradability as complex hydrocarbons need to be converted into monomers before any beneficial use. Similarly, simple sugars as model substrates can easily be calculated as sugar consumed and H_2 produced (Gonzales et al. 2017, Moreno-Garrido 2008). Stoichiometrically, a gram of glucose is converted into 1493 mL $H_{2,}$ and experimental results show 498 mL H_2 per gram of glucose because of competition in microbial species and other unknown reasons (Hallenbeck 2009). Historically, the H_2 production enhancement was focused on parameter optimization, metabolic engineering, and immobilization of microbes (Kumar et al. 2017a, Mishra and Das 2014, Zhang et al. 2011). Recently, the use of nanomaterials has gained success in enhanced bio-hydrogen production (Lin et al. 2016). Several types of nanomaterials and their effects are discussed here. Types of nanoparticles used for the synthesis of biohydrogen are shown in Table 6.

8.6.3.1 Organic nanomaterials

The H_2 production is mainly affected by the presence of volatile fatty acids (VFAs) such as lactic, butyric, propionic, acetic acids, and ethanol as final products. At comparatively higher concentrations, the VFAs negatively affect the H2 production and inhibit the fermentation process. Wimonsong and Nitisoravut have reported using highly porous treated activated carbon (T-AC) that adsorbed the inhibitory butyric acid in aqueous media and caused the enhancement of 73% of H_2 with 33.3 g/L T-AC (Wimonsong and Nitisoravut 2014). In another study, it was observed that nanoporous activated carbon (N-AC) prevented the accumulation of butyric acid and produced 2.6 mol/mol sucrose H_2 (Wimonsong and Nitisoravut 2015). Boshagh et al. reported another way to enhance the H_2 production by immobilizing the *Enterobacter aerogenes* on functionalized multi-walled carbon nanotubes to achieve 47% more H_2 by high biomass accumulation on carbon nanotubes (Boshagh et al. 2019). Liu et al. reported the effect of carbon nanotubes to enhance flocculation and sedimentation capacity of the biomass in a laboratory-scale upflow anaerobic sludge blanket (UASB) reactor. Also, it improved the retention of microbial biomass in the UASB that promoted 24% higher H_2 yields (Liu et al. 2012). The iron oxides enhance the H_2 production by enhancing the electron transfer.

8.6.3.2 Inorganic nanomaterials

Most of the studies have focused on the use of inorganic nanomaterials as different forms of iron. The effect of ferrous iron (Fe^{2+}) was reported to increase 100–105% H_2 using glucose and heat-treated corn starch as substrates, respectively. They suggested the role of ferrous iron in enhancing the metabolic activity using ferridoxin, an electron carrier in hydrogenases (Bao et al. 2013, Nath et al. 2015). Yang et al. observed the decline in the oxidation reduction potential (ORP) of media from –135 to –273 mV that promoted hydrogen-producing bacteria's growth. Moreover, zero-valent iron also promoted cellular respiration by a favorable metabolic pathway that resulted in a 73% increase in H_2 production (Yang and Wang 2018).

Table 6: Nanoparticles used for biohydrogen production.

S. No.	Nanoparticle type	Biomass source	Biohydrogen enhancement (%)	References
		Organic nanomaterials		
1	Treated-activated carbon	Anaerobic sludge	73	(Wimonsong and Nitisoravut 2014)
2	Nanoporous activated carbon	Heat treated anaerobic sludge	62.5	(Wimonsong and Nitisoravut 2015)
3	Functionalized multi-walled carbon nanotubes	Immobilized *Enterobacter aerogenes*	47	(Boshagh et al. 2019)
4	Carbon nanotubes	Anaerobic sludge	24	(Liu et al. 2012)
		Inorganic nanomaterials		
1	Fe^{o}	*Enterobacter* sp. to *Clostridium* sp.	73.1	(Yang and Wang 2018)
2	Fe^{2+}	*Enterobacter cloacae* DH-89	100	(Nath et al. 2015)
3	Fe^{2+}	Mixed culture	105	(Bao et al. 2013)
4	FeO	Anaerobic sludge	7.9	(Engliman et al. 2017)
5	Fe_2O_3	*Enterobacter aerogenes*	63.1	(Lin et al. 2016)
6	Fe_3O_4	Anaerobic sludge	69.6	(Reddy et al. 2017)
7	Hematite (Fe_2O_3)	Mixed culture	32.64	(Han et al. 2011)
8	Maghemite (Fe_3O_4)	*Clostridium* sp. and *Rhodopseudomonas palustris*	57.8	(Nasr et al. 2015)
		Composite nanomaterials		
1	Fe_3O_4/Starch	Sewage sludge	37	(Zhang et al. 2018)
2	Fe^{o}/Activated carbon microelectrolysis	*Clostridium* spp.	50.2	(Zhang et al. 2015)
3	Fe^{2+}/Biochar	Heat treated anaerobic sludge	50	(Zhang et al. 2017)
4	Ni/Graphene	Mixed culture from municipal wastewater	105	(Elreedy et al. 2017)

Iron oxide (FeO) enhances the H_2 production by enhancing the electron transfer without participating in the reaction (Engliman et al. 2017). Hematite (Fe_2O_3), on the other hand, promotes the electron transfer and hydrogenase activity, decreases ORP to promote H_2 production, provides attachment sites for microbes, and slowly releases the ferrous iron into the medium by decreasing hematite's particle size (Han et al. 2011, Lin et al. 2016). Similarly, maghemite (Fe_3O_4) enhanced H_2 production by decreasing lag phase, increasing the hydrogen potential, providing high electron affinity, and reducing protons to H_2 (electron sink) (Nasr et al. 2015, Reddy et al. 2017).

8.6.3.3 Composite nanomaterials

Several composite materials have also been reported for biohydrogen production enhancement. A 200 mg/L Fe_3O_4/starch nanocomposite caused 37% more H_2 by increasing microbial accumulation, electron transfer, and hydrogenase enzymatic activity (Zhang et al. 2018). When zero-valent iron with activated carbon and ferrous iron with biochar was used for the H_2 production, a 50% increase was observed. Unanimously, activated carbon and biochar provided the support and favorable environment to the microbes and provided the buffering capacity to the medium. On the other hand, Fe^{o} and Fe^{2+} components enhanced the hydrogenase activity (Zhang et al. 2017, 2015). A Ni-Graphene nanocomposite promoted a 105% increase in H_2 production from industrial complex mono-ethylene glycol at thermophilic temperature. Ni-Graphene nanocomposite promoted hydrogenase activity, electron transfer and converted H^+ protons to H_2 (Elreedy et al. 2017).

References

Abd Elkodous, M., G.S. El-Sayyad, M.A. Maksoud, R. Kumar, K. Maegawa, G. Kawamura, W.K. Tan, and A. Matsuda. 2021. Nanocomposite matrix conjugated with carbon nanomaterials for photocatalytic wastewater treatment. J. Hazard. Mater. 410: 124657.

Abdullah, F., N.A. Bakar, and M.A. Bakar. 2021. Comparative study of chemically synthesized and low temperature bio-inspired Musa acuminata peel extract mediated zinc oxide nanoparticles for enhanced visible-photocatalytic degradation of organic contaminants in wastewater treatment. J. Hazard. Mater. 406: 124779.

Ahmed, T., M. Noman, M. Shahid, M.B.K. Niazi, S. Hussain, N. Manzoor, X. Wang, and B. Li. 2020. Green synthesis of silver nanoparticles transformed synthetic textile dye into less toxic intermediate molecules through LC-MS analysis and treated the actual wastewater. Environ. Res. 191: 110142.

Akhtar, K., S.A. Khan, S.B. Khan, and A.M. Asiri. 2018. Scanning electron microscopy: Principle and applications in nanomaterials characterization. In Handbook of materials characterization, 113-145: Springer.

Alaei, S., M. Haghighi, J. Toghiani, and B. Rahmani Vahid. 2018. Magnetic and reusable MgO/MgFe2O4 nanocatalyst for biodiesel production from sunflower oil: Influence of fuel ratio in combustion synthesis on catalytic properties and performance. Ind. Crops. Prod. 117: 322-332.

Almomani, F., R. Bhosale, M. Khraisheh, and T. Almomani. 2020. Heavy metal ions removal from industrial wastewater using magnetic nanoparticles (MNP). Appl. Surf. Sci. 506: 144924.

Alshammari, M., M.F. Al Juboury, L.A. Naji, A.A. Faisal, H. Zhu, N. Al-Ansari, and M. Naushad. 2020. Synthesis of a novel composite sorbent coated with siderite nanoparticles and its application for remediation of water contaminated with Congo red dye. Int. J. Environ. Res.: 1-15.

Ambat, I., V. Srivastava, E. Haapaniemi, and M. Sillanpää. 2018. Application of Potassium Ion Impregnated Titanium Dioxide as Nanocatalyst for Transesterification of Linseed Oil. Energy. Fuels. 32: 11645-11655.

Asha, K., C. Hentry, M. Bindhu, A.M. Al-Mohaimeed, M.R. AbdelGawwad, and M.S. Elshikh. 2021. Improved photocatalytic activity for degradation of textile dyeing waste water and thiazine dyes using PbWO4 nanoparticles synthesized by co-precipitation method. Environ. Res.: 111721.

Asri, N.P., R. Saraswati, H. Hindarso, D.A. Puspitasari, and Suprapto. 2021. Synthesis of biodiesel from kesambi (Schleichera oleosa L.) oil using carbon nanotube-supported zinc oxide heterogeneous catalyst. IOP Conference Series: Earth and Environmental Science 749: 012048.

Ban, Y., Z. Xiao, C. Wu, Y. Lv, F. Meng, J. Wang, and Z. Xu. 2021. The positive effects of inoculation using arbuscular mycorrhizal fungi and/or dark septate endophytes on the purification efficiency of CuO-nanoparticles-polluted wastewater in constructed wetland. J. Hazard. Mater.: 126095.

Bannov, A.G., M.V. Popov, and P.B. Kurmashov. 2020. Thermal analysis of carbon nanomaterials: advantages and problems of interpretation. J. Therm. Anal. Calorim. 142: 349-370.

Bao, M.D., H.J. Su, and T.W. Tan. 2013. Dark fermentative bio-hydrogen production: Effects of substrate pre-treatment and addition of metal ions or L-cysteine. Fuel. 112: 38-44.

Barisik, M., S. Atalay, A. Beskok, and S. Qian. 2014. Size dependent surface charge properties of silica nanoparticles. J. Phys. Chem. 118: 1836-1842.

Baskar, G., A. Gurugulladevi, T. Nishanthini, R. Aiswarya, and K. Tamilarasan. 2017. Optimization and kinetics of biodiesel production from Mahua oil using manganese doped zinc oxide nanocatalyst. Renew. Energ. 103: 641-646.

Baskar, G., I.A.E. Selvakumari, and R. Aiswarya. 2018. Biodiesel production from castor oil using heterogeneous Ni doped ZnO nanocatalyst. Bioresour. Technol. 250: 793-798.

Bet-Moushoul, E., K. Farhadi, Y. Mansourpanah, A.M. Nikbakht, R. Molaei, and M. Forough. 2016. Application of CaO-based/Au nanoparticles as heterogeneous nanocatalysts in biodiesel production. Fuel. 164: 119-127.

Boshagh, F., K. Rostami, and N. Moazami. 2019. Biohydrogen production by immobilized *Enterobacter aerogenes* on functionalized multi-walled carbon nanotube. Int. J. Hydrog. Energy. 44: 14395-14405.

Bounab, N., L. Duclaux, L. Reinert, A. Oumedjbeur, C. Boukhalfa, P. Penhoud, and F. Muller. 2021. Improvement of zero valent iron nanoparticles by ultrasound-assisted synthesis, study of Cr (VI) removal and application for the treatment of metal surface processing wastewater. J. Environ. Chem. Eng. 9: 104773.

Bundhoo, M.Z., and R. Mohee. 2016. Inhibition of dark fermentative bio-hydrogen production: a review. Int. J. Hydrog. Energy. 41: 6713-6733.

Chakrabarty, S., S. Bandyopadhyay, M. Pal, and A. Dutta. 2021. Sol-gel derived cobalt containing Ni–Zn ferrite nanoparticles: Dielectric relaxation and enhanced magnetic property study. Mater. Chem. Phys. 259: 124193.

Chatterjee, J., Y. Haik, and C.-J. Chen. 2003. Size dependent magnetic properties of iron oxide nanoparticles. J. Magn. Magn. Mater. 257: 113-118.

Chatterjee, S., S. Mahanty, P. Das, P. Chaudhuri, and S. Das. 2020. Biofabrication of iron oxide nanoparticles using manglicolous fungus Aspergillus niger BSC-1 and removal of Cr (VI) from aqueous solution. Chem. Eng. J. 385: 123790.

Chaturvedi, S., P.N. Dave, and N. Shah. 2012. Applications of nano-catalyst in new era. J. Saudi Chem. Soc. 16: 307-325.

Cherian, E., M. Dharmendirakumar, and G. Baskar. 2015. Immobilization of cellulase onto MnO_2 nanoparticles for bioethanol production by enhanced hydrolysis of agricultural waste. Chinese J. Catal. 36: 1223-1229.

Chethana, D., T. Thanuja, H. Mahesh, M. Kiruba, A. Jose, H. Barshilia, and J. Manjanna. 2021. Synthesis, structural, magnetic and NO_2 gas sensing property of CuO nanoparticles. Ceram. Int. 47: 10381-10387.

Cuong, T.T., H.A. Le, N.M. Khai, P.A. Hung, N.V. Thanh, N.D. Tri, and N.X. Huan. 2021. Renewable energy from biomass surplus resource: potential of power generation from rice straw in Vietnam. Scientific reports 11: 1-10.

Dantas, J., E. Leal, A. Mapossa, D. Cornejo, and A. Costa. 2017. Magnetic nanocatalysts of Ni0. 5Zn0. 5Fe2O4 doped with Cu and performance evaluation in transesterification reaction for biodiesel production. Fuel. 191: 463-471.

Das, P., S. Ghosh, R. Ghosh, S. Dam, and M. Baskey. 2018. Madhuca longifolia plant mediated green synthesis of cupric oxide nanoparticles: a promising environmentally sustainable material for waste water treatment and efficient antibacterial agent. J. Photochem. Photobiol. B, Biol. 189: 66-73.

Deng, X., Z. Fang, Y.-h. Liu, and C.-L. Yu. 2011. Production of biodiesel from Jatropha oil catalyzed by nanosized solid basic catalyst. Energy. 36: 777-784.

Devatha, C., A.K. Thalla, and S.Y. Katte. 2016. Green synthesis of iron nanoparticles using different leaf extracts for treatment of domestic waste water. J. Clean. Prod. 139: 1425-1435.

Dhandapani, P., A.A. Prakash, M.S. AlSalhi, S. Maruthamuthu, S. Devanesan, and A. Rajasekar. 2020. Ureolytic bacteria mediated synthesis of hairy ZnO nanostructure as photocatalyst for decolorization of dyes. Mater. Chem. Phys. 243: 122619.

El-Maraghy, C.M., O.M. El-Borady, and O.A. El-Naem. 2020. Effective removal of levofloxacin from pharmaceutical wastewater using synthesized zinc oxid, graphen oxid nanoparticles compared with their combination. Sci. Rep. 10: 1-13.

Elreedy, A., E. Ibrahim, N. Hassan, A. El-Dissouky, M. Fujii, C. Yoshimura, and A. Tawfik. 2017. Nickel-graphene nanocomposite as a novel supplement for enhancement of biohydrogen production from industrial wastewater containing mono-ethylene glycol. Energy. Convers. Manag. 140: 133-144.

Engliman, N.S., P.M. Abdul, S.-Y. Wu, and J.M. Jahim. 2017. Influence of iron (II) oxide nanoparticle on biohydrogen production in thermophilic mixed fermentation. Int. J. Hydrog. Energy 42: 27482-27493.

Feyzi, M., N. Hosseini, N. Yaghobi, and R. Ezzati. 2017. Preparation, characterization, kinetic and thermodynamic studies of MgO-La2O3 nanocatalysts for biodiesel production from sunflower oil. Chem. Phys. Lett. 677: 19-29.

Fouda, A., S.E.-D. Hassan, E. Saied, and M.S. Azab. 2021. An eco-friendly approach to textile and tannery wastewater treatment using maghemite nanoparticles (γ-Fe2O3-NPs) fabricated by Penicillium expansum strain (Kw). J. Environ. Chem. Eng. 9: 104693.

Gao, J., R. Chen, D. Li, L. Jiang, J. Ye, X. Ma, X. Chen, Q. Xiong, H. Sun, and T. Wu. 2011. UV light emitting transparent conducting tin-doped indium oxide (ITO) nanowires. Nanotechnology. 22: 195706.

Gardy, J., A. Hassanpour, X. Lai, and M. Ahmed. 2016. Synthesis of Ti(SO4)O solid acid nano-catalyst and its application for biodiesel production from used cooking oil. APPL. CATAL A-GEN. 527: 81-95.

Gokhale, A.A., J. Lu, and I. Lee. 2013. Immobilization of cellulase on magnetoresponsive graphene nano-supports. J. Mol. Catal., B Enzym. 90: 76-86.

Gonzales, R.R., G. Kumar, P. Sivagurunathan, and S.-H. Kim. 2017. Enhancement of hydrogen production by optimization of pH adjustment and separation conditions following dilute acid pretreatment of lignocellulosic biomass. Int. J. Hydrog. Energy 42: 27502-27511.

Gowri, A., A. Kumar, and S. Anand. 2021. Recent advances in nanomaterials based biosensors for point of care (PoC) diagnosis of covid-19-A minireview. Trends Analyt Chem: 116205.

Guo, L., Y. Liang, X. Chen, W. Xu, K. Wu, H. Wei, and Y. Xiong. 2016. Effective removal of tetracycline from aqueous solution by organic acid-coated magnetic nanoparticles. J. Nanosci. Nanotechnol. 16: 2218-2226.

Gupta, K., and T.S. Chundawat. 2020. Zinc oxide nanoparticles synthesized using *Fusarium oxysporum* to enhance bioethanol production from rice-straw. Biomass Bioenergy 143: 105840.

Hallenbeck, P.C. 2009. Fermentative hydrogen production: principles, progress, and prognosis. Int. J. Hydrog. Energy 34: 7379-7389.

Han, H., M. Cui, L. Wei, H. Yang, and J. Shen. 2011. Enhancement effect of hematite nanoparticles on fermentative hydrogen production. Bioresour. Technol. 102: 7903-7909.

Hashmi, S., S. Gohar, T. Mahmood, U. Nawaz, and H. Farooqi. 2016. Biodiesel production by using CaO-Al_2O_3 Nano catalyst. Int. j. sci. eng. res. 2: 43-49.

Hethnawi, A., N.N. Nassar, and G. Vitale. 2017. Preparation and characterization of polyethylenimine-functionalized pyroxene nanoparticles and its application in wastewater treatment. Colloids. Surf. A Physicochem. Eng. Asp.525: 20-30.

Holzinger, M., A. Le Goff, and S. Cosnier. 2014. Nanomaterials for biosensing applications: a review. Front. Chem. 2: 63.

Hulwan, D.B., and S.V. Joshi. 2011. Performance, emission and combustion characteristic of a multicylinder DI diesel engine running on diesel–ethanol–biodiesel blends of high ethanol content. Appl. Energy. 88: 5042-5055.

Hyeon, T. 2003. Chemical synthesis of magnetic nanoparticles. Chem.Comm.: 927-934.

Ibrahim, M.L., N.N.A. Nik Abdul Khalil, A. Islam, U. Rashid, S.F. Ibrahim, S.I. Sinar Mashuri, and Y.H. Taufiq-Yap. 2020. Preparation of Na2O supported CNTs nanocatalyst for efficient biodiesel production from waste-oil. Energy Convers. Manag. 205: 112445.

Ikram, M., A. Raza, M. Imran, A. Ul-Hamid, A. Shahbaz, and S. Ali. 2020. Hydrothermal synthesis of silver decorated reduced graphene oxide (rGO) nanoflakes with effective photocatalytic activity for wastewater treatment. Nanoscale. Res. Lett. 15: 1-11.

Ivanova, V., P. Petrova, and J. Hristov. 2011. Application in the ethanol fermentation of immobilized yeast cells in matrix of alginate/magnetic nanoparticles, on chitosan-magnetite microparticles and cellulose-coated magnetic nanoparticles. Int. Rev. Chem. Eng. 3 289 – 299.

Jayabalan, T., M. Matheswaran, T. Radhakrishnan, and S.N. Mohamed. 2021. Influence of Nickel molybdate nanocatalyst for enhancing biohydrogen production in microbial electrolysis cell utilizing sugar industrial effluent. Bioresour. Technol. 320: 124284.

Jordan, J., C.S. Kumar, and C. Theegala. 2011. Preparation and characterization of cellulase-bound magnetite nanoparticles. J. Mol. Catal., B Enzym. 68: 139-146.

Kakaei, K., A. Rahimi, S. Husseindoost, M. Hamidi, H. Javan, and A. Balavandi. 2016. Fabrication of Pt–CeO_2 nanoparticles supported sulfonated reduced graphene oxide as an efficient electrocatalyst for ethanol oxidation. Int. J. Hydrog. Energy 41: 3861-3869.

Kaliva, M., and M. Vamvakaki. 2020. Nanomaterials characterization. In Nanotechnology, 401-433: Elsevier.

Kaur, M., M. Kaur, and V.K. Sharma. 2018. Nitrogen-doped graphene and graphene quantum dots: A review onsynthesis and applications in energy, sensors and environment. Adv. Colloid Interface Sci. 259: 44-64.

Khan, M., H.H. Ngo, W. Guo, Y. Liu, L.D. Nghiem, F.I. Hai, L. Deng, J. Wang, and Y. Wu. 2016. Optimization of process parameters for production of volatile fatty acid, biohydrogen and methane from anaerobic digestion. Bioresour. Technol. 219: 738-748.

Khoo, K.S., W.Y. Chia, D.Y.Y. Tang, P.L. Show, K.W. Chew, and W.-H. Chen. 2020. Nanomaterials Utilization in Biomass for Biofuel and Bioenergy Production. 13: 892.

Kim, J.H., C.-H. Ryu, M. Ji, Y. Choi, and Y.-I. Lee. 2021. Synthesis and Optical Property of (GaN) 1-x (ZnO) x Nanoparticles Using an Ultrasonic Spray Pyrolysis Process and Subsequent Chemical Transformation. J. Korean Inst. Met. Mater. 28: 143-149.

Kim, Y.-K., and H. Lee. 2016. Use of magnetic nanoparticles to enhance bioethanol production in syngas fermentation. Bioresour. Technol. 204: 139-144.

Kim, Y.-K., S.E. Park, H. Lee, and J.Y. Yun. 2014. Enhancement of bioethanol production in syngas fermentation with Clostridium ljungdahlii using nanoparticles. Bioresour. Technol. 159: 446-450.

Kothari, R., D. Singh, V. Tyagi, and S. Tyagi. 2012. Fermentative hydrogen production–An alternative clean energy source. Renew. Sust. Energ. Rev. 16: 2337-2346.

Krishnadasan, S., R. Brown, A. Demello, and J. Demello. 2007. Intelligent routes to the controlled synthesis of nanoparticles. Lab. Chip. 7: 1434-1441.

Kumar, G., P. Sivagurunathan, B. Sen, A. Mudhoo, G. Davila-Vazquez, G. Wang, and S.-H. Kim. 2017b. Research and development perspectives of lignocellulose-based biohydrogen production. Int. Biodeterior. Biodegradation. 119: 225-238.

Kumar, G., P. Sivagurunathan, B. Sen, S.-H. Kim, and C.-Y. Lin. 2017a. Mesophilic continuous fermentative hydrogen production from acid pretreated de-oiled jatropha waste hydrolysate using immobilized microorganisms. Bioresour. Technol. 240: 137-143.

Li, J.F., E.J. Rupa, J. Hurh, Y. Huo, L. Chen, Y. Han, J. chan Ahn, J.K. Park, H.A. Lee, and R. Mathiyalagan. 2019. Cordyceps militaris fungus mediated Zinc Oxide nanoparticles for the photocatalytic degradation of Methylene blue dye. Optik. 183: 691-697.

Lin, R., J. Cheng, L. Ding, W. Song, M. Liu, J. Zhou, and K. Cen. 2016. Enhanced dark hydrogen fermentation by addition of ferric oxide nanoparticles using Enterobacter aerogenes. Bioresour. Technol. 207: 213-219.

Liu, J., Z. Chen, F. Zhang, G. Ji, S. Zhong, Y. Wu, M. Wang, G. Sun, and H. Wang. 2019. Microstructural evolution and mechanical property of nanoparticles reinforced Al matrix composites during accumulative orthogonal extrusion process. Mater. Charact. 155: 109790.

Liu, T., C. Wang, W. Wang, G. Yang, Z. Lu, P. Xu, X. Sun, and J. Zhang. 2021. The enhanced properties in photocatalytic wastewater treatment: Sulfanilamide (SAM) photodegradation and Cr6+ photoreduction on magnetic Ag/ZnFe2O4 nanoarchitectures. J. Alloys Compd. 867: 159085.

Liu, Z., F. Lv, H. Zheng, C. Zhang, F. Wei, and X.-H. Xing. 2012. Enhanced hydrogen production in a UASB reactor by retaining microbial consortium onto carbon nanotubes (CNTs). Int. J. Hydrog. Energy. 37: 10619-10626.

López-Tinoco, J., J. Lara-Romero, R. Rangel, J. Apolinar-Cortés, F. Paraguay-Delgado, S. Jiménez-Sandoval, L. Bazán-Díaz, and R. Mendoza-Cruz. 2021. Microwave-assisted synthesis of ceria nanoparticles on carbon nanotubes and their dye-removal assesstment. J. Mater. Res. Technol. 13: 70-82.

Lu, C., L. Han, J. Wang, J. Wan, G. Song, and J. Rao. 2021. Engineering of magnetic nanoparticles as magnetic particle imaging tracers. Chem. Soc. Rev.

Madhuvilakku, R., and S. Piraman. 2013. Biodiesel synthesis by TiO2–ZnO mixed oxide nanocatalyst catalyzed palm oil transesterification process. Bioresour. Technol. 150: 55-59.

Malek, A., A. Ganta, G. Divyapriya, I.M. Nambi, and T. Thomas. 2021. Hydrogen production from human and cow urine using in situ synthesized aluminium nanoparticles. Int. J. Hydrog. Energy.

Manivasagan, P., K.-H. Kang, D.G. Kim, and S.-K. Kim. 2015. Production of polysaccharide-based bioflocculant for the synthesis of silver nanoparticles by Streptomyces sp. Int. J. Biol. Macromol. 77: 159-167.

Markeb, A.A., J. Llimós-Turet, I. Ferrer, P. Blánquez, A. Alonso, A. Sánchez, J. Moral-Vico, and X. Font. 2019. The use of magnetic iron oxide based nanoparticles to improve microalgae harvesting in real wastewater. Water. Res. 159: 490-500.

Mathur, P., S. Saini, E. Paul, C. Sharma, and P. Mehtani. 2021. Endophytic fungi mediated synthesis of iron nanoparticles: characterization and application in methylene blue decolorization. Curr. Opin. Green Sustain. Chem. 4: 100053.

Mazaheri, H., H.C. Ong, H.H. Masjuki, Z. Amini, M.D. Harrison, C.-T. Wang, F. Kusumo, and A. Alwi. 2018. Rice bran oil based biodiesel production using calcium oxide catalyst derived from Chicoreus brunneus shell. Energy. 144: 10-19.

Mehwish, H.M., M.S.R. Rajoka, Y. Xiong, H. Cai, R.M. Aadil, Q. Mahmood, Z. He, and Q. Zhu. 2021. Green synthesis of a silver nanoparticle using Moringa oleifera seed and its applications for antimicrobial and sun-light mediated photocatalytic water detoxification. J. Environ. Chem. Eng. 9: 105290.

Miletić, N., A. Nastasović, and K. Loos. 2012. Immobilization of biocatalysts for enzymatic polymerizations: possibilities, advantages, applications. Bioresour. Technol. 115: 126-135.

Mishra, P., and D. Das. 2014. Biohydrogen production from Enterobacter cloacae IIT-BT 08 using distillery effluent. Int. J. Hydrog. Energy. 39: 7496-7507.

Mishra, S., N.G. Shimpi, and T. Sen. 2013. The effect of PEG encapsulated silver nanoparticles on the thermal and electrical property of sonochemically synthesized polyaniline/silver nanocomposite. J. Polym. Res. 20: 1-10.

Moradnia, F., S.T. Fardood, A. Ramazani, B.-k. Min, S.W. Joo, and R.S. Varma. 2021. Magnetic Mg0. 5Zn0. 5FeMnO4 nanoparticles: green sol-gel synthesis, characterization, and photocatalytic applications. J. Clean. Prod. 288: 125632.

Moreno-Garrido, I. 2008. Microalgae immobilization: current techniques and uses. Bioresour. Technol. 99: 3949-3964.

Nasr, M., A. Tawfik, S. Ookawara, M. Suzuki, S. Kumari, and F. Bux. 2015. Continuous biohydrogen production from starch wastewater via sequential dark-photo fermentation with emphasize on maghemite nanoparticles. J .Ind. Eng. Chem. 21: 500-506.

Nath, D., A.K. Manhar, K. Gupta, D. Saikia, S.K. Das, and M. Mandal. 2015. Phytosynthesized iron nanoparticles: effects on fermentative hydrogen production by Enterobacter cloacae DH-89. Bull. Mater. Sci. 38: 1533-1538.

Nayebzadeh, H., N. Saghatoleslami, and M. Tabasizadeh. 2017. Application of microwave irradiation for preparation of a KOH/calcium aluminate nanocatalyst and biodiesel. Chem. Eng. Technol. 40: 1826-1834.

Nigam Ahuja, N., A. Ansari, R. Rajput, and P. Singh. 2020. Synthesis and Characterization of Zero Valent Iron Nanoparticles for Textile Wastewater Treatment. Pollution. 6: 773-783.

Pandey, R.R., M. Fukumori, A. TermehYousefi, M. Eguchi, D. Tanaka, T. Ogawa, and H. Tanaka. 2017. Tuning the electrical property of a single layer graphene nanoribbon by adsorption of planar molecular nanoparticles. Nanotechnology. 28: 175704.

Pena, L., M. Ikenberry, K. Hohn, and D. Wang. 2012. Acid-functionalized nanoparticles for pretreatment of wheat straw. J Biomater Nanobiotechnol 3: 342–352.

Prakash, L.V., A. Gopinath, R. Gandhimathi, S. Velmathi, S. Ramesh, and P. Nidheesh. 2021. Ultrasound aided heterogeneous Fenton degradation of Acid Blue 15 over green synthesized magnetite nanoparticles. Sep. Purif. Technol. 266: 118230.

Qiu, F., Y. Li, D. Yang, X. Li, and P. Sun. 2011. Heterogeneous solid base nanocatalyst: preparation, characterization and application in biodiesel production. Bioresour. Technol. 102: 4150-4156.

Radwan, M., M.A. Rashad, M. Sadek, and H.A. Elazab. 2019. SYNTHESIS, CHARACTERIZATION AND SELECTED APPLICATION OF CHITOSAN-COATED MAGNETIC IRON OXIDE NANOPARTICLES. J. Chem. Technol. 54.

Rahman, M., T. Sarmah, P. Dihingia, R. Verma, S. Sharma, D.N. Srivastava, L.M. Pandey, and M. Kakati. 2021. Bulk synthesis of tungsten-oxide nanomaterials by a novel, plasma chemical reactor configuration, studies on their performance for waste-water treatment and hydrogen evolution reactions. Chem. Eng. J.: 131111.

Rahmani Vahid, B., M. Haghighi, S. Alaei, and J. Toghiani. 2017. Reusability enhancement of combustion synthesized MgO/MgAl2O4 nanocatalyst in biodiesel production by glow discharge plasma treatment. Energy. Convers. Manag. 143: 23-32.

Rane, A.V., K. Kanny, V. Abitha, and S. Thomas. 2018. Methods for synthesis of nanoparticles and fabrication of nanocomposites. In Synthesis of inorganic nanomaterials, 121-139: Elsevier.

Razack, S.A., S. Duraiarasan, and V. Mani. 2016. Biosynthesis of silver nanoparticle and its application in cell wall disruption to release carbohydrate and lipid from C. vulgaris for biofuel production. Biotechnol. Rep. 11: 70-76.

Reddy, K., M. Nasr, S. Kumari, S. Kumar, S.K. Gupta, A.M. Enitan, and F. Bux. 2017. Biohydrogen production from sugarcane bagasse hydrolysate: effects of pH, S/X, Fe 2+, and magnetite nanoparticles. Environ. Sci. Pollut. Res. 24: 8790-8804.

Ren, D., H. Yang, L. Zhou, Y. Yang, W. Liu, X. Hao, and P. Pan. 2021. The Land-Water-Food-Environment nexus in the context of China's soybean import. Adv. Water. Resour. 151: 103892.

Rodríguez-Couto, S. 2019. Green nanotechnology for biofuel production. In Sustainable Approaches for Biofuels Production Technologies, 73-82: Springer.

Sandhya, R., R. Velavan, and J. Ravichandran. 2020. Biodiesel production from waste cooking oil using copper doped zinc oxide nanocatalyst – process optimisation and economic analysis. Int. J. Oil, Gas Coal Technol. 25: 488-497.

Sano, N., K. Yamada, S. Tsunauchi, and H. Tamon. 2017. A novel solid base catalyst for transesterification of triglycerides toward biodiesel production: Carbon nanohorn dispersed with calcium ferrite. Chem. Eng. J. 307: 135-142.

Sanusi, I.A., F.D. Faloye, and E.G. Kana. 2019. Impact of various metallic oxide nanoparticles on ethanol production by Saccharomyces cerevisiae BY4743: screening, kinetic study and validation on potato waste. Catal. Lett. 149: 2015-2031.

Sanusi, I.A., T.N. Suinyuy, A. Lateef, and G.E. Kana. 2020. Effect of nickel oxide nanoparticles on bioethanol production: process optimization, kinetic and metabolic studies. Process. Biochem. 92: 386-400.

Sekoai, P.T., C.N.M. Ouma, S.P. Du Preez, P. Modisha, N. Engelbrecht, D.G. Bessarabov, and A. Ghimire. 2019. Application of nanoparticles in biofuels: an overview. Fuel. 237: 380-397.

Seo, Y.I., K.H. Hong, S.H. Kim, D. Chang, K.H. Lee, and Y. Do Kim. 2012. Removal of bacterial pathogen from wastewater using Al filter with Ag-containing nanocomposite film by in situ dispersion involving polyol process. J. Hazard. Mater. 227: 469-473.

Shameer, P.M., R. Kasimani, S. Rajamohan, and P. Ramakrishnan. 2017. Effects of fuel injection parameters on emission characteristics of diesel engines operating on various biodiesel: A review. 67: 1267-1281.

Shukla, A.K., A.K. Upadhyay, and L. Singh. 2021. Algae-Mediated Biological Synthesis of Nanoparticles: Applications and Prospects. In Algae., 325-338: Springer.

Sikkema, R., S. Proskurina, M. Banja, and E. Vakkilainen. 2021. How can solid biomass contribute to the EU's renewable energy targets in 2020, 2030 and what are the GHG drivers and safeguards in energy-and forestry sectors? Renew. Energ. 165: 758-772.

Singh, A., S. Pandey, M.A. Siddiqui, and N. Kumar. 2021. A brief review on Nanoparticles: Type of platforms, biological synthesis & evaluation. Int. j. indig. herbs. drugs.: 23-28.

Singh, D., D. Sharma, S.L. Soni, S. Sharma, P. Kumar Sharma, and A. Jhalani. 2020. A review on feedstocks, production processes, and yield for different generations of biodiesel. Fuel. 262: 116553.

Siva, S., and M. Muthu. 2015. Production of biodiesel by transesterification of algae oil with an assistance of nano-CaO catalyst derived from egg shell. Int. J. Chemtech. Res. 7: 2112-2116.

Sivagurunathan, P., G. Kumar, A. Mudhoo, E.R. Rene, G.D. Saratale, T. Kobayashi, K. Xu, S.-H. Kim, and D.-H. Kim. 2017. Fermentative hydrogen production using lignocellulose biomass: an overview of pre-treatment methods, inhibitor effects and detoxification experiences. Renew. Sust. Energ. Rev. 77: 28-42.

Srivastava, G., S. Roy, and A.M. Kayastha. 2015. Immobilisation of Fenugreek β-amylase on chitosan/PVP blend and chitosan coated PVC beads: A comparative study. Food. Chem. 172: 844-851.

Srivastava, S., Z. Usmani, A.G. Atanasov, V.K. Singh, N.P. Singh, A.M. Abdel-Azeem, R. Prasad, G. Gupta, M. Sharma, and A. Bhargava. 2021. Biological nanofactories: using living forms for metal nanoparticle synthesis. Mini. Rev. Med. Chem. 21: 245-265.

Teo, S.H., A. Islam, and Y.H. Taufiq-Yap. 2016. Algae derived biodiesel using nanocatalytic transesterification process. Chem. Eng. Res. Des. 111: 362-370.

Thangaraj, B., and S. Piraman. 2016. Heteropoly acid coated ZnO nanocatalyst for Madhuca indica biodiesel synthesis. Biofuels. 7: 13-20.

Tripathi, R.M., R.K. Gupta, A.S. Bhadwal, P. Singh, A. Shrivastav, and B. Shrivastav. 2015. Fungal biomolecules assisted biosynthesis of Au–Ag alloy nanoparticles and evaluation of their catalytic property. IET. Nanobiotechnol. 9: 178-183.

Tsade, H., S.T. Anshebo, and F.K. Sabir. 2021. Preparation and characterization of functionalized cellulose nanomaterials (CNMs) for Pb (II) ions removal from wastewater. J. Chem. 2021.

Tsegay, M., H. Gebretinsae, and Z. Nuru. 2021. Structural and optical properties of green synthesized Cr_2O_3 nanoparticles. Mater. Today. 36: 587-590.

Vandenabeele, C.R., and S. Lucas. 2020. Technological challenges and progress in nanomaterials plasma surface modification–a review. Mater. Sci. Eng. R Rep. 139: 100521.

Vasistha, S., A. Khanra, and M.P. Rai. 2021. Influence of microalgae-ZnO nanoparticle association on sewage wastewater towards efficient nutrient removal and improved biodiesel application: An integrated approach. J. Water.Process. Eng. 39: 101711.

Vu, D.K.N., and D.K.V. Nguyen. 2021. Gamma Irradiation-Assisted Synthesis of Silver Nanoparticle-Embedded Graphene Oxide-TiO2 Nanotube Nanocomposite for Organic Dye Photodegradation. J. Nanomater. 2021.

Wang, C., D. Chen, Q. Wang, and R. Tan. 2017. Kanamycin detection based on the catalytic ability enhancement of gold nanoparticles. Biosens. Bioelectron. 91: 262-267.

Wang, H., X. Liang, J. Wang, S. Jiao, and D. Xue. 2020. Multifunctional inorganic nanomaterials for energy applications. Nanoscale. 12: 14-42.

Wei, L., Z. Li, J. Li, Y. Zhang, B. Yao, Y. Liu, W. Song, X. Fu, X. Wu, and S. Huang. 2020. An approach for mechanical property optimization of cell-laden alginate–gelatin composite bioink with bioactive glass nanoparticles. J. Mater. Sci. Mater. Med. 31: 1-12.

Wen, L., Y. Wang, D. Lu, S. Hu, and H. Han. 2010. Preparation of KF/CaO nanocatalyst and its application in biodiesel production from Chinese tallow seed oil. Fuel. 89: 2267-2271.

Wimonsong, P., and R. Nitisoravut. 2014. Biohydrogen enhancement using highly porous activated carbon. Energy. Fuels. 28: 4554-4559.

Wimonsong, P., and R. Nitisoravut. 2015. Comparison of different catalyst for fermentative hydrogen production. J. Clean. Energy. Techno. 3: 128-131.

Xia, A., A. Jacob, C. Herrmann, M.R. Tabassum, and J.D. Murphy. 2015. Production of hydrogen, ethanol and volatile fatty acids from the seaweed carbohydrate mannitol. Bioresour. Technol. 193: 488-497.

Yadav, V.K., D. Ali, S.H. Khan, G. Gnanamoorthy, N. Choudhary, K.K. Yadav, V.N. Thai, S.A. Hussain, and S. Manhrdas. 2020. Synthesis and characterization of amorphous iron oxide nanoparticles by the sonochemical method and their application for the remediation of heavy metals from wastewater. Nanomaterials. 10: 1551.

Yang, G., and J. Wang. 2018. Improving mechanisms of biohydrogen production from grass using zero-valent iron nanoparticles. Bioresour. Technol. 266: 413-420.

Yaqoob, A.A., T. Parveen, K. Umar, and M.N. Mohamad Ibrahim. 2020. Role of nanomaterials in the treatment of wastewater: A review. Water. 12: 495.

Yazdi, M.E.T., M.S. Amiri, S. Akbari, M. Sharifalhoseini, F. Nourbakhsh, M. Mashreghi, M.R. Abbasi, M. Modarres, and A. Es-haghi. 2020. Green synthesis of silver nanoparticles using helichrysum graveolens for biomedical applications and wastewater treatment. Bionanoscience. 10: 1121-1127.

Yeganeh, M.S., A.R. Kazemizadeh, A. Ramazani, P. Eskandari, and H.R. Angourani. 2020. Plant-mediated synthesis of Cu0. 5Zn0. 5Fe2O4 nanoparticles using Minidium leavigatum and their applications as an adsorbent for removal of reactive blue 222 dye. Mater. Res. Express. 6: 1250f1254.

Yu, G., X. Wang, J. Liu, P. Jiang, S. You, N. Ding, Q. Guo, and F. Lin. 2021. Applications of nanomaterials for heavy metal removal from water and soil: a review. Sustainability. 13: 713.

Zhang, C., F.-X. Lv, and X.-H. Xing. 2011. Bioengineering of the Enterobacter aerogenes strain for biohydrogen production. Bioresour. Technol. 102: 8344-8349.

Zhang, J., C. Fan, and L. Zang. 2017. Improvement of hydrogen production from glucose by ferrous iron and biochar. Bioresour. Technol. 245: 98-105.

Zhang, J., C. Fan, H. Zhang, Z. Wang, J. Zhang, and M. Song. 2018. Ferric oxide/carbon nanoparticles enhanced bio-hydrogen production from glucose. International Journal of Hydrogen Energy 43: 8729-8738.

Zhang, L., L. Zhang, and D. Li. 2015. Enhanced dark fermentative hydrogen production by zero-valent iron activated carbon micro-electrolysis. Int. J. Hydrog. Energy. 40: 12201-12208.

Zhou, X.-h., B.-c. Huang, T. Zhou, Y.-c. Liu, and H.-c. Shi. 2015. Aggregation behavior of engineered nanoparticles and their impact on activated sludge in wastewater treatment. Chemosphere. 119: 568-576.

Zou, C.-D., Y.-L. Gao, Y. Bin, and Q.-J. Zhai. 2010. Size-dependent melting properties of Sn nanoparticles by chemical reduction synthesis. T. NONFERR. METAL. SOC. 20: 248-253.

9

Oil Recovery from Mining Effluents Using Nanotechnology

Sougata Ghosh[1,*] and *Bishwarup Sarkar*[2]

9.1 Introduction

Growing population at an alarming rate with large scale industrialization and urbanization would increase the worldwide energy demand by 50% by the end of 2030. Renewable energy being insufficient to meet this demand, oil will contribute the major share of the total energy requirement at least for next few decades. Hereby, exploration of new oil fields or enhanced recovery of oil from the existing oil fields has attracted attention of researchers. Unfortunately, the available conventional methods for oil recovery can only recover one-third of the total oil from the reservoir (Kamal et al. 2017). Various primary, secondary and tertiary oil recovery techniques are used as depicted in Fig. 1. The primary oil recovery techniques are employed where the hydrocarbons naturally rise to the surface due to the existing displacement energy in a reservoir, such as gas-cap drive, solution-gas drive, and natural water drive, etc. On the other hand, water or gas is commonly injected in secondary techniques to provide energy to sweep the residual oil out of the wellbore. Tertiary processes include the application of chemicals, miscible gases, and/or thermal energy to mobilize the additional oil after the secondary recovery process (Adil et al. 2018). The oil recovery efficiency in primary techniques depends on the effective maintenance of pressure while the efficiency in tertiary recovery processes depends upon the interaction of the injected fluid with the reservoir rock/oil system. Various factors influencing such interactions critical for enhanced oil recovery (EOR) include lowering of interfacial tensions (IFT), reduction in oil viscosity, oil swelling, and alteration of wettability and/or favourable phase behaviour.

[1] Department of Microbiology, School of Science, RK University, Rajkot, Gujarat, India.
[2] College of Science, Northeastern University, Boston, MA, USA.
* Corresponding author: ghoshsibb@gmail.com

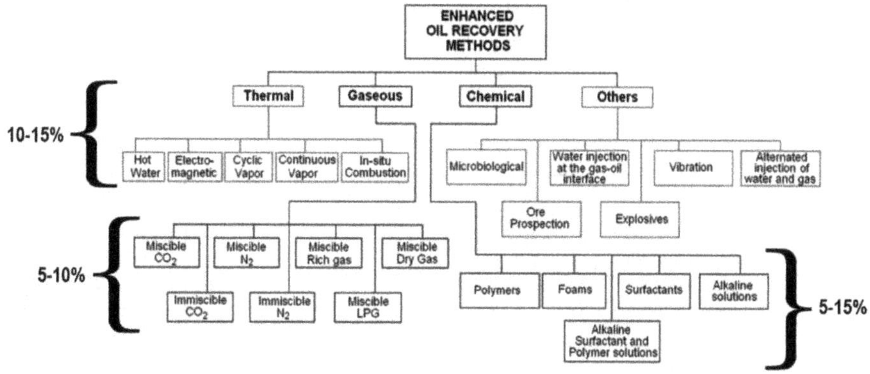

Fig. 1: Various available EOR methods, with their typical percentage incremental recovery. Reprinted from Adil, M., Lee, K., Mohd Zaid, H., Ahmad Latiff, N.R. and Alnarabiji, M.S. 2018. Experimental study on electromagnetic-assisted ZnO nanofluid flooding for enhanced oil recovery (EOR). PLoS ONE. 13(2): e0193518. (Open Access).

The major challenge in EOR is the location of the oil reservoirs that are mostly concealed in deep water. Under the extreme environmental conditions in such location, often the property of the surfactant used for flooding and gas used for EOR drastically change (Chang et al. 2006). Also, these methods are costly and sometimes hazardous. Similarly, the limitation of the thermal methods includes involvement of the high energy that makes the process less economically viable. Moreover, the large amount of heat loss in this method is also undesired (Adil et al. 2018). In view of the background, novel, inexpensive, efficient, and environmentally benign techniques have been developed recently. This chapter highlights various nanotechnology driven solutions for EOR that include nanofluids, nanoparticles, nanocomposites, carbon-nanotubes, and nanomembranes as listed in Table 1. Further, the mechanism of the enhancement of the EOR is discussed for each type. Eventually, the scope of improvement and development of large scale nanotechnological process for EOR is also presented.

9.2 Nanofluids

Oil recovery can be enhanced significantly by altering the wettability at different stages by transition of an oil-wet to a strongly water-wet condition. Recently, nanofluids have gained wide attention as promising wettability modifiers. Alumina-based nanofluids were used by Giraldo et al. (2013) to alter the wettability conditions of sandstone cores. Contact angle measurements of untreated sandstone cores were found to vary between 104° and 142° that were reduced to practically zero upon nanofluid treatment indicating alteration of rock wettability from an oil-wetting to a strongly water-wetting condition. Imbibition tests revealed that a concentration of 100 ppm of alumina nanoparticles created the optimum synergistic effect on wettability alteration. Treatment with 1000 ppm of alumina nanoparticles or above resulted in inhibition of imbibition process which was speculated to be due to aggregation and particle coalescence that can negatively affect both coating and cleaning mechanisms of nanofluid treatments. Core displacement tests highlighted the shift of wettability

Table 1: Use of various nanomaterials for EOR.

Nanoparticle	Additive	Ultimate Oil Recovery (%)	References
Alumina-based nanofluids	-	-	Giraldo et al. 2013
Fluorescent carbon nanofluids	Brine	39.1%	Li et al. 2017
LHP silica nanofluid	Brine	62.23%	Hendraningrat et al. 2013
Polymeric SiO_2 nanofluids	Polyethylene Glycol	80.6%	Zhang et al. 2016
TiO_2 nanofluids	-	61.9%	Bayat et al. 2014
TiO_2 nanofluids	Brine	80%	Ehtesabi et al. 2014
ZnO nanofluids	Brine and SDBS	10.27%	Adil et al. 2018
ZrO_2	Cetrimonium bromide surfactant	40%	Karimi et al. 2012
Al_2O_3, Fe_2O_3 and SiO_2 nanofluids	Propanol	89.5%, 87.2%, and 91.7%	Joonaki and Ghanaatian 2014
Ultadispersed multimetallic nanofluids	Vacuum gas oil	30%	Hashemi et al. 2013
AlOOH nanoparticles	Deionised water with SDS	-	Yang et al. 2017
Citric acid coated- Fe_3O_4 nanoparticles	Brine	22%	Divandari et al. 2019
Si nanoparticles	Guar Gum solution	44.28%	Bera et al. 2020
Polymer coated- silica nanoparticles	Synthetic seawater	71.5%	Bila et al. 2019
SnO_2 nanoparticles	Brine	61%	Jafarnezhad et al. 2017
Fe_3O_4@chitosan nanocomposites	-	67.5%	Rezvani et al. 2018
LoSal-ZnO/SiO_2 nanocomposites	Diluted Seawater	66.24%	Ali et al. 2019
Multi-wall carbon nanotubes	Brine	31.8%	Alnarabiji et al. 2016
P(NIPAAm-co-NMA)/ ChNWs NF membranes	-	-	Wu et al. 2018
SNB membranes	-	-	Li et al. 2021

due to nanofluid treatment wherein, oil effective permeability at a given water saturation increased from 521.6 mD to 696.2 mD.

Li et al. (2017) reported development of a novel nanofluid composed of fluorescent carbon nanoparticles that could be potentially used for enhanced oil recovery (EOR). Fluorescent carbon nanoparticles were observed to be spherical with approximately 10 nm in size that may be useful in efficient mobility in a reduced permeability medium. FTIR analysis revealed presence of large amount of carboxylic groups providing excellent ability to disperse in water. A light yellow colour was observed when fluorescent carbon nanoparticles (0.005 wt. %) were dispersed in brine that turned to bright green under UV excitation at 365 nm. DLS analysis indicated a narrow size distribution of particles between 10 and 30 nm. Improvement in electrostatic repulsion among the particles was observed to be the result of a negative zeta potential of –35 mV. The IFT was found to be reduced down to 13.4 nM/m in presence of 0.1 wt. % of nanofluid. The wettability of oil-wet surface was changed to water-wet in presence of 0.1 wt. % of nanofluid with an increase in contact angle of oil on the surface followed by gradual displacement of oil. The stability of nanofluid was demonstrated with storage at 90°C for 30 days resulting in no change in transmittance. Similarly, stability of nanofluid was observed to be unchanged in presence of increasing concentrations of NaCl. Ultra-low permeable sandstone cores were taken for verification of EOR abilities of nanofluid. Recovery of oil in presence of 0.05 wt. % nanofluid was found to be 39.1% as compared to higher concentrations of nanofluid. Furthermore, fluorescence microscopy was used for confirmation of seeping of nanoparticles into the pores of ultra-low permeable cores.

In another study, a lipophobic and hydrophilic nanoparticle (LHP) composed of SiO_2 was reported by Hendraningrat et al. (2013) for its application in EOR. Berea sandstone core plugs were used as a porous medium having a permeability range of 9–400 mD and a degassed crude oil from a field in North Sea had been employed. LHP silica nanoparticles were dispersed in brine and the nanofluids were injected as tertiary recovery mode after flooding at room temperature. It was observed that the IFT value upon introduction of 0.01 wt. % LHP silica nanoparticles was reduced to half as compared to brine-crude oil system that acted as a reference. On the other hand, pH value was decreased from 6 to 5 upon increasing the concentration of nanofluid from 0.05 to 0.1 wt. %. Contact angles of aqueous phase were also found to reduce upon introduction of nanoparticles thus increasing water wetness. It was also observed that increase in concentration of nanofluid led to subsequent increase in the porosity and permeability impairment. Differential pressure was also found to increase upon injection of nanofluid into the core plug which could be due to adsorption of nanoparticles leading to blockage of pore throats, which subsequently led to 5–6% final increase in oil recovery. The ultimate oil recovery in presence of nanofluid was increased as well with simultaneous increase of nanoparticles from 0.01 to 0.05 wt. % as compared to brine injection, thus implying effective EOR strategies by applying LHP silica nanofluids.

Zhang et al. (2016) reported EOR from sintered bead pack using polymeric SiO_2 nanofluids containing polyethylene glycol and dispersed in 0.25 wt. % brine solution. Dynamic light scattering (DLS) method was used for characterization of nanoparticles and the average diameter was found to be 9.5 ± 0.5 nm. The viscosity of nanofluid was observed to be 1.097 mPa at 20°C with a density of 1.0 ± 0.01 g/mL and pH of 6.5 ± 0.5 at 1 atm pressure. Sintered bead packs were made using a mixture of 1 mm and 250–300 μm borosilicate glass beads. Flooding experiments indicated displacement of around 65% of oil by brine injection and an additional 14.6% recovery was obtained with 0.277 wt. % nanofluid injections as the tertiary process. Interestingly, it was noted that there was an increase in pressure drop of about 0.1 to 0.15 psi during nanofluid injection. Also, flooding with 1.35 mM of sodium dodecyl sulphate solution (SDS) was performed to exclude oil recovery induced by interfacial tension. It was observed that 2% of additional oil recovery was purely due to interfacial tension. Distribution of oil and nanofluid in the sintered bead pack was determined using a high-resolution X-ray microtomography (X-ray micro-CT). It was observed that the oil porous medium was predominantly infused with oil after brine flooding and due to gravitational effect; majority of oil was displaced from bottom and retained on top. Furthermore, injection of nanofluid was observed to cause an apparent reduction in gravitational effect leading to uniform distribution of oil along the porous medium. Brine flooding was found to retain trapped oil ganglia on the upper part of porous medium while nanofluid flooding induced breakage of these large clusters into small oil blobs. Increase in capillary numbers was observed to decrease additional oil recovery by nanofluid. It was proposed that such decrease in EOR by nanofluid could be due to insufficient time for advancement of nanofluid film through structural disjoining pressure in order to displace oil. Also, it was suggested that mobilization by Ca enhanced brine flooding may provide less oil availability for nanofluid to displace. Thus, it was concluded that nanofluids are effective as compared to high Ca brine flooding promoting better EOR.

TiO_2 nanoparticle mediated EOR from limestone media was reported by Bayat et al. (2014). Initially, sedimentation experiments were performed in order to determine the suspension stability of TiO_2 nanoparticle dispersed in deionized water. Time-resolved optical absorbance indicated negligible sedimentation for the first 60 min. of experiment. In order to determine the tendency of TiO_2 nanoparticles to adsorb on the surface of limestone, transportation tests were performed that showed presence of TiO_2 nanoparticles in the effluent after 1.3 pore volumes of injection in the limestone. Recovery of 72.2% of TiO_2 nanoparticles were seen while 27.8% remained in the porous media. It was proposed that due to difference between electric surface charge signs of TiO_2 nanoparticles and limestone grains, there was considerable adsorption observed on limestone surface. Figure 2 shows the field emission scanning electron microscope (FESEM) image of limestone after TiO_2 flooding, indicating adsorption of nanoparticles along with Energy Dispersive X-Ray (EDX) analysis. Displacement test results showed 50.9% oil recovery in presence of TiO_2 nanofluid as compared to 47.3% in presence of 0.3 wt. % NaCl as control. The oil recovery was further enhanced up to 61.9% when temperature was raised to 60°C.

Ehtesabi et al. (2014) also fabricated TiO_2 based nanofluids for application in EOR from sandstone cores. Nanoparticles were synthesised using titanium tetraisopropoxide (TITP) that were suspended in various salts such as NaCl, Na_2SO_4, KCl and $CaCl_2$. TiO_2 anatase nanoparticles showed proper stability in presence of NaCl and Na_2SO_4 without any sediment formation after 24 h of incubation. However, there was weak stability in presence of KCl and no stability in presence of $CaCl_2$. In presence of 10000 ppm NaCl, the sizes of nanoparticles were observed to increase from 40 nm to 52 nm after 24 h and 48 h, respectively. Oil recovery was found to increase up to 80% in presence of 0.01% TiO_2 anatase nanoparticles as compared to 49% recovery factor upon flushing the core with 5000 ppm brine solution. Contact angle of water droplets were found to reduce from $125 \pm 3°$ to $90 \pm 3°$ upon flooding with 0.01% of TiO_2 nanofluids. It was proposed that the hydroxyl groups present at the surface of TiO_2 nanoparticles were responsible for wettability alteration from oil-wet to water-wet. Scanning electron microscopy (SEM) and energy-dispersive

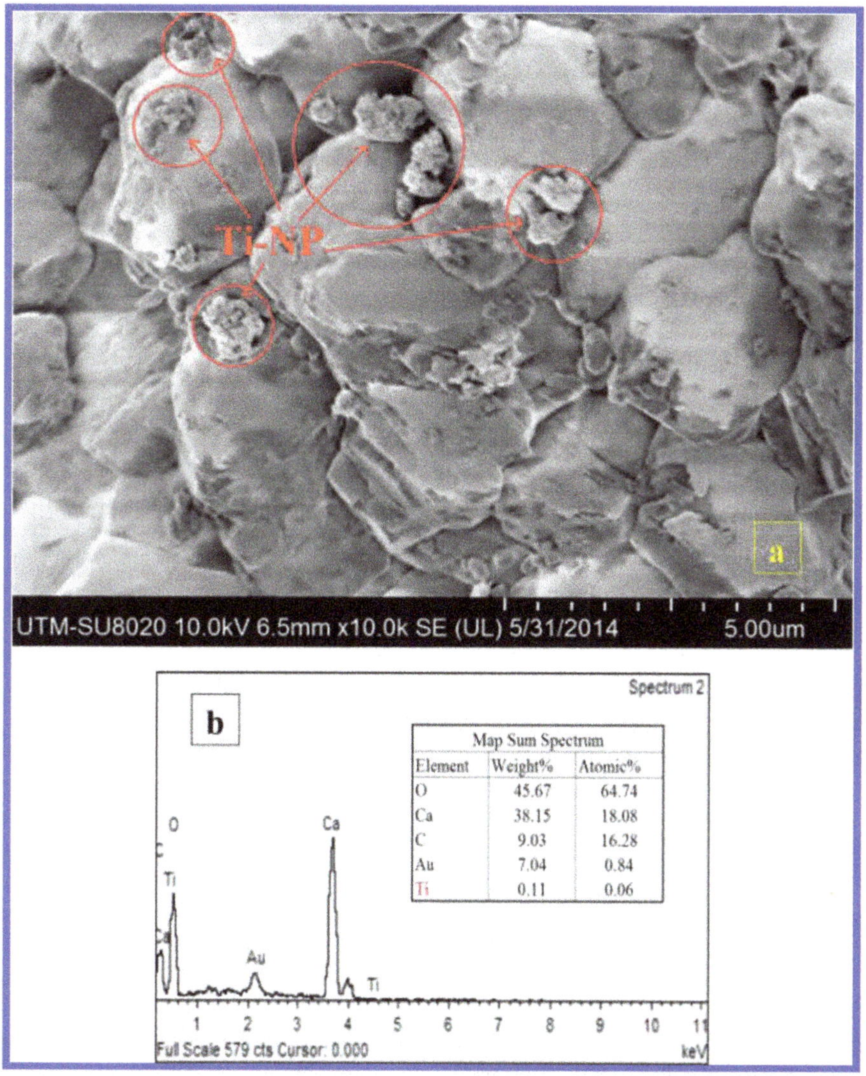

Fig. 2: (a) FESEM image from limestone grains after the flooding of TiO_2 nanofluid. (b) EDX measurement after the flooding of TiO_2 nanofluid from the entrance side of the column. Reprinted with permission from Bayat, A.E., Junin, R., Samsuri, A., Piroozian, A. and Hokmabadi, M. 2014. Impact of metal oxide nanoparticles on enhanced oil recovery from limestone media at several temperatures. Energy Fuels. 28(10): 6255–6266. Copyright © 2014, American Chemical Society.

X-ray spectroscopy (EDS) measurements indicated homogenous deposition of TiO_2 nanoparticles on the surface of the core. SEM images highlighted formation of TiO_2 nanorods having a diameter of 60 nm. It was found that amorphous TiO_2 nanoparticles formed quantitatively higher sediment on the core plug as compared to anatase TiO_2 nanoparticles along with higher pressure drop and amplified nanorods formation that was assumed to plug the pores of sandstone core. Therefore, TiO_2 nanoparticles were proposed to be potentially useful in EOR applications as an inexpensive and eco-friendly additive under low concentrations.

Adil et al. (2018) reported application of electromagnetic-assisted ZnO nanofluid mediated EOR. ZnO nanoparticles were synthesised via sol-gel auto-combustion method and were found to have an average size of 55.7 and 117.1 nm when calcined at 500°C and 800°C, respectively. These nanoparticles were then dispersed in brine (3 wt. % NaCl) followed by supplementation of an anionic surfactant, sodium dodecylbenzene sulfonate (SDBS) as a stabilizer. Sandpack flooding experiments were performed to check efficiency of oil recovery. It was observed that 0.1 wt. % of ZnO nanoparticles, when dispersed in 0.025 wt. % of SDBS, was able to substantially reduce residual oil saturation, thus providing greater oil recovery as compared to SDBS alone with additional recovery of 8.5–10.2% of original oil in place (OOIP). However, a certain incubation time was required for this additional oil displacement which was proposed to be due to physiochemical interactions between the nanoparticles and the sandpack. IFT analysis showed that nanoparticles were responsible for decrease in IFT between oleic phase and aqueous phase. The contact angle was also reduced as the size of the nanoparticle was decreased. The viscosity of oil was also observed with increase in temperature from 25°C to 95°C, thus resulting in reduced mobility ratio that would facilitate movement of nanofluids providing better oil recovery. The total oil recovery after re-injection of brine was found to increase from 64.9 to 68.5% of OOIP in case of ZnO nanoparticles calcined at 500°C, while for particles calcined at 800°C, the increase was from 64.6 to 71.3% of OOIP. In addition, electromagnetic (EM)-assisted nanoflooding was observed to further increase the additional oil recovery to 10.4 and 9% OOIP after water injection in case of 500°C and 800°C calcined nanoparticles, respectively. Three mechanisms were proposed for increment in oil recovery in presence of EM waves. First, it was suggested that the dielectric polarization of nanoparticles may cause deformation of the shape of oil drops that can lead to increase in surface area, thus resulting in higher particles adsorption and reduction in interfacial tension. Next, an increase in free energy of surface was observed due to increase in rate of wettability alteration. Finally, improvement of mobility ratio as a consequence of electro-rheological (ER) effect could increase the viscosity of nanofluids.

Alteration of wettability of a carbonate reservoir rock in order to increase oil recovery was reported by Karimi et al. (2012) using ZrO_2-based nanofluids. Sol-gel method was used to synthesise ZrO_2 nanoparticles using zirconium oxychloride ($ZrOCl_2 \cdot 8H_2O$) as the source of Zr and Cetrimonium bromide (CTAB) as surfactant. Contact angle measurements were taken using a flat and relatively smooth calcite plates. Water-phase contact angle of clean calcite plates was measured and found to be about 32° and 35° in presence of air and *n*-heptane, respectively. Calcite plates that were aged in crude oil sample at 80°C for one week were observed to have a strong change from water-wet to oil-wet condition with a water-phase contact angle of 118° and 180° in presence of air and *n*-heptane, respectively. As shown in Fig. 3, this change in wettability was confirmed by SEM images. It was found that cleaned substrates have sharp-edged carbonate particles while plates aged in oil were observed to have a relatively smooth surface. Alteration in wettability from oil-wet to strongly water-wet was observed upon introduction of ZrO_2 nanofluid. This change was proposed to be due to adsorption of nanoparticles leading to subsequent change in free energy of the surface. It was observed that a minimum of 48 h was required

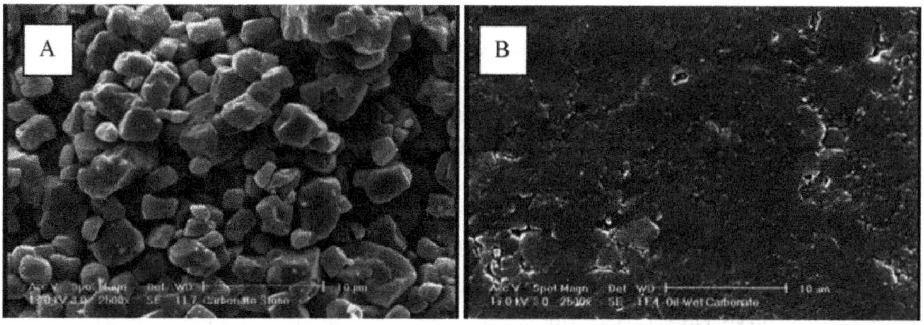

Fig. 3: SEM images of a clean carbonate rock (A) and a carbonate rock aged in oil (B). Reprinted with permission from Karimi, A., Fakhroueian, Z., Bahramian, A., Khiabani, N.P., Darabad, J.B., Azin, R. and Arya, S. 2012. Wettability alteration in carbonates using zirconium oxide nanofluids: EOR implications. Energy Fuels. 26(2): 1028–1036. Copyright © 2012 American Chemical Society.

Fig. 4: SEM images of an oil-wet carbonate rock aged in fluid 1 (A), fluid 2 (B), and fluid 3 (C). Reprinted with permission from Karimi, A., Fakhroueian, Z., Bahramian, A., Khiabani, N.P., Darabad, J.B., Azin, R. and Arya, S. 2012. Wettability alteration in carbonates using zirconium oxide nanofluids: EOR implications. Energy Fuels. 26(2): 1028–1036. Copyright © 2012 American Chemical Society.

for alteration of wettability. EDX analysis showed that no Zr element was present in the original reservoir rock and it was made up of Ca, hence supporting the notion of adsorption of Zr onto the surface of aged rock. SEM images, as depicted in Fig. 4, showed that compared to fluid 1 (absence of nanoparticles), fluids 2 and 3 facilitated formation of nanosized ribbons and nanoflower like morphologies in aged oil-wet plates. Spontaneous imbibition tests revealed that a maximum recovery rate of around 40% was seen shortly after contact between nanofluids and core plugs saturated with oil. Two mechanisms were speculated based on these quick fluid imbibition results which involve either formation of nanoparticle- surfactant complexes which could subsequently interact with oil and rock or else formation of emulsions that could be stabilized by nanoparticles.

Application of aluminium oxide, iron oxide, and silicon oxide nanoparticles treated by silane for EOR was reported by Joonaki and Ghanaatian (2014). All three kinds of nanoparticles were dispersed in propanol to form nanofluids. Contact angle measurements after nanofluid applications suggested wettability alteration. SiO_2 nanofluids showed the best results by shifting the contact angle value from 134° to 82° while Al_2O_3 showed better alteration activity as compared to Fe_2O_3 nanoparticles. A concentration of 1.5 g/L was found to be ideal for reduction in contact angle values. Likewise, IFT between oil-water interphase was found to decrease from 38.5 dyne/cm to 2.25, 2.75 and 1.45 dyne/cm in presence of Al_2O_3, Fe_2O_3 and SiO_2 nanofluids, respectively. Thus, it suggested the fact that these nanoparticles are efficient in mobilization of oil. Sandstone cores were used for EOR analysis of nanofluids. Initially, three pore volumes of synthetic brine solution were injected into the core followed by equal volumes of three different nanofluids. Oil recovery due to water flooding was found to be 56.6%, followed by a 20.2%, 17.3%, and 22.5% increase due to Al_2O_3, Fe_2O_3 and SiO_2 nanofluids injection, respectively. Hence, the total recovery due to Al_2O_3, Fe_2O_3 and SiO_2 nanofluids was observed to be 76.8%, 73.9%, and 79.1%, respectively. This recovery was further enhanced to 89.5%, 87.2%, and 91.7% of total oil recovery in presence of Al_2O_3, Fe_2O_3 and SiO_2 nanofluids, respectively, when initial flooding was done with propanol instead of brine solution. Reduction of IFT and alteration of wettability from water-wet to neutral-wet state was proposed to be the two main reasons for EOR in presence of nanofluids that help in mobilization of oil droplets toward the flow line due to reduction in capillary pressure.

In another study, Hashemi et al. (2013) reported EOR via ultradispersed multimetallic nanoparticles. A water-in-vacuum gas oil microemulsion method was employed in order to prepare nanocatalyst suspension wherein, vacuum gas oil (VGO) was mixed with HLB-8 under shaking conditions (700 rpm) at 60°C. Oil sand packed bed columns were saturated with Athabasca bitumen. A steam injection test was performed in order to evaluate porous media in which introduction of steam was followed by VGO at 3.5 MPa and two different temperatures of 320 and 340°C. Steam-assisted gravity drainage (SAGD) was shown to recover about 50% OOIP. It was also found that presence of 15 wt. % of light hydrocarbon such as pentane was able to increase the efficiency of oil recovery which was proposed to be due to reduction of bitumen viscosity in order to increase the mobility that could result in higher recovery. It was also speculated that pentane could act as a vaporising agent as well as diluent for bitumen. Recovery performance of VGO injection was found to enhance in the presence of ultra-dispersed nanocatalysts providing around 30% bitumen recovery at 340°C.

9.3 Nanoparticles

Stabilization of foams by *in situ* modified AlOOH nanoparticles was reported by Yang et al. (2017) for its application in EOR. Dispersion of 1.0 wt. % of AlOOH nanoparticles were prepared in deionised water along with addition of varying concentrations of SDS. Adsorption of SDS molecules onto the surface of AlOOH nanoparticles was analysed by zeta potential measurements wherein, zeta potential values of nanoparticles decreased from +55.4 mV to 0 mV with subsequent increase in SDS. The zeta potential was reduced to -52.1 mV in presence of 30 mmol/L of SDS. This adsorption was confirmed by adsorption isotherm analysis. Foam generation was observed in SDS/AlOOH dispersions; however, a slight decrease in foamability was observed with increasing concentrations of nanoparticles as seen in Fig. 5. At a relative SDS concentration of 5, the half-life of foam in presence of 1.0 wt. % AlOOH nanoparticles was observed to be 71.4 h. The contact angle was found to rapidly increase from 18.5° to 55.5° in presence of increasing concentrations of SDS. Formation of a nanoparticle network within the foam film was observed to reduce drainage, coalescence, and gas diffusion, hence increasing the stability of foams. Sandpack flooding experiments revealed effective pore blocking ability of SDS/AlOOH foam. The high foam stability also enables it to withstand water flooding and other external pressures, hence, making it a potential tool for EOR.

Divandari et al. (2019) demonstrated application of citric-acid coated magnetite nanoparticles for EOR. Nanoparticles were synthesised using 1mmol of citric acid trisodium salt dehydrate and 2M $FeSO_4.7H_2O$. Spherical Fe_3O_4 nanoparticles coated with citric acid on the outer surface were formed where the size of nanoparticles were found to be less than 80 nm. It was speculated that since these citric acid-coated magnetite nanoparticles are water soluble, the charged carboxyl groups present in every citrate ion may act as a surfactant and the repulsive forces between the charges will make the nanoparticles more stable and dispersive in water. The magnetizing power and sedimentation rate of 0.8 wt. % nanoparticles suspended in 5000 ppm NaCl were demonstrated by simple experiments. Instantaneous reaction and agglomeration was observed when a hand-held magnet was placed near the solution that was not

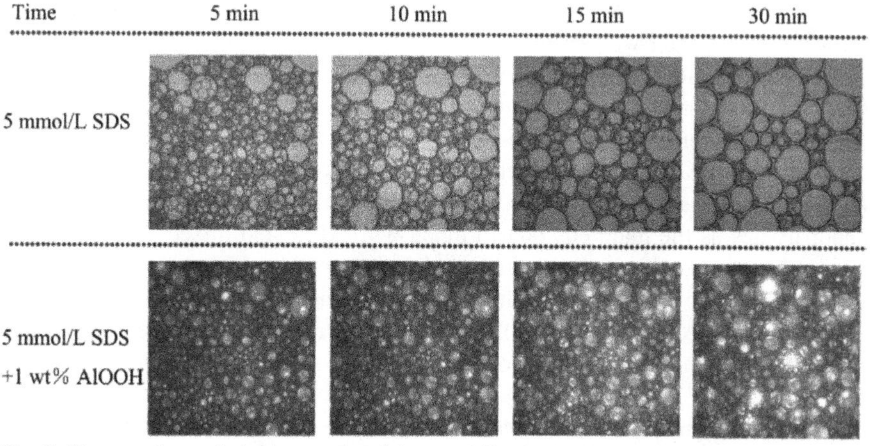

Fig. 5: Time-varying optical micrographs of foam stabilized solely by SDS and foam stabilized by AlOOH nanoparticles, in synergy with SDS. Reprinted with permission from Yang, W., Wang, T., Fan, Z., Miao, Q., Deng, Z. and Zhu, Y. 2017. Foams stabilized by in situ-modified nanoparticles and anionic surfactants for enhanced oil recovery. Energy Fuels. 31(5): 4721–4730. Copyright © 2017 American Chemical Society.

seen for non-coated nanoparticles. The sedimentation rate was found to be lowered in case of surface-coated magnetite nanoparticles. Contact angle measurements suggested that citric acid coated Fe_3O_4 nanoparticles could alter the wettability of surface making it strongly water-wet. IFT values were also found to decrease from 37 mN/m to 16 mN/m when 2 wt. % of nanoparticles was added to the brine solution. A maximum ultimate oil recovery factor of around 83% was reached in the presence of magnetic field with an intensity of 2750 Gauss. A flow rate of 0.1 cc./min. was found to recover almost 100% of oil. As compared to hematite nanoparticles, citric acid coated magnetite nanoparticles provided a better enhanced oil recovery with a 22% increase and 60% reduction of water ratio indicating plugging efficiency.

In another study, Bera et al. (2020) found an enhanced oil recovery (EOR) when 4000 ppm of guar gum solution was mixed with 0.2% of silica nanoparticles. Initially, X-ray diffraction (XRD) analysis was used to determine the porosity and permeability of rock samples and was found to be in the range of 15–17% and 200–205 mD, respectively. The lowest shear rate and the highest viscosity of crude oil were observed to be at 30°C. XRD analysis also revealed that the major component of the core sample is silica. Viscosity of guar gum solution was observed to decrease with increase in solution temperature and shear rate, respectively. It was suggested that average speed of molecules increases with subsequent rise in temperature, therefore resulting in an overall decrease in the intermolecular forces. On the other hand, addition of silica nanoparticles showed an opposite effect wherein, supplementation of nanoparticles resulted in increase of viscosity of guar gum solutions. It was speculated that interlink bonds formed between silica nanoparticles and guar gum solution may lead to formation of a stable polymer slug hence, increasing the viscosity even at higher temperatures. This synergistic effect of nanoparticle and polymer resulted in maximum oil recovery of 44.28% when 4000 ppm guar gum solution was mixed with 0.2 wt. % silica nanoparticle. Furthermore, it was found that the mixture of nano-silica and polymer solution changed the contact angle value from 73° to 115° indicating effective alteration from oil-wet rock to water-wet. Promotion of adsorption of silica nanoparticles onto the sandstone rock may be one of the reasons for alteration of wettability. Phase behaviour study indicated that nano-silica and polymer solution was able to form emulsion layer only under low temperature conditions that were broken at around 80°C.

Polymer-coated silica nanoparticles were investigated by Bila et al. (2019) for EOR. Commercialized nanoparticles were used in this study that comprised of SiO_2 nanoparticles as the main component along with Al_2O_3 and mixed oxides as well. The surfaces of these particles were modified through polymer chain attachment for enhancing the long-term stability of these nanoparticles when dispersed in fluid such as synthetic seawater (SSW). Neutral-wet Berea sandstone rock cores were flooded with nanofluids as a secondary recovery scheme that gave a maximum oil recovery of 71.5% OOIP as compared to 56.05% OOIP from reference water flooding as illustrated in Fig. 6. The differential pressure across the core was found to remain lower as compared to water flooding differential pressure. In tertiary recovery scheme, around 2.6% to 5.2% of OOIP was recovered from the cores. IFT was found to decrease from 10.6 mN/m to 6.8 mN/m between crude oil samples and SSW in presence of polymer-coated silica nanoparticles within 2 h of incubation.

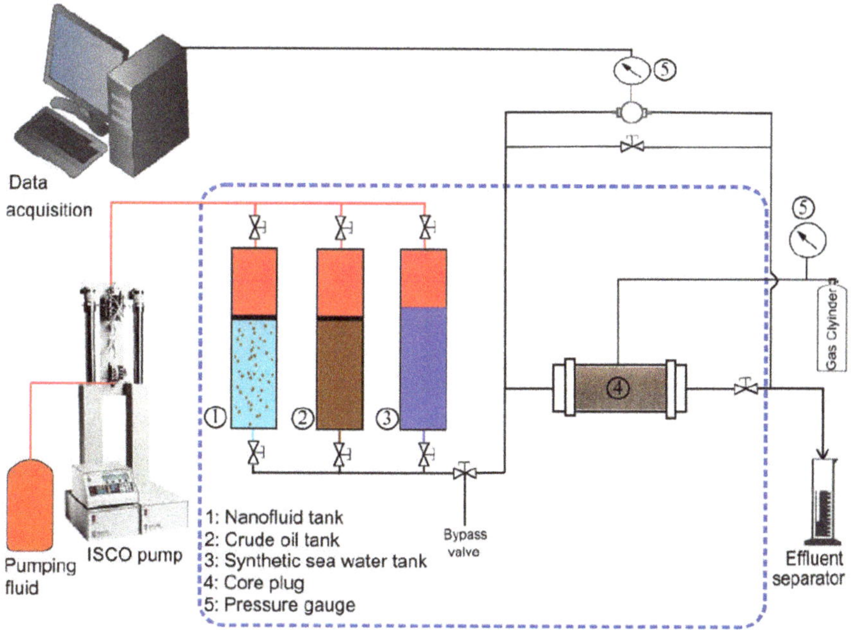

Fig. 6: Schematic diagram of core flooding experiment. Reprinted from Bila, A., Stensen, J.Å. and Torsæter, O. 2019. Experimental investigation of polymer-coated silica nanoparticles for enhanced oil recovery. Nanomaterials (Basel). 9(6): 822 (Open access).

Core permeability was found to be impaired upon secondary treatment with nanoparticles while tertiary injection resulted in both impairment and improvement in the permeability. Finally, it was confirmed that the wettability was altered from neutral-wet to water-wet that could be the mechanism for recovery of oil.

Jafarnezhad et al. (2017) reported application of SnO_2 nanoparticles on EOR from porous carbonate media. Nanoparticles were synthesised using sol-gel method from $SnCl_2.2H_2O$. Characterization of nanoparticles were performed using XRD analysis that indicated presence of tetragonal rutile phase SnO_2 nanoparticles. These nanoparticles were then dispersed in brine solution made up of mixture of salts. A degassed crude oil sample was used for the oil recovery studies. The initial contact angles were found to be around 124° suggesting weakly oil-wet phase that significantly changed to 0° after flooding with nanofluid indicating alteration of wettability. SEM images of carbonate cores were taken before and after flooding with nanofluid. IFT values between the oleic phase and aqueous phase were found to be decreased in presence of nanoparticles. The recovery factor was found to be 44% after flooding the medium with brine solution. Further increase of 24% was observed when 0.1 wt. % of SnO_2 nanoparticles was introduced into the system. However, it was noted that further increase in nanoparticle concentration to 0.5% resulted in decrease in recovery factor.

Multi-wall carbon nanotubes (MWCNTs) were used by Alnarabiji et al. (2016) for EOR. Glass micromodels were used as porous media having a diameter in the range of 30–0 μM. MWCNTs were dispersed in 0.3 wt. % of NaCl for preparation of nanofluid. The porous medium was initially flooded with brine solution followed by saturating with Heavy Arab crude oil sample at an injection rate of 0.8 mL/min. The morphology of MWCNTs were characterised using High Resolution Transmission Electron Microscope (HRTEM) that confirmed the presence of nanotubes having a diameter of 20 nm. The hydrophobic nature of MWCNTs was confirmed with a contact angle measurement of 136° along with a surface area of 30 m^2/g. Tertiary stage of oil recovery was studied using three different concentrations of MWCNTs fluid as an EOR agent. Highest oil recovery of 31.8% OOIP with a mobility retention factor of 10 was obtained using 0.05 wt. % concentration of MWCNTs. Increment in oil recovery was proposed to be mainly because of increase in fluid velocity through the oil-trapped channels along with reduction in IFT between oil and water.

9.4 Nanocomposites

Rezvani et al. (2018) reported investigation of Fe_3O_4@chitosan nanocomposites for potential EOR. The nanocomposites were synthesised using sol-gel method and the surface area was observed to be 84.96 BET m^2/g. XRD analysis was performed showing a diffraction pattern similar to that of pure nanocomposites of Fe_3O_4@chitosan without any presence of iron oxide. Fourier-Transform Infrared (FTIR) spectroscopy indicated the presence of O-H stretching vibration on nanocomposite surface along with presence of N-H bending vibration and C-O stretching vibration as well. Contact angle measurements were observed to be reduced over 24 h of incubation with the nanocomposite indicating wettability alteration to neutral condition that could potentially increase oil recovery. Interfacial tension (IFT) reduction was observed when the concentration was increased from 30 to 17.29 $mN.m^{-1}$ at 25°C. IFT reduction between oil and water was proposed to be due to adsorption of asphaltene on the surface of nanocomposites. A carbonate sand pack was used to investigate effect of Fe_3O_4@chitosan nanocomposites on oil recovery. The sand pack was initially saturated with formation water followed by injection of oil with a flow rate of 1 mL/min. to maintain a constant and stable pressure difference between inlet and outlet of the sand pack. Upon introduction of increased concentrations of nanocomposites, reduction in pressure difference was observed suggesting reduced oil trapping and blockage of pore throats in the porous sand pack. It was further seen that Fe_3O_4 nanoparticles had the potential to improve oil relative permeability via strong reduction in viscosity of oil. The reduction in viscosity was also considered to be due to breakdown of asphaltene molecules present in the crude oil sample. Addition of 0.03 wt. % of nanocomposites resulted in 103 cp reduction in viscosity. Soaking time of nanofluids on the porous media was also evaluated by injecting seawater along with nanofluids. While only 4.9% increase in oil recovery was observed upon seawater injection, the increase reached up to 9.3% upon nanofluid injection which could be due to reduction in IFT and alteration of wettability to neutral condition. Therefore, an ultimate recovery of 67.5% was obtained using 0.03 wt. % of Fe_3O_4@chitosan nanocomposites making it a potential tool for EOR.

Since physical and chemical methods involve toxic and corrosive chemical and hazardous reaction conditions, biological methods for synthesis of nanoparticles have come up as powerful alternatives. Several bacteria can synthesize copper, platinum and palladium nanoparticles (Ghosh 2018, Bloch et al. 2021, Ranpariya et al. 2021). Likewise, medicinal plants are considered as rich source of diverse phytochemicals that can reduce the metal ions to corresponding nanoparticles and also stabilize them (Ghosh et al. 2016a, Ghosh et al. 2016b, Ghosh et al. 2016c, Ghosh et al. 2016d). The reducing sugar, ascorbic acid, citric acid, saponins, poly phenols, and flavonoids play a critical role for shape evolution of the phytogenic nanoparticles (Ghosh et al. 2015a-d). Numerous medicinal plants like *Gloriosa superba, Barleria prionitis, Gnidia glauca, Plumbago zeylanica, Litchi chinensis,* and *Platanus orientalis* are reported to synthesize biocompatible nanoparticles (Rokade et al. 2018, 2017, Jamdade et al. 2019, Shende et al. 2017, 2018, Bhagwat et al. 2018, Shinde et al. 2018). Among several nanoparticles, zinc oxide nanoparticles, either defect rich or rod shaped, were reported to have attractive surface properties, radical scavenging potential and antidiabetic activities (Adersh et al. 2015, Robkhob et al. 2020, Karmakar et al. 2020). ZnO/SiO_2 nanocomposites were synthesised using pomegranate seed extract by Ali et al. (2019) and demonstrated for their potential use in EOR. The natural polymer-coated nanocomposites were dispersed in diluted seawater to prepare low salinity (LoSal)-polymeric nanofluids (1:20 seawater dilution). Stability of various concentrations of nanocomposites within the nanofluid solutions were observed with 2000 ppm of nanocomposites identified to be optimal for IFT and core-flooding tests. Ultrasonic waves were utilised to prevent sedimentation and to maintain proper stability of the fluids. IFT measurements indicated a 93.6% decrease with a minimum value of 2.016 mN/m between a crude oil sample and LoSal-polymeric nanofluid having 2000 ppm of nanocomposites at 70°C and 1500 psi. A carbonate core was used for determination of oil displacement, wherein, 46.96% OOIP was recovered by brine injection. This cumulative oil recovery was further enhanced by 19.28% OOIP upon LoSal-polymeric nanofluid injection, indicating the change in wettability system to a stronger water-wet condition. Hence, application of such LoSal-polymeric nanofluids could be effective for improving oil recovery process from carbonate reservoirs.

9.5 Nanomembranes

An eco-friendly approach for synthesis of hydrophilic nanofibrous (NF) membranes was reported by Wu et al. (2018) that could separate oil-in-water emulsion and could be potentially applicable for EOR. A conventional single-spinneret electrospinning method was utilised to synthesise NF membranes using a superhydrophilic copolymer poly(N-isopropylacrylamide-co-N-methylolacrylamide), that is, P(NIPAAm-co-NMA) made from N,N-isopropylacrylamide (NIPAAm) along with chitin nanowhiskers (ChNWs) that acted as a reinforcement and "co-crosslinking" hub. This co-crosslinking was characterised by FTIR spectroscopy wherein, a peak at 1073 cm^{-1} was indicative of presence of –C-O-C- bond between the –OH group of NMA units and ChNWs. Morphology of NF membranes was found to be uniform, smooth, continuous and bead-free with no adjacent nanofiber adhesions. Hence, a randomly oriented 3D nonwoven porous network was formed. Wettability of P(NIPAAm-co-NMA)/ChNWs NF membranes was analysed using water contact angle and underwater oil contact angle measurements. A water contact angle of around 0° was observed when a water droplet was placed on the surface of NF membrane in absence of ChNWs suggesting a high hydrophilicity. Thus, it indicates the hydrophilic nature of NF membranes. The ChNWs-10% NF membranes showed a contact angle of 143° when immersed in water highlighting the oleophobic characteristics of ChNWs-10% NF membranes. Oil-in-water emulsion separations were analysed in which a better separation efficiency of 99.8% was obtained with NF membranes without ChNWs; however, the relative separation flux was found to be quite low. The P(NIPAAm-co-NMA)/ChNWs NF membranes were found to have a better separation flux highlighting the fact that co-cross linking contributes in stability of membranes. Hence, ChNWs-10% NF membranes were further investigated for separation of oil-in-water emulsions containing various oils. Separation efficiencies of different oils such as hexadecane, petroleum ether, isooctane, gasoline, and edible oil were found to be in the range of 99.1–99.9%, therefore making it a promising applicant for EOR from mining effluents as well.

Li et al. (2021) reported synthesis of silica nanofibre/nanobead (SNB) membranes that could be used for separation of oil from wastewater samples. Electro-spinning and electro-spraying techniques were combined to form SNB membranes using different precursors. FTIR analysis revealed complete decomposition of organic components upon calcination. The average diameter of nanomembrane was found to be around 185 nm with a smooth surface. Wetting properties indicated that the membranes were super-amphiphilic in air while it became super-oleophobic when kept inside water. Oil contact angle was gradually increased from 148° to 162° with decrease in ratio of polymer and silica solution. The optimised adhesive properties of oil droplets were evaluated by the underwater sliding angle, adhesion force, and oil contact angle hysteresis (OCAH) that were found to be 2.5°, 0.4 mN and 3.7°, respectively. The robustness of SNB membranes were observed with a maximum tensile strength of 3.2 MPa and 6% strain highlighting its use in EOR. In addition, the large aspect ratio along with the amorphous atomic structure with some defects in microstructure provided outstanding flexibility to the membrane. Surfactant-free (SFEs) and surfactant-stabilized (SSEs) oil-in-water emulsions were used for separation performance analysis. A high separation efficiency of 99.98 and 98.8% was observed with different SFEs and SSEs such as petroleum ether, nhexane, cyclohexane, and diesel oil, along with a flux value of 4958 and 2237 L/m^2/h, respectively. Separation efficiency was also evaluated by total organic content (TOC) measurements and was found to be significant with a filtrate TOC content of 41 mg/L from hot SSEs, indicating high thermal stability as well. Therefore, it was proposed that SNB membranes could be useful for oil-in-water emulsion separation.

9.6 Conclusions and future perspectives

In this chapter, various nanomaterials are discussed which play a promising role for oil recovery. Nanofluids based on various metal and metal oxide nanoparticles can alter the wettability of the surfaces leading to enhancement of the oil displacement and recovery. Similarly, several monometallic, bimetallic or functionalized nanoparticles and even carbon nanotubes can modulate the IFT which is very critical for oil extraction. Another noteworthy development is nanocomposites which are often polymer coated nanoparticles like chitosan coated magnetic nanocomposites. Such coating gives the nanoparticles stability, reusability and recyclability. Nanoparticles impregnated polymeric nanomembranes also have potential applications in EOR.

However, it should be noted that physical and chemical process of synthesis of nanoparticles often involves hazardous reaction conditions and toxic chemicals. Hence, exposing the natural oil reservoirs to these nanoparticles may pose a serious environmental threat as the nanoparticles may persist in the nature for longer duration. This might adversely affect the flora and fauna in the vicinity of the oil reservoirs. Thus, there is a need to think of alternative source of nanoparticles which are environmentally benign and biocompatible. Various bacteria, algae, fungi and medicinal plants are reported to synthesize nanoparticles of exotic shape and size. Such biogenic nanoparticles should be explored for EOR. Biologically synthesized nanoparticles can be incorporated in the nanofluids, nanocomposites or nanomembranes.

Naturally occurring biopolymers can be used for developing coated nanoparticles. It would be interesting to evaluate the relationship between the IFT (crude oil/NF), contact angle (NF/quartz), and viscosity ratio (crude oil/NF) with the biomaterial coated nanocomposites. Again, optimization of the synthesis parameters is very critical to ensure the tuneable size of the resulting nanoparticles as oil displacement mechanism is dependent on the particle size. Similarly, blocking the high permeable water channels by the nanoparticles is also dependent upon the physicochemical properties of the nanoparticles. In view of the background, nanotechnology driven solution can be considered as a powerful tool for EOR.

References

Adersh, A., Kulkarni, A.R., Ghosh, S., More, P., Chopade, B.A. and Gandhi, M.N. 2015. Surface defect rich ZnO quantum dots as antioxidant inhibiting α-amylase and α-glucosidase: A potential anti-diabetic nanomedicine. J. Mater. Chem. B. 3: 4597–4606.

Adil, M., Lee, K., Mohd Zaid, H., Ahmad Latiff, N.R. and Alnarabiji, M.S. 2018. Experimental study on electromagnetic-assisted ZnO nanofluid flooding for enhanced oil recovery (EOR). PLoS ONE. 13(2): e0193518.

Ali, J.A., Kolo, K., Manshad, A.K. and Stephen, K.D. 2019. Low-salinity polymeric nanofluid-enhanced oil recovery using green polymer-coated ZnO/SiO_2 nanocomposites in the Upper Qamchuqa Formation in Kurdistan Region, Iraq. Energy Fuels. 33(2): 927–937.

Alnarabiji, M.S., Yahya, N., Shafie, A., Solemani, H., Chandran, K., Hamid, S.B.A. et al. 2016. The influence of hydrophobic multiwall carbon nanotubes concentration on enhanced oil recovery. Procedia Engineering. 148: 1137–1140.

Bayat, A.E., Junin, R., Samsuri, A., Piroozian, A. and Hokmabadi, M. 2014. Impact of metal oxide nanoparticles on enhanced oil recovery from limestone media at several temperatures. Energy Fuels. 28(10): 6255–6266.

Bera, A., Shah, S., Shah, M., Agarwal, J. and Vij, R.K. 2020. Mechanistic study on silica nanoparticles-assisted guar gum polymer flooding for enhanced oil recovery in sandstone reservoirs. Colloids Surf. A Physicochem. Eng. Asp. 598: 124833.

Bhagwat, T.R., Joshi, K.A., Parihar, V.S., Asok, A., Bellare, J. and Ghosh, S., 2018. Biogenic copper nanoparticles from medicinal plants as novel antidiabetic nanomedicine.World J. Pharm. Res. 7(4): 183–196.

Bila, A., Stensen, J.Å. and Torsæter, O. 2019. Experimental investigation of polymer-coated silica nanoparticles for enhanced oil recovery. Nanomaterials (Basel). 9(6): 822.

Bloch, K., Pardesi, K., Satriano, C. and Ghosh, S. 2021. Bacteriogenic platinum nanoparticles for application in nanomedicine. Front. Chem. 9: 624344.

Chang, H.L., Zhang, Z.Q., Wang, Q.M., Xu, Z.S., Guo, Z.D., Sun, H.Q. et al. 2006. Advances in polymer flooding an alkaline/surfactant/polymer processes as developed and applied in the People's Republic of China. J. Pet. Technol. Society of Petroleum Engineers. 58: 84–89.

Divandari, H., Hemmati-Sarapardeh, A., Schaffie, M. and Ranjbar, M. 2019. Integrating synthesized citric acid-coated magnetite nanoparticles with magnetic fields for enhanced oil recovery: Experimental study and mechanistic understanding. J. Pet. Sci. Eng. 174: 425–436.

Ehtesabi, H., Ahadian, M.M., Taghikhani, V. and Ghazanfari, M.H. 2014. Enhanced heavy oil recovery in sandstone cores using TiO_2 nanofluids. Energy Fuels. 28(1): 423–430.

Ghosh, S., More, P., Nitnavare, R., Jagtap, S., Chippalkatti, R., Derle, A. et al. 2015a. Antidiabetic and antioxidant properties of copper nanoparticles synthesized by medicinal plant *Dioscorea bulbifera*. J. Nanomed. Nanotechnol. S6: 007.

Ghosh, S., Jagtap, S., More, P., Shete, U.J., Maheshwari, N.O., Rao, S.J. et al. 2015b. *Dioscorea bulbifera* mediated synthesis of novel $Au_{core}Ag_{shell}$ nanoparticles with potent antibiofilm and antileishmanial activity. J. Nanomater. 2015, Article ID 562938.

Ghosh, S., Nitnavare, R., Dewle, A., Tomar, G.B., Chippalkatti, R., More, P. et al. 2015c. Novel platinum-palladium bimetallic nanoparticles synthesized by *Dioscorea bulbifera* : Anticancer and antioxidant activities. Int. J. Nanomedicine. 10(1): 7477–7490.

Ghosh, S., More, P., Derle, A., Kitture, R., Kale, T., Gorain, M. et al. 2015d. Diosgenin functionalized iron oxide nanoparticles as novel nanomaterial against breast cancer. J. Nanosci. Nanotechnol. 15(12): 9464–9472.

Ghosh, S., Patil, S., Chopade, N.B., Luikham, S., Kitture, R., Gurav, D.D. et al. 2016a. *Gnidia glauca* leaf and stem extract mediated synthesis of gold nanocatalysts with free radical scavenging potential. J. Nanomed. Nanotechnol.7: 358.

Ghosh, S., Chacko, M.J., Harke, A.N., Gurav, S.P., Joshi, K.A., Dhepe, A. et al. 2016b. *Barleria prionitis* leaf mediated synthesis of silver and gold nanocatalysts. J. Nanomed. Nanotechnol. 7: 4.

Ghosh, S., Gurav, S.P., Harke, A.N., Chacko, M.J., Joshi, K.A., Dhepe, A. et al. 2016d. *Dioscorea oppositifolia* mediated synthesis of gold and silver nanoparticles with catalytic activity. J. Nanomed. Nanotechnol. 7: 5.

Ghosh, S., Harke, A.N., Chacko, M.J., Gurav, S.P., Joshi, K.A., Dhepe, A. et al. 2016c. *Gloriosa superba* mediated synthesis of silver and gold nanoparticles for anticancer applications. J. Nanomed. Nanotechnol. 7: 4.

Ghosh, S. 2018. Copper and palladium nanostructures: a bacteriogenic approach. Appl. Microbiol. Biotechnol. 101(18): 7693–7701.

Giraldo, J., Benjumea, P., Lopera, S., Cortés, F.B. and Ruiz, M.A. 2013. Wettability alteration of sandstone cores by alumina-based nanofluids. Energy Fuels. 27(7): 3659–3665.

Hashemi, R., Nassar, N.N. and Almao, P.P. 2013. Enhanced heavy oil recovery by *in situ* prepared ultradispersed multimetallic nanoparticles: A study of hot fluid flooding for Athabasca bitumen recovery. Energy Fuels. 27(4): 2194–2201.

Hendraningrat, L., Li, S. and Torsæter, O. 2013. A coreflood investigation of nanofluid enhanced oil recovery. J. Pet. Sci. Eng., 111: 128–138.

Jafarnezhad, M., Giri, M.S. and Alizadeh, M. 2017. Impact of SnO_2 nanoparticles on enhanced oil recovery from carbonate media. Energy Sources, Part A: Recovery, Utilization, and Environmental Effects. 39(1): 121–128.

Jamdade, D.A., Rajpali, D., Joshi, K.A., Kitture, R., Kulkarni, A.S., Shinde, V.S. et al. 2019. *Gnidia glauca* and *Plumbago zeylanica* mediated synthesis of novel copper nanoparticles as promising antidiabetic agents. Adv. Pharmacol. Sci., 9080279.

Joonaki, E. and Ghanaatian, S.J.P.S. 2014. The application of nanofluids for enhanced oil recovery: effects on interfacial tension and coreflooding process. Pet. Sci. Technol. 32(21): 2599–2607.

Kamal, M.S., Adewunmi, A.A., Sultan, A.S., Al-Hamad, M.F. and Mehmood, U. 2017. Recent advances in nanoparticles enhanced oil recovery: Rheology, interfacial tension, oil recovery, and wettability alteration. J. Nanomater., 2473175.

Karimi, A., Fakhroueian, Z., Bahramian, A., Khiabani, N.P., Darabad, J.B., Azin, R. et al. 2012. Wettability alteration in carbonates using zirconium oxide nanofluids: EOR implications. Energy Fuels. 26(2): 1028–1036.

Karmakar, S., Ghosh, S. and Kumbhakar, P. 2020. Enhanced sunlight driven photocatalytic and antibacterial activity of flower-like ZnO@MoS_2 nanocomposite. J. Nanopart. Res. 22: 11.

Li, M., Gao, X., Wang, X., Chen, S. and Yu, J. 2021. Wettable and flexible silica nanofiber/bead-based membranes for separation of oily wastewater. ACS Appl. Nano. Mater. 4: 2952–2962.

Li, Y., Dai, C., Zhou, H., Wang, X., Lv, W., Wu, Y. et al. 2017. A novel nanofluid based on fluorescent carbon nanoparticles for enhanced oil recovery. Ind. Eng. Chem. Res. 56(44): 12464–12470.

Ranpariya, B., Salunke, G., Karmakar, S., Babiya, K., Sutar, S., Kadoo, N. et al. 2021. Antimicrobial synergy of silver-platinum nanohybrids with antibiotics. Front. Microbiol. 11: 610968.

Rezvani, H., Riazi, M., Tabaei, M., Kazemzadeh, Y. and Sharifi, M. 2018. Experimental investigation of interfacial properties in the EOR mechanisms by the novel synthesized Fe_3O_4@ Chitosan nanocomposites. Coll. Surf. A Physicochem. Eng. Aspec. 544: 15–27.

Robkhob, P., Ghosh, S., Bellare, J., Jamdade, D., Tang, I.M. and Thongmee, S. 2020. Effect of silver doping on antidiabetic and antioxidant potential of ZnO nanorods. J. Trace Elem. Med. Biol. 58: 126448.

Rokade, S., Joshi, K., Mahajan, K., Patil, S., Tomar, G., Dubal, D. et al. 2018. *Gloriosa superba* mediated synthesis of platinum and palladium nanoparticles for induction of apoptosis in breast cancer. Bioinorg. Chem. Appl. 4924186.

Rokade, S.S., Joshi, K.A., Mahajan, K., Tomar, G., Dubal, D.S., Parihar, V.S. et al. 2017. Novel anticancer platinum and palladium nanoparticles from *Barleria prionitis*. Glob. J. Nanomedicine, 2(5): 555600.

Shende, S., Joshi, K.A., Kulkarni, A.S., Charolkar, C., Shinde, V.S., Parihar, V.S. et al. 2018. *Platanus orientalis* leaf mediated rapid synthesis of catalytic gold and silver nanoparticles. J. Nanomed. Nanotechnol. 9: 2.

Shende, S., Joshi, K.A., Kulkarni, A.S., Shinde, V.S., Parihar, V.S., Kitture, R. et al. 2017. *Litchi chinensis* peel : A novel source for synthesis of gold and silver nanocatalysts. Glob. J. Nanomedicine 3(1): 555603.

Shinde, S.S., Joshi, K.A., Patil, S., Singh, S., Kitture, R., Bellare, J. et al. 2018. Green synthesis of silver nanoparticles using *Gnidia glauca* and computational evaluation of synergistic potential with antimicrobial drugs. World J. Pharm. Res. 7(4): 156–171.

Wu, J.X., Zhang, J., Kang, Y.L., Wu, G., Chen, S.C. and Wang, Y.Z. 2018. Reusable and recyclable superhydrophilic electrospun nanofibrous membranes with *in situ* co-cross-linked polymer–chitin nanowhisker network for robust oil-in-water emulsion separation. ACS Sustainable Chem. Eng. 6(2): 1753–1762.

Yang, W., Wang, T., Fan, Z., Miao, Q., Deng, Z. and Zhu, Y. 2017. Foams stabilized by *in situ*-modified nanoparticles and anionic surfactants for enhanced oil recovery. Energy Fuels. 31(5): 4721–4730.

Zhang, H., Ramakrishnan, T.S., Nikolov, A. and Wasan, D. 2016. Enhanced oil recovery driven by nanofilm structural disjoining pressure: flooding experiments and microvisualization. Energy Fuels. 30(4): 2771–2779.

10

Nanomaterials for Recovery of Lead Heavy Metal from Wastewater

Papiya Dhara[1,*] and *Kenneth T.V. Grattan*[2]

10.1 Aquatic environment pollution by heavy metals: Problems and context

Water is "life-blood of the biosphere" and has important role in growth of civilization. The most abundant natural resource of our Earth is water, but only about 1% of that resource is accessible for human consumption (Abou-Shanab et al. 2013, Adeleye et al. 2016). Aquatic life is getting deteriorated with pollution which is generating due to rapid industrialization. Industrial wastage from various industries, such as mining, alloy industry, metal plating industries, electroplating and battery manufacturing, sewage effluent, mining waste, and agricultural run-off are contaminated with a numerous variety of toxic chemicals, with heavy metals and organics being the most significant and hazardous contaminants. Discharging these wastes without treatment into the aquatic bodies causes prime contamination of heavy metals in different water resources, such as ground water, pond, river, marine etc. (Afzal et al. 2018). Cost effective recycling and re-use of water is remarkably less in low-income, developing countries, only 3% of all water available is fresh water and only 8% of industrial and municipal wastewater is under any sort of treatment (Roy and Bhattacharya 2015). Contamination of water resources by heavy metals is a life-threatening issue which adversely affects plants, animals, and human health. Mercury (Hg(II)), molybdenum (Mo(II)), nickel (Ni(II)), chromium (Cr(III)), cadmium (Cd(II)), copper (Cu(II)), cobalt (Co(II)), iron (Fe(II)), lead (Pb(II)), selenium (Se(II)), and zinc (Zn(II)) can

[1] Department of Physics, Adamas University, Kolkata, India, 700126.
ORCID.org/0000-0002-3030-4481
[2] Department of Electrical and Electronics Engineering, City, University of London, EC1V 0HB, UK.
Email: T.Sun@city.ac.uk ; K.T.V.Grattan@city.ac.uk
ORCID.org/0000-0003-2250-3832
* Corresponding author: papiyadharaismpolito@gmail.com

get accumulated in the food chain and circulate great eco-toxicological hazard to living organisms, including plants, animals and humans.

The heavy metals are accumulated in the environment first, then they start to accumulate in living organisms and are transferred from one trophic level to another in the food chains. The accessibility of accumulation of heavy metals in biota completely depends on the rate of consumption in living element and their rate of elimination from the living body.

The preservation of heavy metals in the body of living elements depends on some factors, such as the biological evolution of the metal concerned and the physiological mechanisms developed by the organism for the regulation, homeostasis and detoxification of the heavy metal (Ali and Khan 2018).

Heavy metal transfer from the soil to plant is a very important step in the trophic transfer of heavy metals in food chains.

These metals taken by plants from polluted soil are afterward transferred to herbivorous animals (Nica et al. 2012). Human food chain will be contaminated by consuming the crops such as cereals and vegetables which is a very serious issue.

In a developing country, people are doing irrigation of agricultural soils with wastewater. Higher concentration of heavy metal has been reported in leafy vegetables grown with wastewaters compared to those grown with groundwater. The transformation of heavy metals from roots of the rice plant to stem, leaves and rice grains is of human health concern as it requires water during most of its growth period. It has been reported that people consume Cd in the highest amount through consumption of rice (Cai et al. 2015).

Quantifying the accumulation rate should be very important in this scenario and certain terms have been used to quantify the degree or extent of accumulation of heavy metals in biota. Some of these quantitative terms are bioconcentration factor (BCF), bioaccumulation factor (BAF), bioaccumulation coefficient (BAC), etc., as mentioned by Ali et al. (Ali et al. 2019).

Freshwater fish are also consuming different poisonous heavy metals released to freshwater bodies from different natural and anthropogenic sources (Rahman et al. 2012). Physiological changes due to the effect of heavy metal on fish Labeo rohita has been reported by Alkesh (Alkesh 2017).

Recently, many researchers from different parts of world have been reporting on hazardous aquatic status. In spite of satisfactory protection and mitigation measures, large parts of the Nile Delta of Egypt suffer from severe coastal erosion. Most of the coastal lakes are suffering from the excessive discharge of industrial, agricultural and domestic sewage flow (UNEP 1999). Heavy metals concentrations, such as that of Fe, Cu, Cd, Pb, Zn, Cr, Mn, Hg and Ni, have been determined in water collected from six areas at Assiut Governorate on river Nile, Egypt (Omar 2013). It is revealed that Zn, Cu and Fe concentrations are the highest in water followed by Mn, Cr, Pb, Cd, Ni and Hg in areas under investigation.

A team of researchers led by the National Center for Polar and Ocean Research (NCPOR) has conducted a study around three popular urban beaches in Tamil Nadu, India, namely, the Marina Beach, Edward Elliott Beach, and Silver Beach. They have analysed sediment samples for concentrations of heavy metals like iron (Fe), zinc (Zn), copper (Cu), manganese (Mn), nickel (Ni), and lead (Pb). According to Bureau of Indian Standards, the concentration level of these heavy metals has been found to be under the safe limit, but the researchers are concerned that long term exposure to these heavy metals would have an adverse effect on the marine biota (Downtoearth 2020).

10.2 Sources of lead ions in the aquatic environment and toxicity issues

Lead is a subject of significant concern and attention due to its toxic nature and its extensive occurrence in aquatic environments. It has a melting point 327.4°C and boiling point 1725°C. It can be combined with sulphur to give PbS, and $PbS0_4$. Oxygen can combine with lead to give $PbC0_3$ and ranges from 10 to 30 mg kg^{-1} in the Earth's crust. It is naturally available in the Earth's crust, but majorly the high level of lead found throughout the environment mainly comes from human activities. Lead occurs in the aquatic ecosystem from both natural and anthropogenic sources. Direct exposure of lead ions occurs into both fresh and marine ecosystems, as well as indirect routes of exposure happens through dry and wet deposition and land run-off (Biney et al. 1994). Important natural sources are volcanic activity, continental weathering and forest fires (FAO 1992). The anthropogenic sources are as follows: batteries and electrical devices; pigments and paints; alloys and solders; pesticides; glass; fertilizer; refiners; fuel; and plastic (Fifield and Haines 2000).

Lead is used as solders, bearings, cable covers, ammunition, plumbing, pigments and caulking. Ionic lead, Pb (II), lead oxides and hydroxides and lead-metal oxyanion complexes are the general forms of Pb that are released into the soil, ground water and surface waters. Very small lead oxide (PbO) layer over the surface of the metal can react with oxygen and water, and be converted into lead hydroxyl. Most stable forms of lead are Pb(II) and lead hydroxyl complexes. In soil matrix, it has been found that lead hydroxides, lead sulphide are the most stable solid (Rajeswari and Namburu 2014). Several basic salts of lead such as $Pb(OH)_2$, and $2PbCO_3$ are most widely used in white paints' pigment. Lead dioxide and lead sulphate are present in the reversible reaction that occurs during the charge and discharge of lead storage battery.

The primary drinking water standards for lead vary among different organizations. The EPA sets a discharge limit of 0.015 mg/L (USEPA 1999) for lead in water released into the environment, while the National Environment Regulations fix the limit at 0.1 mg/L for drinking water (BIS 1999). The World Health Organization provides a risk-based guideline of 10 μg/L (WHO 1993) for lead in drinking water to protect public health. These variations in standards may stem from different methodologies, data, and considerations, and local authorities must adopt appropriate standards based on scientific evidence and regional factors to ensure safe drinking water.

Drinking water gets contaminated with lead when plumbing materials that contain lead corrode, especially where the water has high acidity or low mineral content that corrodes pipes and fixtures. The most common sources of lead in drinking water are lead pipes, faucets and fixtures. In 2020, USEPA published the final regulation "Use of Lead-Free Pipes, Fittings, Fixtures, Solder, and Flux for Drinking Water" as amendment of existing regulations based on the Reduction of Lead in Drinking Water Act (RLDWA) (USEPA 2020).

Humans can be exposed to heavy metals in the work place and in the environment. Workers of mining and industrial operations may inhale dust and matter containing metal particles. Ullah et al. has reported that prolonged exposure to welding fumes contaminates significant amount of the heavy metals including Pb in blood which results in increased oxidative stress (Mahmood et al. 2015).

Antioxidants and enzymes have the thiol group and Pb metal may increase the generation of reactive oxygen species (ROS) like hydroxyl radical, superoxide radical, and hydrogen peroxide. ROS can destroy the inherent antioxidant immunity of cells and it results in "oxidative stress" (Ercal et al. 2001).

Heavy metal ions interact with various cell components such as DNA and nuclear proteins and produce damage in DNA that may lead to cell cycle modulation, carcinogenesis or apoptosis (Kasprzak 2002, Beyersmann and Hartwig 2008). High degree of toxicity in human health has been found for lead even at lower levels of exposure as several studies have demonstrated that ROS production by lead and oxidative stress play an important role in the toxicity and carcinogenicity (Tchounwou et al. 2004).

The Agency for Toxic Substances and Disease Registry (ATSDR) toxicological profile briefly characterizes the toxicology and adverse health effects information of lead (ATSDR 2020). It is reported by experts that lead is a cumulative toxicant that affects multiple body systems (gastrointestinal tract, kidney, central nervous system) and it is particularly harmful to young children (WHO 2021).

10.3 Technology development for lead ion removal

The permissible limit of Pb (II) that has been declared by EPA in wastewater is 0.015 mgL^{-1}. The lead-ion concentrations in industrial wastewaters are measured as 200–500 mgL^{-1} which is high in comparison to water quality standards. Hence, suitable waste water treatment technology is required for removal of lead ion from the aquatic environment to reduce the lead level of 0.05–0.10 mgL^{-1} (Goel et al. 2005, Ucun et al. 2003, Özacar et al. 2008, Vilar et al. 2005).

Several conventional methods for removing heavy metals from waste water are used in practice, such as chemical precipitation, flotation, coagulation and flocculation, membrane filtration, cementation, adsorption, ion exchange, and electrochemical deposition.

Chemical precipitation technique is an effective technique for removing of lead from wastewater. Chemicals react with the heavy metals available in wastewater and insoluble precipitates form. Afterwards, these precipitates are removed using sedimentation technique and finally it is followed by generation of a high-water content sludge, the disposal of which is cost intensive (Barakat 2011). Lime added with fly ash has been used as precipitant which is low cost. Bisulphite is also frequently used for precipitation and these precipitants have been found inefficient in removal of the lead ions at low concentration (Chen et al. 2009, Ahluwalia and Goyal 2007). Magnesium hydroxy carbonate has been used as precipitating agent and the chemical precipitation of heavy metal is found in noticeable amount at 99.9% (Zhang and Duan 2020).

Flotation has nowadays found widespread use in wastewater treatment (Fu and Wang 2011). The main flotation processes for the removal of metal ions from solution are Dissolved air flotation (DAF), ion flotation and precipitation flotation. In the beginning, DAF had been widely applied to remove heavy metal (Tassel et al. 1997, Waters 1990).

The effectiveness of ion flotation to remove cadmium, lead and copper from dilute aqueous solution has been investigated by using biosurfactant tea saponin (Yuan et al. 2008). The maximum removal of Pb(II), Cu(II), and Cd(II) can reach 89.95%, 81.13% and 71.17%, respectively, when the ratio of collector to metal has been 3:1. It is reported recently that Salicylhydroxamic acid (SHA) is an efficient collector and it can separate oxide minerals by utilizing flotation method. During flotation, it is combined with Pb(II) to enhance the floatability of targeted minerals (Zhao et al. 2020). Flotation is established as an effective process for the removal of Pb(II) from wastewater by using graphene oxide (GO). The experimental results have proposed more than 99% of Pb(II) removal by flotation and the initial solution pH and dosage of GO significantly controlled the removal of Pb(II) (Peng et al. 2018).

Ion exchange is another well-known process in industry waste water management while handling a relatively large volume which is basically the exchange of cations or anions between the exchange medium and the wastewater. Generally, people use matrices for ion exchange which are synthetic organic ion exchange resins (Dorfner 1991). A very cost-effective approach of lead ion removal from binary mixtures and solid phase extraction using Purolite C100 resin has been reported having 99.71% lead ion separation efficiency (Badawy et al. 2009, Liang et al. 2006). Zewail and Yousef has proposed lead ion removal from waste water using spouted bed with AMBERJET 1200 Na resin with 98% efficiency (Zewail and Yousef 2015). Even though the ion exchange is very efficient in lead removal from waste water, it is not widely used as ion-exchange resins which is affected by certain parameters such as pH, temperature, initial metal concentration and contact time (Gode and Pehlivan 2006). Lalmi and Bouhidel has proposed cation exchange by usual resin Purolite C100E in Ca(II) form for removal of Pb(II) from toxic waters and regeneration using $Ca(NO_3)_2$ (Lalmi and Bouhidel 2018).

Chang and Wang have proposed that coagulation/flocculation could not be treated as a complete process of removal of the heavy metal from wastewater (Chang and Wang 2007). Generally, coagulation and flocculation are followed by sedimentation and filtration for removal of heavy metal from wastewater (Akbal and Camcı 2010, Chang et al. 2009). Pang et al. has proposed methodology to remove lead from wastewater with aluminium sulphate (alum), polyaluminium chloride (PACL), and magnesium chloride ($MgCl_2$) as coagulants and korat PA3230 as the polyelectrolyte (Pang et al. 2011). Author has reported that maximum removal of lead is 99% at an optimum pH of 6.2–7.8 for alum, 8–9.3 for PACL and 8.7–10.9 for $MgCl_{2.}$ Iwuozor has reviewed different aspects of coagulation and flocculation for wastewater treatment (Iwuozor 2019).

Membrane filtration is one of the best emerging technique for heavy metal filtration from wastewater. Depending on the size of the particle retainable by membrane, various types of membrane filtrations are there, such as ultrafiltration, nanofiltration and reverse osmosis which are applicable for lead removal. Micellar enhanced ultrafiltration (MEUF) is an advance ultrafiltration technique having high removal efficiency of lead ions. Samper et al. have reported the use of MEUF to remove Pb(II) from synthetic water using two anionic surfactants, SDS and linear alkylbenzene sulfonate (LAS), in a laboratory-scale membrane system having enhanced removal efficiency (Samper et al. 2009). A semi-permeable membrane is utilized in the reverse osmosis (RO) process permitting the water through the membrane to be purified and rejecting the heavy metals by the use of pressure. Dialynas and Diamadopoulos have reported 100% lead removal efficiency by using RO (Dialynas and Diamadopoulos 2009). Nanofiltration (NF) is as an intermediate process between ultrafiltration (UF) and reverse osmosis (RO), where the membrane pore size ranging in between UF and RO (Mehdipour et al. 2015). A membrane has been fabricated by simultaneous co-extrusion of polybenzimidazole (PBI) and polyethersulfone (PES)/polyvinylpyrolidone (PVP) dopes through a triple orifice spinneret, and has given 93% rejection for lead (Zhu et al. 2014). Gholami et al. have improved the membrane by combining it with the nanosized ferric oxides in membrane matrix (Gholami et al. 2014). Mehdipour et al. have proposed polyamide NF membrane having increased efficiency of lead removal (97.5%) (Mehdipour et al. 2015). Natural Acacia gum (AG) has been used to enhance the removal of lead from contaminated water using polymer-enhanced membrane filtration (PEMF). The highest binding capacity of AG with lead has been found to be about 12.2 $mg.g^{-1}$, while1000 ppm AG is added to 35 ppm lead solution at pH 7. The lead removal efficiencies of this matrix have been reported up to 99.9% with a dosage of 1000 $mg.L^{-1}$ of AG at pH 7 (Manawi et al. 2018). The superior permeates quality of membrane filtration processes have enhanced highwater recovery and open up their applications for wastewater remediation, meanwhile increasing the concentration of feed water (Hube et al. 2020). Advance chemical precipitation enabled membrane filtration (CPEMF) technique has been established for the removal of Pb(II) ions. Two different membranes are used, such as microfiltration (MF) and ultrafiltration (UF) for experiment at three different pH levels, i.e., pH 7.5, pH 8.5, and pH 10. The result has been proposed as the CPEMF method which can achieve detection ability of lead below the drinking water standard at 1 mM dissolved inorganic carbon (DIC) and pH 8.5 (Xu et al. 2021).

Cementation is sort of another precipitation method applying an electrochemical mechanism. In the electrochemical mechanism metal of higher oxidation potential replaces lower oxidation metal ion and higher oxidation potential metal ion goes into electrochemical solution. Mostly, copper is separated by cementation along with noble metals, such as Ag, Au and Pb. Investigation on lead-ion cementation system has been done using an iron sphere-packed bed and Lead-ion Percentage Removal (LPR) has been reported as 99.52%. The efficiency of lead removal has been tested using wastewater from the manufacture of storage batteries (Agclidis 1988). The kinetics of the cementation of lead from acidic solutions onto rotating iron discs have been studied over a large range of initial lead ion concentrations, disc rotation speeds and temperatures. It has been observed that the rate of the reaction is affected by the increase in the electrode surface area with increasing cementation time and initial lead ion concentrations (Makhloufi et al. 2000). Lead ions removal from wastewater consisting of lead nitrate by cementation on a gas-sparged zinc rod has been investigated. Activation energy of the process, and the rate controlling step rates of cementation have been studied and are expressed in terms of the mass transfer coefficient (Nosier and Sallam 2000). The galvanic cell containing parallel copper and zinc plates is used for wastewater treatment for remediation of various heavy metal ions from solution efficiently (Youssef et al. 2021).

Electrochemical deposition is another improving technology for remediation of wastewater. This technology has a relatively large capital investment and the electricity supply is expensive, so it is not applied on a large scale. There are several different types of electrochemical treatment processes available such as electrocoagulation (EC), electroflotation (EF), electrochemical precipitation (EP) and so on. Belkacem et al. have investigated the treatment of wastewater using the EF technique with aluminium electrodes (Belkacem et al. 2008). Optimized parameters of the study have been utilized on the separation of some heavy metal ions such as iron, nickel, copper, zinc, lead and cadmium and metal removal rate reached up to 99%. A cell containing aluminium electrode as anode and stainless-steel electrode as cathode has been used in EC process to study lead removal efficiency from wastewater. Complete lead removal of 100% has been achieved at best operating conditions of cell (Shakir and Husein 2009). Utilizing high density of surface functional groups of graphene-oxide, low-concentration and high-concentration heavy-metal pollutants can be removed by electrochemical method with graphene-oxide-modified carbon felt electrode (CF-GO) (Liu at el. 2019). Iron anode composition has been modified by addition of zinc to remove chemical contaminants up to 100% including lead with other toxicants, such as nitrates, fluoride, arsenic, beryllium, chromium, copper, etc. (Hussain et al. 2021).

Compared to other conventional technologies for lead ion removal from wastewater, adsorption has some remarkable advantages, such as relatively low-cost process, less volume of chemical and biological sludge, low cost, high efficiency, regeneration of adsorbents and possibility of metal recovery. Adsorption is basically a surface phenomenon involving binding molecules or particles in a solution onto a surface and there are two types of adsorptions: physical adsorption and chemical adsorption. In physical adsorption interactions, it is found that fully reversible enabling desorption occurs at the same temperature and is not site specific. However, chemical sorption is an irreversible process, and chemical-like bonding onto the surface of adsorbents and the chemisorbed molecules are fixed at specific sites (Roy and Bhattacharya 2015). Wadhawan et al. have reviewed function of nanomaterials as adsorbents in various heavy metal ion removal from waste water (Wadhawan et al. 2020). Activated carbon (AC) adsorbents are powerful treatment for domestic and industrial waste water in the removal of lead metal because of its large micropore and mesopore volumes and the resulting high surface area (Jabbari et al. 2016, Abbas and Alalwan 2019). Anirudhan and Sreekumari have synthesised activated carbon using waste coconut button for removal of Pb(II) and more than 90% efficiency has been reported (Anirudhan and Sreekumari 2011). It has been found that adsorbent agricultural waste materials such as peanut shells, nut shells, plum seeds, eucalyptus bark, olive pips, peach stones, and pine sawdust are considered as economic and eco-friendly (Wang et al. 2007). Adsorption capacities of Cd(II), Zn(II) and Pb(II) metal ions from water have been investigated using maize cob and husk, the maximum adsorptions have occurred at 495.9 $mg.g^{-1}$ for Zn(II) ion, 456.7 $mg.g^{-1}$ for Pb(II) ion, and 493.7$mg.g^{-1}$ for Cd(II) ion (Igwe et al. 2005). The ability of biological wastes, such as two non-living (dried) fresh water algae (*Oedogonium* sp. and *Nostoc* sp.) under varying range of pH (2.99–7.04), contact time (5–300 min), biosorbent dose (0.1–0.8 g/L) and initial metal ion concentrations (100 and 200 mg/L) has been examined for Pb (II) ion removal. *Oedogonium* sp. (145 mg/g) has been found more suitable than *Nostoc* sp. (93.5 mg/g) for the advance of an efficient biosorbent for the removal of Pb(II) from aqueous solution (Gupta and Rastogi 2008). Numerous other agricultural by-products for Pb removal have been reviewed and the adsorption capacity has reported in the range of 8.3 mg/g for chitosan and 263 mg/g for coir pith waste of coconut (Alalwan et al. 2020a). Industrial by-products include fly ash, waste iron, iron slags, and hydrous titanium oxide; these materials can modify to upgrade its Pb (II) removal capacity from wastewater. Feng et al. have investigated Pb(II) removal using iron slag (Feng et al. 2004). Fly ash is the most available industrial waste and it is reported that fly ash from coal-burning has been used for removal of Cu(II) and Pb(II) ions (Alinnor 2007). Natural zeolites have their significant properties as ion exchange capability and it is gaining a significant interest as adsorbents of heavy metals from waste water. Pan et al. have reported Pb(II) ion removal by zirconium phosphate having the high adsorption capacity of 398 $mg.g^{-1}$ (Pan et al. 2007). Crosslinked starch gel is modified biopolymers for heavy metals removal and high adsorption capacity of 433 $mg.g^{-1}$ have been reported by Crini (Crini 2005). The adsorption efficiency always depends on the type of adsorbents and its adsorption capacity in single and multi-component systems (Lin et al. 2000).

The bonding in adsorption is raised up due to the two types of force of attraction present between the adsorbate and the adsorbent, such as weak van der Waals of attraction and strong chemical bond. The process of adsorption is constituted by following successive steps, such as diffusion of the adsorbate from the bulk solution to the surface of adsorbent, and diffusion of adsorbate through the boundary layer. Afterwards, intraparticle diffusion which is adsorbate diffusion from the surface into the adsorbent pores and finally binding of adsorbate to the active sites on the adsorbent surface happens (Lee et al. 2019). It has been described that the adsorption mechanisms of heavy metal ions onto porous nanomaterials occur through ion exchange and π-cation electrostatic interactions (Holmberg 2006, Flores-Cano 2013).

10.4 Models of adsorption isotherms and kinetics

Adsorption rate investigation is significant to optimize the best operating conditions of the process and Alalwan et al. have reported the important parameters, such as residence time, solution pH, temperature and concentration which affect the adsorption rate (Alalwan et al. 2018).

In adsorption system, determination of kinetic is very important as it effects the configuration of adsorption systems and the reaction rate regulatory step of the chemical reaction. Adsorption kinetics basically describe the relations between the amount of adsorbates adsorbed on the adsorbents (q_t) and the contact time (t) (Matouq 2015). Thomas, Yoon-Nelson, Bed Depth Service Time (BDST), Clark models, Adams-Bohart, and Wolborska are various kinetic models that have been introduced to describe the dynamic process of adsorption systems (Alalwan et al. 2020b). There are several kinetic models based on solution concentration and the capacity of the adsorbent which are sufficiently informative to access the data for the dynamics of the biosorption process, reaction pathways and associated mechanisms. The first and second order reversible and irreversible models, in addition to pseudo first and second orders models, are commonly used kinetic models based on concentration of solution (Alalwan et al. 2020a). Mostly, used models based on the adsorption capacity are Lagergren's first order equation, the Redlich Peterson model and the BET model (Sud et al. 2008).

In adsorption equilibrium, the relation between the amount of the adsorbates on the adsorbents and the equilibrium concentrations of the adsorbates under constant temperatures is called the adsorption isotherm (Theodore and Ricci 2011). Numerous isotherm models are established including Temkin, Jovanovic, Halsey, Langmuir, Freundlich, Dubinin–Radushkevich, Sips model, etc., and Langmuir, and Freundlich models are broadly used for isotherms. These isotherms models describe the relationship between the amount adsorbed by a unit weight of solid adsorbent, and the amount of non-adsorbed solute at equilibrium.

10.4.1 Adsorption equilibria

The equilibrium adsorption capacity, q_e (mgg^{-1}) of heavy metal ions aqueous has been calculated using the following mass balance equation:

$$q_e = \frac{(C_0 - C_e) \times V}{m} \tag{10.1}$$

where C_0 and C_e (mg L^{-1}) are initial and equilibrium concentrations of metal ions, respectively, V (L) is the sample volume, m (g) is the mass of adsorbent.

10.4.2 Langmuir isotherm model

The Langmuir Adsorption Isotherm is considered for single layer uniform adsorption onto one particular surface containing a finite number of identical sites and afterwards there is no movement of the adsorbate in the plane of the surface (Langmuir 1916). In Langmuir isotherm model, it is considered that altogether the active sites of adsorption have the unique binding energy and a single adsorbate can bind with each site (Sha 2009). To determine the maximum capacity of the adsorbent from overall single layer coverage of the adsorbent surface, Langmuir isotherm model can be utilized. The Langmuir's isotherm non-linear equation is as follows:

$$q_e = \frac{q_m \times b \times C_e}{(1 + b \times C_e)} \tag{10.2}$$

The linear form of the Langmuir model is represented as:

$$\frac{C_e}{q_e} = \frac{1}{q_{m \times b}} + \frac{C_e}{q_m} \tag{10.3}$$

where q_e is the equilibrium adsorption capacity of adsorbent in mg (heavy metal ions) g^{-1} (adsorbent), C_e is the equilibrium concentration of heavy metal ions in $mg\ L^{-1}$, q_m is the maximum amount of metal ions adsorbed in *mg* (heavy metal ions) g^{-1} (adsorbent), and *b* is the constant that refers to the bonding energy of adsorption in $L\ mg^{-1}$.

10.4.3 Freundlich isotherm model

The Freundlich model is based on the multilayer adsorption, used for heterogeneous surface energies of adsorption, unlike Langmuir Adsorption Isotherm. This isotherm presents the ratio of the quantity of the solute adsorbed on a known mass of adsorbent to the concentration of the solute in the solution, which differs at different concentrations (Freundlich 1906). The Freundlich equation is stated as:

$$q_e = K_f C_e^{\frac{1}{n}} \tag{10.4}$$

The linearized Freundlich isotherm is given as:

$$\log q_e = \log K_f + \frac{1}{n} \log C_e \tag{10.5}$$

Here, q_e is the equilibrium adsorption capacity of the adsorbent in mg (heavy metal ions) g^{-1} (adsorbent), *Ce* is the equilibrium concentration of heavy metal ions in mg L^{-1}, K_f (Freundlich constant) is the constant related to the adsorption capacity of the adsorbent in mg L^{-1} and *n* (Freundlich constant) is the constant related to the adsorption intensity or index of surface heterogeneity.

10.4.4 Sips Model

Repo et al. have described the Sips isotherm as a hybrid model of the Langmuir and the Freundlich isotherms (Repo et al. 2010). The Sips isotherm effectively converts to the Freundlich isotherm at low adsorbate concentrations and this model assumes a monolayer adsorption capacity characteristic of the Langmuir isotherm at high adsorbate concentrations (Ho et al. 2002).
The Sips model is expressed as (Repo et al. 2011):

$$q_e = \frac{q_m (K_S C_e)^{n_S}}{1 + (K_S C_e)^{n_S}} \tag{10.6}$$

where q_e is the equilibrium adsorption capacity of the adsorbent *mg* (heavy metal ions). g^{-1} (adsorbent), C_e is the equilibrium concentration of the adsorbate ($mg.L^{-1}$), q_m is the Sips maximum adsorption capacity ($mg.g^{-1}$), K_S ($L.mg^{-1}$) is the Langmuir equilibrium constant and n_S is comparable to the Freundlich heterogeneity factor n_F ($n_S = 1/n_F$).

The isotherm models deliver vital information about the interactive behaviour between absorbent and heavy metal at constant temperature and equilibrium solute concentration. It will help to estimate the amount of the absorbent required to isolate a suitable concentration of heavy metal from the aqueous solution.

The pseudo-first-order and pseudo-second-order kinetic models constitute the adsorption kinetics. The pseudo-first order model represents the adsorption of liquid–solid phase systems based on the adsorption capacity (ÇAvu and GuRdag 2016) and the pseudo-second order model is based on the adsorption capacity of the solid phases.

The pseudo-first order model of Lagrange can be generally expressed as follows (Farhan et al. 2012):

$$\frac{dq_t}{dt} = K(q_e - q_t) \tag{10.7}$$

The integral form of this equation is presented as follows:

$$\log (q_e - q_t) = \log q_e - \frac{K_1}{2.303} t \tag{10.8}$$

where q_e is the equilibrium adsorption capacity of the adsorbent ($mg.g^{-1}$), q_t is the adsorption capacity ($mg.g^{-1}$) when the contact time is t, K and k_1 are the rate constants of the pseudo first-order adsorption model.

The pseudo-second-order model can be expressed as follows (Matouq et al. 2015):

$$\frac{dq_t}{dt} = k_2 (q_e - q_t)^2 \tag{10.9}$$

The integral form of this equation is presented as follows:

$$\frac{t}{q_t} = \frac{1}{k_2 q_e^2} + \frac{t}{q_e} t \tag{10.10}$$

where q_e is the equilibrium adsorption capacity of the adsorbent ($mg.g^{-1}$), q_t is the adsorption capacity ($mg.g^{-1}$) while the contact time is t, and k_2 is the rate constant of the pseudo second-order adsorption model.

10.5 Nanomaterials for removing heavy metals: Illustration with lead ions

Nanomaterials are a form of nano-science having the smallest structures developed by humans (Chaturvedi et al. 2012). Nanoparticles have structure components with one dimension at least less than 100 nm and the normal rules of physics and chemistry often no longer being applied (Amin et al. 2014). Nanowires, nanotubes, films, particles, quantum dots and colloids are the variety of nanomaterials that have been developed (Lubick and Betts 2008).

Due to higher surface to volume ratio, nanomaterials possess size dependent properties such as high reactivity, higher adsorption capacity and higher dissolution activity which are utilized in wastewater treatment. There are some more attractive properties, including superparamagnetism semiconducting and quantum confinement effect which are beneficial for the treatments.

Recently, many advance developments have been found in water treatment technology by using different nanomaterials, such as nanostructured catalytic membranes, nanosorbents, nanocatalysts, bioactive nanoparticles, biomimetic membrane and molecularly imprinted polymers (MIPs) for removing toxic metal ions (specially lead ion), disease causing microbes, organic and inorganic solutes from water. Recently, various nanomaterials have been reviewed particularly for heavy metals' remediation (Tahoon et al. 2020, Singh et al. 2021).

There are various wastewater treatment methods for removal of lead as discussed, including precipitation, flotation, coagulation and flocculation, membrane filtration, cementation, adsorption, ion exchange, and electrochemical deposition, while adsorption is considered quite attractive in terms of its efficiency of removal of lead from dilute solutions, economics and handling for many researchers (Agrawal et al. 2005, Vukojevi´c et al. 2006). In spite of various absorbents' availability for lead removal, nano-absorbents have been cited mostly, in different literatures, as one of the most advanced processes for lead ion separation from wastewater (Zhang et al. 2014a,b, Shamsizadeh et al. 2014, Kyzas and Matis 2015).

Nano-adsorbents are roughly classified into various groups depending on their role in adsorption process. They can be listed as metallic nano-particles, nanostructured mixed oxides, magnetic NPs and metallic oxide NPs. In addition, a recent development on carbonaceous nano-materials (CNMs) included carbon nanotubes, carbon nano-particles and carbon nanosheets. Various types of silicon nano-material are also used as nano-adsorbents such as silicon nanotubes, silicon nanoparticles and silicon nanosheets. In addition, nanoclays, polymer-based nanomaterials, nanofibres and aerogels are some of the nano-materials used for adsorption of heavy metals from wastewater (Anjum et al. 2019)

10.5.1 Carbon-Based Nanomaterials

Here, the major carbon-based nano absorbents are mainly presented for lead ion removal, such as activated carbon-based, carbon nanotubes-based and graphene-based nanomaterials. Synthesis and functionalization of each class of carbon-based nanomaterials have been critically reviewed, emphasizing on adsorption behaviour of carbon-based nanomaterials, interaction type between the sorbent and Pb(II) ions, and effect of pH of samples as a critical factor in the adsorption of Pb(II)ions (Ghorbani et al. 2020).

10.5.1.1 Activated carbon (AC)

AC has attractive properties to capture heavy metal ions, such as high porosity, and high surface area. In addition, it can be made from easily available, cheap carbonaceous precursors in environment such as coal, wood, coconut shells and agricultural wastes (Chatterjee et al. 2010, 2011, Bina et al. 2012, Rodrı´guez et al. 2010). Specifically, Chatterjee et al. 2011 have synthesized Powdered activated carbon (PAC) from Eucalyptus camaldulensis Dehn bark and adsorption capacity has been reported as 0.89 $mmol.g^{-1}$ at 60°C for Pb(II). Activated carbon is also prepared from various agricultural wastes by other groups of researchers, for example, moso and ma bamboo (Lo et al. 2012), grape (Sardella et al. 2015), olive stones (Bohli et al. 2015), lignin (Gonzalez-Serrano et al. 2004) and dust (Bohli et al. 2015, Sardella et al. 2015). 99.9% lead removal has been achieved by using moso and ma bamboo activated carbons (Lo et al. 2012). AC from grape is also an emerging absorbent having removal of lead 1.93 $mmol.g^{-1}$ as reported by Sardella et al. (Sardella et al. 2015).

10.5.1.2 Carbon Nanotubes (CNT)

Carbon nanotubes (CNT) were discovered by Lijima in 1999 and they can be either a single-wall (SWCNTs), which is layered rolled up graphene, or multiwall carbon nanotubes (MWCNTs) formed by multi-layered rolled up graphene (Meng et al. 2010). Utilization of carbon nanotubes in treating heavy metal wastewater is expected due to many superiorities of CNT including the high adsorption capacity, large specific surface area and fast adsorption kinetics (Yan et al. 2008).

Carbon nanotubes have been reported to have outstanding adsorption effects towards Pb (II) (Kabbashi et al. 2009). A molecular dynamic simulation of bare SWCNTs and their functionalized counterparts, such as SWCNTs-OH, SWCNTs-NH_2, and SWCNTs-COOH has been conducted for the adsorption capacities of heavy metal ions (Cd(II), Cu(II), Pb(II), and Hg(II)) from aqueous media (Anitha et al. 2015). SWCNTs-COOH have much adsorption capacities, for example, 96.02 mg/g for Pb(II) which is about 150–230% higher compared to bare SWCNTs, while SWCNTs-OH and SWCNTs-NH have been found to be weak in adsorption, 10–47% higher adsorption compared to SWCNTs, as reported by author. El-Sheikh et al. has proposed that oxidants such as NaOCl, HNO_3, $KMnO_4$ and H_2SO_4 could modify the surface of CNTs which advance the adsorption capacity of CNTs (El-Sheikh et al. 2011). SWCNTs-polysulfone nanocomposite-based membrane is applied to enhance Pb (II) ion removal capability with high rejection value of 94.2% (Gupta 2015). Kosa et al. have modified MWCNTs using eight-hydroxyquinoline for removal of lead and the removal efficiency reported more than 80% (Kosa et al. 2012). Moreover, it is less than the other proposed value and authors have explained that this is due to the lower specific area of the pristine MWCNTs and 8-HQ-MWCNTs, which are 69.1 $m^2.g^{-1}$ and 76.2 $m^2.g^{-1}$, respectively. Oxidized MWCNTs have been studied as efficient adsorbent of Pb(II) ions from aqueous solution (Xu et al. 2008). Carbon nanotubes/metal oxide (CNT/MO) composites have interesting properties as a consequence of interaction between CNTs and metal oxides nanoparticles (Chu et al. 2010). Synthesisation of CNT/MO composites can be done using various techniques such as pressure-less sintering, *in situ* chemical vapour deposition, ultra-sonication, or sol-gel method with spark plasma sintering.

MWCNT/alumina nanocomposite has been found as one of the efficient adsorbent for remediation of Pb(II) ions from aqueous solution, as approximately 100% removal was achieved with 50 mg of the coated MWCNTs while 85% removal was achieved with 50 mg of uncoated MWCNTs (Gupta et al. 2011). The cause of increase in adsorption has been explained as the electrostatic attraction between the pairs of electrons on the oxygen atoms of alumina and the cationic lead. The removal of Pb(II) from water by MnO_2/CNTs has been investigated and the metal removal varied from 77 to 98% with a change in pH from 2 to 4. This result is evidence of high dependency of absorption capacity of absorbents on the solution pH value.

Pb(II) adsorption via MnO2/CNTs shows about 300% increase over the adsorption by CNTs (Wang et al. 2007). CNT-iron oxides magnetic composite has also been utilized as adsorbent for removal of Pb(II) from water (Peng et al. 2005).

Bankole et al. have investigated the removal of heavy metals via batch adsorption process from industrial electroplating wastewater utilizing purified carbon nanotubes (P-CNTs) and polyhydroxylbutyrate functionalized carbon nanotubes (PHB-CNTs) (Bankole et al. 2019). However, PHB-CNTs have been studied more efficiently than P-CNTs for heavy metals' removal based on ion exchange and electrostatic forces mechanism. Raw value of lead availability in industrial waste water is 4.94 $mg.L^{-1}$; after batch adsorption by P-CNTS, it has been reduced to 0.0160 $mg.L^{-1}$, while by PHB-CNTS it is 0.0395 $mg.L^{-1}$ (WHO permissible value 0.01/0.05 $mg.L^{-1}$). Investigation on the adsorption of Pb(II) from aqueous solution onto a sugarcane bagasse/MWCNT composite has been performed and the Langmuir adsorption isotherm presented the best fit to the data and Elovich kinetics model has been found to be suitable for kinetics studies (Hamza et al. 2013). The ethylenediamine-grafted multi-walled carbon nanotubes (MWCNTs-EDA-I and MWCNTs-EDA-II) have been presented as nano-absorbent having Pb (II) absorption capacity 157.19 $mg.g^{-1}$, where MWCNTs-EDA-I has a higher maximum adsorption capacity than MWCNTs-EDA-II (Hu et al. 2012). Atieh et al. have studied the adsorption of Pb(II) from water by using COOH functionalized CNTs. It is reported that there will be 100% of Pb(II) removed at pH 7 (Atieh et al. 2010).

Although there are many advantages of CNTs to remove heavy metals from wastewater, it still has some weaknesses. The high costs of CNTs are obstacles for their commercial use. After adsorption of heavy metal ions, it is very challenging to separate CNTs from waste water and ultimately the treatment process will be costly. Moreover, there is the risk of pollution and toxicological study of CNTs is required (Kumar et al. 2014). Recently, CNT has been reviewed for heavy metals removal including lead from aqueous environment (Baby et al. 2019).

10.5.1.3 Graphene-based nanomaterials

Graphene was discovered in 2004 by Sir Andre Geim and Sir Konstantin Novoselov and they received a Nobel Prize for their discovery in 2010. Graphene (2-D) is being used widely in almost every field. Additionally, these nano-absorbents are getting tremendous attention in water treatment due to their unique physicochemical characteristics, namely electronic properties, high surface area, thermal mobility, high mechanical strength and tuneable surface chemistry (Mohan et al. 2018, Tabish et al. 2018). Porous graphene has been designed and is utilized as an adsorbent for the removal of heavy metal ions as well as other pollutants from water by Tabish et al. (2018). Researchers have reported many modified graphene used as successful adsorbents such as GO, EDTA-graphene oxide (EDTAGO), and chitosan-GO for removal of lead ions from water (Zhao et al. 2011, Madadrang et al. 2012, Sitko et al. 2013).

The role of several functional groups present in GO makes it soluble in polar and non-polar solvents as the oxygen functionality makes GO hydrophilic and the graphene domain makes GO hydrophobic. As all the functional groups located on the edges and on the basal plane of GO, it helps in removal of heavy metal ions (Bao et al. 2013). Aggregation is a major disadvantage of graphene layers, and researchers have proposed a way to prevent it partially through its composite formation (Yusuf 2015, Peng et al. 2017). Removal of lead using functionalized GO/RGO/GO-aerogel, GO-polymer composites, and GO-magnetites has been established as an efficient process of waste water treatment (Kumar et al. 2019).

The improved adsorption for Pb(II) ions from the aqueous solution is observed with an EDTA-GO adduct in comparison with GO and EDTA-RGO adduct. The maximum adsorption capacities are reported for the EDTA-GO adduct and GO for Pb(II) removal as 479 ± 46 and 328 ± 39 $mg.g^{-1}$, respectively (Madadrang 2012). The reason for enhanced performance of EDTA-GO is correlated with the chelating characteristics of EDTA with functionalized graphene sheets. Madadrang et al. have explained mainly two adsorption process: EDTA-GO adduct, i.e., the ion-exchange reaction process between Pb(II) and carboxylic and hydroxyl function group responsible for chelate complex formation between Pb (II) and EDTA over GO surface. In the absorption process, Pb(II) equilibrium concentration reached lower than admissible value, i.e., < 10 ppm. In addition, higher adsorption capacity is attributed to the higher stability constant (Log K ≈ 18.0) of Pb(II)-EDTA complex. As GO is dispersed nicely in water, separation of GO from water is very tedious, which basically reduces flexible use of GO. This problem has been solved by graphene modified absorbent. Calcium alginate (CA) beads embedded graphene oxide is reduced by polyethylenimine to increase the adsorption capacity towards heavy metals. The experiment for adsorption is carried out batch-wise in a shaker bath. The highest adsorption capacities of this GO-based nanocomposite towards Pb (II) is reported as 602 $mg.g^{-1}$ (Arshad et al. 2018).

Fan et al. have investigated magnetic chitosan/graphene oxide (MCGO) as sorbents for the removal of Pb(II) ions from aqueous solutions (Fan et al. 2012). It is reported by author that adsorption of Pb(II) ions on MCGO surface is strongly dependent on pH value of the solution; additionally, MCGO is stable and easily recovered. The plentiful functional groups on the surfaces of MCGO played an important role in Pb(II) ion adsorption. The maximum adsorption capacity for Pb(II) is estimated to be 76.94 mg/g. Kong et al. have studied the adsorption capacity of graphene sheets before and after modification with 1-pyrenebutyric acid (PBA) via the stacking interaction for heavy metal ions, such as Cu(II), Cd(II), and Pb(II) at different pH and contact time (Kong et al. 2013). PBA modified graphene (GS-PBA) is found to be a better adsorbent for heavy metal ions than graphene. The results are reported as the adsorption capacity of graphene 99.3 mg/g for Pb(II), while that of graphene/PBA 124.2 mg/g for Pb(II) at pH of 5. Vilela et al. have designed a novel graphene oxide-based microbots (GO, Ni, and Pt composed the structure of microbots, and GOx-microbots) which could be an automotive system for binding heavy metals, as well as transferring and removing it (Vilela et al. 2016).

The results are presented as the mobile GOx-microbots could remove Pb (II) ten times more efficiently, i.e., reduced from 1000 ppb down to below 50 ppb in 60 min in comparison with nonmobile GOx-microbots. The reusability of the GOx-microbots is also proposed after eliminating the lead ions from the surface of the microbots. Graphene oxide (GO) and two goethite (α-FeOOH)/reduced graphene oxide (rGO) composites (composite 1: 0.10 g GO: 22.22 g α-FeOOH and composite 2: 0.10 g GO:

5.56 g α-FeOOH) have been studied for adsorption of Pb(II) from aqueous medium (Gordon-Nuñez et al. 2019).

Fullerenes were invented in 1985 from interstellar dust, and they have a closed-caged pentagonal and hexagonal ring structure, having a hydrophobic character, high electron affinity and high surface to volume ratio, and surface defects (Kroto et al. 1985). Fullerenes adsorbed species by the penetration of adsorbates in the gap between the carbon nanoclusters. Moreover, it has lower aggregation tendency and large surface area which is beneficial for adsorption of heavy metal ions from water (Lucena et al. 2011). Ciotta et al. have performed the atomic absorbance spectroscopy of a Pb(II) contaminated aquas solution owing to the adsorption by unfolded-fullerene nanoparticles (UFNPs) to investigate efficiency of removal of Pb(II) ions (Ciotta et al. 2019). The adsorption isotherms have been studied and the Freundlich model provides a better fit for the Pb(II) adsorption by UFNPs than the Langmuir one.

10.5.2 Silica-Based Nanomaterials

Excellent surface characteristics, i.e., pore size, high surface area, non-toxicity and adsorption capacity are the properties possessed by silica-based nanomaterial which are excellent for removing heavy metals from waste water (Mahmoud et al. 2016). Basically, functionalized or composite form of silica is employed for heavy metal ions' removal (Kegl et al. 2020). Different research groups have employed several functional group for adsorption study, such as silica functionalized nanomaterials (NMs), silica/polymer nano compounds (NCs), functionalized silica/compounds NCs and silica core@shell (NPs). Kegl et al. have proposed a novel γ-Fe_2O_3–NH_4OH@SiO_2(APTMS) nanoparticles and adsorption of dysprosium ($Dy^{3+)}$ ions by this absorbent (Kegl et al. 2019). The adsorption result fits the Temkin isotherm, and the process is endothermic and spontaneous. Three silica-based nanomaterials, including the amino functionalized silica gel (NH2-SG), amino functionalized silica nano hollow sphere (NH2-SNHS) and non-functionalized silica nano hollow sphere (SNHS) have been investigated for removal of Cd (II), Ni (II), and Pb (II) in a batch mode. NH2-SNHS is reported as most efficient absorbent for Pb (II), having maximum absorbance capacity of 96.79 mg. g^{-1}. The adsorption isotherms are similar to the Langmuir-Freundlich (Sips) isotherm well and the kinetic data fit the pseudo-second-order well. Non-metallic nano silica spheres (NSHS) with calcium carbonate ($CaCO_3$) synthesis and characterization have been performed for removal of Pb(II) ions from industrial wastewater. Temperature dependency on the rate of adsorption and adsorption capacity are established by results. In addition, it is concluded that the adsorption efficiency is decreased with an associated increase in initial metal ion concentration. The optimum adsorption capacity of NSHS for Pb(II) ions is shown as 266.89 $mg.g^{-1}$ (Manyangadze et al. 2020). Huang et al. have investigated adsorption of Pb(II), Zn(II) and Cu(II) ions from aqueous solution by fabricating magnesium silicate hierarchical nanostructures (Huang et al. 2017). It has been established by the investigation that adsorption of Pb(II), Zn(II) and Cu(II) ions are best fitted by Langmuir isotherm (R^2 = 0.999) with highest adsorption capacity of 436.68 mg.g^{-1} for Pb(II) ions. Mahmoud et al. have carried out study on nanopolyaniline and crosslinked nanopolyaniline, immobilized onto the nanosilica to make nanocomposites, Sil-Phy-NPANI and Sil-Phy-CrossNPANI (Mahmoud et al. 2016). The highest adsorption capacities of Sil-Phy-NPANI and Sil-Phy-CrossNPANI towards Pb (II) by using a batch technique reach up to 900 $mol.g^{-1}$ and 1450 $mol.g^{-1}$, respectively, as calculated from Langmuir isotherm.

10.5.3 Zero-valent metal-based nanomaterials

Recently, zero-valent metal nanoparticles have received attention of researchers due to their potential in water treatment and remediation. Nanoscale zero valent iron (nZVI) is a composite consisting of Fe (0) and ferric oxide coating and it has adaptability and simple synthesis that can encourage to remove metallic and metal contaminants (O'Carroll et al. 2013). Zhang et al. have investigated 20% (w/w) nanoscale zero-valent iron (nZVI) loaded onto kaolinite as a support material (K-nZVI) (Zhang et al. 2011). It has been found that more than 96% of Pb(II) is removed from aqueous solution using K-nZVI as 500 mg/L Pb(II) is isolated within 30 min under the conditions of 10 g/L of K-nZVI, pH 5.10 and a temperature of 30°C. Huang et al. have proposed that the high reducing capacity and large specific surface area is attributed to the extraordinary performance of nZVI in removing heavy metals from wastewater (Huang et al. 2013). Authors have also indicated that nZVI particles could remove Cd(II), Cr(II) and Pb(II) from aqueous solutions. nZVI is reported as it is oxidized with oxygen and water within aqueous solution and aggregate easily which results in decreasing reduction process and reacting surface area, respectively (Tratnyek et al. 2008). To overcome the difficulties, a novel nZVI-modified material by combing nZVI with an anionic surfactant sodium dodecyl sulfate (SDS) has been synthesized (Huang et al. 2015). Zarime et al. have proposed a new nZVI-based nanocomposite by using low-cost bentonite to treat Pb(II), Cu(II), Cd(II), Co(II), Ni(II) and Zn(II) in water (Zarime et al. 2017). Bentonite-nZVI composite possesses a higher removal capacity towards these heavy metals in comparison with the bentonite. Moazeni et al. have proposed a high removal of lead with nZVI 80%–100% at pH 9, contact time of 5 minutes, nZVI dosage of 0.1 g/L and lead initial concentration of 10 mg/L (Moazeni et al. 2017).

Numerous reports based on the interaction between Ag nanoparticles and Hg(II) ions have been presented (Morris et al. 2002, Fan et al. 2009). Silver nanoparticles have been used as colorimetric sensors for water pollutants, focussing on several heavy metals ions, such as Hg(II), Ni(II),Cu(II), Fe(III), Mn(II), Cr(III/V) Co(II), Cd(II), and Pb(II). Selectivity of Pb(II) ions in aqueous samples has been developed using 1-(2-mercaptoethyl)-1,3,5-triazinane-2,4,6-trione-functionalized silver NPs. The Pb (II) ion acts as the metal centre of the coordination complex, which forms N–Pb(II)–O coordination bonds with the AgNPs. The selectivity and sensitivity noticeably improved in the pH range of 7–8 (Prosposito et al. 2020). The sensor has displayed a linear correlation with Pb (II) ion concentrations within the linear range of 0.1–0.6 μg/mL and the limits of detection in tap and pond water as 0.02 and 0.06 μg/mL, respectively (Noh et al. 2015).

Simultaneous optimization of the important variables for synthesis of Ag-NPs has been done by Central composite design (CCD) under response surface methodology (RSM) to remove Pb(II) ion. The study has been done with some variable parameters such as contact time, pH, and the amount of reagent. The maximum of Pb(II) ion removal has been obtained while the concentration of sodium citrate, silver nitrate, lead nitrate and L-cysteine has been fixed as 0.05 M, 0.005 M, 27.78 mg L^{-1} and 0.25 M, respectively. The maximum removal of Pb(II) ions by optimized Ag-NP has been reported as 105.4 mg.g^{-1} from samples with high Pb(II) ion concentration in short reasonable time (Hosseini et al. 2016).

Attatsi and Nsiah have tested the efficiency of AgNPs using standard solutions of cobalt and lead. A bathochromic shift in the absorption wavelength is observed from 406 to 445 nm and from 406 to 458 nm for Co(II) and Pb(II), respectively. Basically, this shift indicates an adsorption of these heavy metals. Adsorption characteristics of the nanoparticles have been depicted at various incubation periods and there is a reduction in co(II) and Pb(II) from 33.13 to 53.34% and 79.9–92.92%, respectively from day 1–14 (Attatsi and Nsiah 2020).

Gold nanoparticles (AuNPs) have some unique properties, for example high surface area per volume ratios, ease of surface modification and high stability and biocompatibility (Qian et al. 2013). Basically, these properties are encouraging enhanced removal capacity and high selectivity towards a variety of target species including heavy metal ions. AuNPs can be used as colorimetric probes, which is based on their highly sensitive surface plasmon resonance (SPR); there will be colour change while the aggregation of AuNPs is induced by heavy metal ions Hg(II), Pb(II) and Cu(II) (Lin et al. 2011). Annadhasan et al. have proposed that the –OH and –NH2 groups are factors for the reduction of $[AuCl4]^-$ and the $-COO^-$ groups are responsible for the stabilization of AuNPs formed. It is also reported that L-tyrosine capped AuNPs is very sensitive towards lead ions with the limit of detection as low as 16 nM (Annadhasan et al. 2014). Synthesis of AuNPs, followed by their functionalization and characterization and their application as heavy metal detector as well as for removal, has been reviewed by Nitti (Nitti 2014). Kim et al. have first reported the use of 11-mercaptoundecanoic acid (MUA) capped Au NPs (d ~ 13 nm) for the detection of heavy metal ions Hg(II), Pb(II) and Cd(II). The drawback of this method is the inability to separate between the different metal ions (Kim et al. 2001).

DNA-coated Au NPs can aggregate due to DNA binding with the DNAzyme, finally producing a blue-coloured sol. Pb(II) ions can disaggregate the nanoparticles after the DNAzyme cleaves the DNA strands which results in a red-coloured sol. The Pb(II) detection limit has been reported much less than the EPA limit for drinking water (0.015 mg L^{-1}) as 0.62 μgL^{-1} (0.00062mgL^{-1}) (Wang et al. 2008). Numerous recent research work on treatment of pollutant water by using gold nanoparticles has been highlighted to stimulate removal and detection of pollutants (Huifeng et al. 2013). Two types of bovine serum albumin (BSA)-capped 14.2 nm diameter Au NPs have been introduced for the separate detection of Hg(II) and Pb(II) ions in highly saline media. Gel-based membrane of BSA-capped AuNPs is also designed for removal of Hg(II) or Pb(II) from aqueous solutions having recoveries of 95–106% (Lee et al. 2012).

10.5.4 Metal oxide-based nanomaterials

Metal oxide nanoparticles have exceptionally high removal capacity and these properties make them selective towards heavy metals as efficient adsorbents. Metal oxides-based nanomaterials include manganese oxides, nanoscale iron oxides, zinc oxides, aluminium oxides, titanium oxides, zirconium oxides, magnesium oxides and cerium oxides (Yang et al. 2019).

10.5.4.1 Iron oxide-based nanomaterials

Iron oxide-based nanomaterial is more attractive for removal of heavy metals from the water because of their attractive features, such as small size, high surface area and magnetic property. Magnetic property of iron oxide nanoparticles is very significant as it accelerates the separation of heavy metals from the system (Xu et al. 2012, Yang et al. 2012). Versatility of iron oxides-based nanoparticle is its reusability which could make the water treatment more economic. Here in, the application of important iron oxides nanoadsorbents such as magnetite (Fe_3O_4), maghemite (γ-Fe_2O_3), and hematite (α-Fe_2O_3) are discussed for removal of Pb (II) from wastewater.

Investigation on hematite performance has been done in different functional factors, including multiple metal species, the dosage of adsorbents and temperature. The affinity between the Pb(II) and hematite has been observed maximum in comparison to other heavy metals. Superparamagnetic hematite nanoparticles are synthesized and are very effective to treat the acid mine drainage (AMD) containing Ni(II), Al(III), Mg(II), Mn(II), etc., in a batch mode (Kefeni et al. 2018).

Maghemite nanoparticles with a surface area of 79.35 $m^2.g^{-1}$ has been applied to remove Pb(II) and Cu(II) in wastewater through the batch adsorption. The maximum Langmuir adsorption capacities have been reported as 68.9 and 34 $mg.g^{-1}$ for Pb(II) and Cu(II), respectively. Here, the adsorption of metal ions is a consequence of electrostatic interactions. The surface of maghemite is functionalized with FeOH groups in water which can constitute positive Fe-+OH2 or negative FeO^- groups with the change of pH. Fe $(III)O^-$ or Fe (III)OH can easily incorporate with metal ions. Fe $(III)O^-$ or Fe (III)OH sites are generated more while pH increases. Finally, there will be improvement of the adsorption capabilities for Pb (II) and Cu (II). In addition, there will be excess Fe-+OH2 sites produced with the decrease of pH, which will keep away Pb (II) and Cu (II) on the surface and it will decrease the removal capacity (Rajput et al. 2017). Maghemite (γ-Fe2O3) nanoparticles have been tested for the selective remediation of poisonous heavy metals from electroplating wastewater. The adsorption of Pb(II) has increased with pH and the equilibrium is achieved within 15 min Cheng et al. 2012). Polymer-modified maghemite nanomaterials can utilize advantages of both polymers and maghemite to attract heavy metals. Mercaptoethylamino monomer-modified maghemite nanomaterial (MAMNPs) has been synthesized having specific surface area of MAMNPs 92.41 $m^2.g^{-1}$ and the maximum removal capacity of Pb (II) is reported as 118.51 $mg.g^{-1}$ by using the Sips isotherm (Madrakian et al. 2015).

Maghemite nanotubes are synthesized by microwave irradiation method and it has been found that the surface area of the nanoadsorbaent is 321.638m^2-g^{-1}. The maximum adsorption capacities of the maghemite nanotubes' adsorbents towards

Cu(II), Pb(II), and Zn(II) are 111.11, 71.42, and 84.95 $mg.g^{-1}$, respectively, as reported from the fitted Langmuir isotherms. The kinetic data of adsorption of Pb(II) ions onto the synthesized nanoadsorbents is followed by a pseudosecond-order equation indicating their chemical adsorption (Roy and Bhattacharya 2012).

Magnetite (Fe_3O_4) based nanomaterials are one of the widely used nanometer adsorbent as it is low cost, simplicity of use, easy availability and eco-friendly. Nessar has reported the adsorption of Pb(II) ions by magnetite with increasing temperature which indicates endothermic adsorption and the adsorption equilibrium data is fitted with the Langmuir and Freundlich adsorption isotherm models. The maximum adsorption capacity of Pb is reported as 36$mg.g^{-1}$ (Nassar 2010). Hydrothermal method has been used to synthesize water soluble magnetite nanoparticles with high solubility and stability. It has shown high affinity for Pb(II) and Cr(VI) than water insoluble magnetite nanoparticles. These water soluble magnetic nanoparticles have the capacity to remove 90% of Pb(II) in 2 min (Wang et al. 2012). Giraldo et al. have synthesized magnetite nanoparticles by using a co-precipitation method and it has been used to treat Pb(II), Cu(II), Zn(II), and Mn(II) in a batch mode (Giraldo et al. 2013). The best adsorption capacity of 0.180 $mmol.g^{-1}$ for Pb(II) ions has been found for nanosized magnetite. The reason for the highest attraction towards Pb(II) is attributed to the various electrostatic interactions between the heavy metal ions and the adsorbent sites.

It has been reported by Singh et al. that the removal of toxic metal ions from wastewater can be done using carboxyl-, amine- and thiol-functionalized magnetite nanoparticles (succinic acid, ethylenediamine and 2,3-dimercaptosuccinic acid, respectively). The functionalized magnetic nanoadsorbents captured various metal ions including Pb(II), either by forming chelate complexes, by ion exchange process or through electrostatic interaction based on their surface functionality (Singh et al. 2011). Magnetite can be combined with multiwall carbon nanotube as magnetite nanocomposites, where thiol functional group is attached on the surface of CNTs/Fe_3O_4 using 3-mercaptopropyltriethoxysilane. The maximum Langmuir adsorption capacity for removal of Pb (II) is 65.40 $mg.g^{-1}$ (Zhang et al. 2012). Magnetite can be functionalized with amine by hydrothermal method in which iron salt $FeCl_3{\cdot}6H_2O$ is reacted with ethylene glycol followed by sodium acetate and ethanediamine. The mesoporous morphology of amine functionalized Fe_3O_4 is found with suitable pore volume. The adsorption rate of amino functionalized Fe_3O_4 for Pb (II) is fitted with pseudosecond-order kinetic model (Xin et al. 2012). Huang et al. have designed a novel core-shell adsorbent by coating the organodisulfide polymer (PTMT) onto the amino-functionalized magnetite nanoparticles (Huang et al. 2018). The adsorbents have high adsorption capacities toward high-concentration Pb(II), Hg(II) and Cd(II) in a batch adsorption experiment. The adsorption capacity of Pb(II) has been reported as 533.13 $mg.g^{-1}$ at 600 $mg.L^{-1}$ concentration. Reusability of PTMT has been demonstrated as extraordinary as re-adsorption of Pb(II) remained constant after five cycles. Polymer modified magnetite nanoparticles have been used for almost 100% removal rate of Pb(II) ions from water at pH 8 (Ge et al. 2012, Warner et al. 2010). The adsorption process is found to be highly dependent on the amount, surface functionality and pH of the medium. It is reported that ethylenediaminetetraacetic acid-functionalized (EDTA) Fe_3O_4 nanomagnetic chelators (NMCs) has affinity

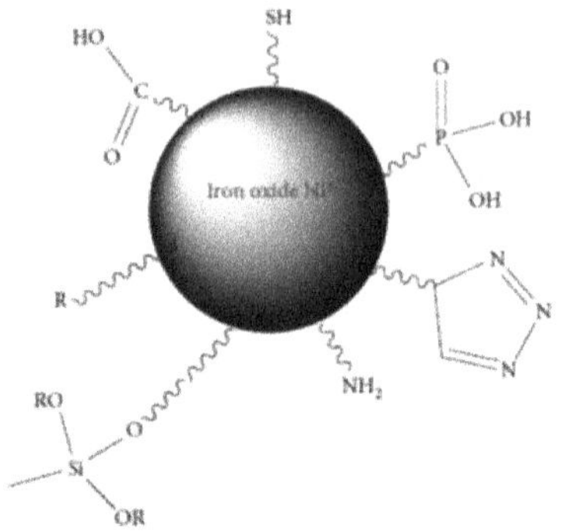

Fig. 1: Schematic representation of surface modification of iron oxide nanoparticles. Reproduced with permission from (Dave and Lakhan et al. 2014).

towards the adsorption of Pb(II) from wastewater (Visa et al. 2009). Recently, Sadak et al. have demonstrated polyacrylic acid (PAA) conjugated ferric oxide (Fe_3O_4) magnetic NPs (MNPs), further functionalized with CR azo dye (PAA-CR/MNPs). Adsorption affinity of PAA-CR/MNPs structure for various cations such as Fe^{2+}, Fe^{3+}, Cd^{2+}, Cu^{2+}, and Pb^{2+} have been investigated at various pHs, temperatures, reaction conditions, and times. A special prominence is given to Pb(II) having maximum removal efficiency at 6.5 pH and 45 min of reaction time (Sadak et al. 2020). Dave et al. have reviewed various other methods to fabricate different iron-oxide based nanoparticles for removal of the Pb(II), as well as other heavy metal ions (Dave and Lakhan 2014). Surface modification of iron oxide nanoparticles for heavy metal incorporation has been shown in Fig. 1. Recently, magnetite nanomaterial was used effectively that could easily separate lead ion with 85% treatment efficiency from aqueous environment at pH 6 (Stoian et al. 2021). An innovative idea of synthesis of magnetic iron oxide–tea waste nanocomposite is also reported for adsorption of lead ions from aqueous solutions and magnetic elimination of the adsorbent has been described afterwards (Khanna et al. 2020).

10.5.4.2 Manganese oxides (MnO) nanoparticles

Manganese oxide (MnOs) NPs possess a high adsorption ability due to their high BET surface area and polymorphic structure (Luo et al. 2010). Modified MnOs include nanoporous/nanotunnel manganese oxides and hydrous manganese oxide (HMO) (Gupta et al. 2015). Various heavy metals including Pb(II) adsorptions have been reported onto HMOs, which happened due to the inner-sphere formation mechanism. Recently, HMO-BC nanocomposite has been fabricated by infusing the HMO nanoparticles with the biochar (BC). HMO-BC nanocomposite exhibited an enhanced removal effect towards Pb(II) and Cd(II) in a wide range of pH (Wan et al. 2018). The novel dumbbell-like manganese dioxide/gelatin has been fabricated and its adsorption performances have been studied specifically for Pb(II) and Cd(II) ions. The batch adsorption study has disclosed that the maximum adsorption capacities towards Pb(II) and Cd(II) are 318.7 and 105.1 $mg.g^{-1}$, respectively, as deduced from the Langmuir mode l(Wang et al. 2018).

10.5.4.3 Zinc oxide (ZnO) nanoparticles

Zinc oxide nanoparticles have achieved their popularity as adsorbents for heavy metals as they have high surface area, extraordinary removal capacity and low cost (Kumar et al. 2013a). Kumar et al. have reported the high removal efficiency of Pb(II) and Cd(II) from wastewater by using mesoporous hierarchical ZnO nano-rods (Kumar et al. 2013b). Novel ZnO nanoparticles (NPs) are synthesized in zerumbone solution by a green approach and assessed for their ability to absorb Pb(II) ions from aqueous solution. The formation of the nanoabsorbents is established by UV–visible studies, X-ray diffraction (XRD), and Transmission Electron Microscopy (TEM). Investigation on Pb(II) removal by Zno NPs through batch experiments, have been done and impact of various parameters, such as adsorbent mass, Pb(II) concentration, solution temperature, pH of solution, and contact time variations, on lead removal efficiency has been carried out. The surface of nano-adsorbent can do protonation depending on the pH of solution. While the pH of solution increases, the number of protons on the surface of nano-adsorbent starts decreasing and excess negative groups will produce for complexation of metal cations. It is reported that the biosynthesized nanoparticles adsorption for the Pb(II) ions is fitted with Langmuir adsorption model and pseudo-second-order equation. The maximum adsorption capacity of Pb(II) is found to be 19.65 $mg.g^{-1}$ under pH of 5 and temperature of 70°C in aqueous solution (Azizi et al. 2017). Special rod-like shaped ZnO particles are synthesized using solid precipitation technique. The average length and diameter of ZnO particle has been reported as 497.34 ± 15.55 and 75.78 ± 10.39 nm, respectively. The removal efficiency of Pb(II) ions is demonstrated as more than 85% under exposure of 1 hour of UV light (Le et al. 2019). Zinc oxide nanoparticle (ZnO-NP) has been combined with the layers of montmorillonite by green simple heat method using zinc nitrate and sodium alginate, zinc oxide precursor and stabilizer, respectively. Adsorption of heavy metal ions, such as Pb(II), and Cu(II) by ZnO/MMT has been increased with ZnO during the use of the ZnO/MMT nanocomposite for the removal of copper and lead ions from aqueous solutions. Mechanisms of the adsorption of Cu(II) and Pb(II) onto ZnO/MMT has been shown in Fig. 2. Reusability of the nanocomposite has been confirmed after utilizing it for at least three times. The percentage removal of Cu(II) and Pb(II) ions is 89.5% and 97.2% using the prepared ZnO/MMT nanocomposite. It has been tabulated that adsorption capacity of Pb(II) is 88.50 $mg.g^{-1}$ (Sani et al. 2017).

10.5.4.4 Magnesium oxide (MgO) nanoparticles

There are numerous reasons, such as high adsorption capacity, economic, nontoxic, plenty and eco-friendly character of magnesium oxide nanoparticles which make them attractive nano adsorbents for heavy metals including Pb(II) (Cai et al. 2016). Modification of NPs morphology can enhance the adsorption capacity of MgO. Cai et al. have demonstrated that Cd(II), Pb(II), and *Escherichia coli* could be removed from water by utilizing MgO nanoparticles produced by the sol–gel method. The adsorption capacities of MgO nanoparticles towards Cd(II), Cu(II), Ni(II) and Pb(II) from aqueous solutions have been reported as 135, 149.1, 149.9 and 148.6 $mg.g^{-1}$, respectively (Mahdavi et al. 2013). In a batch adsorption study, Xiong et al. have

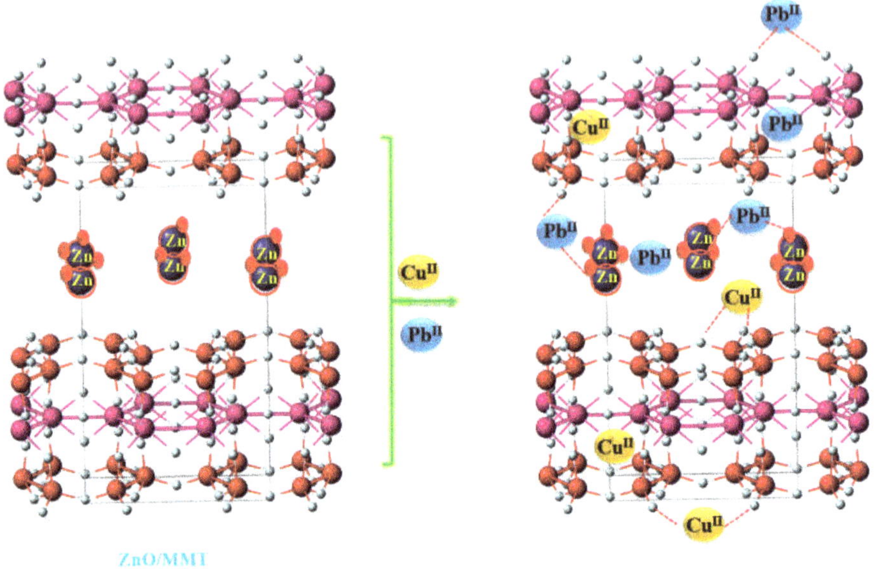

Fig. 2: Proposed mechanisms for the adsorption of Cu(III) and Pb(II) onto ZnO/MMT. Reproduced with permission from (Sani et al. 2017) .

reported high adsorption capacities of MgO nanoparticles towards Cd (II) and Pb (II) as 2294 and 2614 mg.g^{-1}, respectively, by calculating from the Langmuir equation. The high adsorption originated by external mass transfer and intraparticle diffusion. The generation of OH^- dissociated from Mg (OH)2 and the collaborative effects of adsorption facilitated the achievement of this adsorption capacity for Pb(II) ions (Xiong et al. 2015). A novel structure of MgO microsphere has been introduced having improved adsorption affinity for the removal of heavy metals (Gupta et al. 2015).

Flowerlike mesoporous MgO has been reported for efficient adsorption of Pb(II) and Cd(II) (Li et al. 2003). A nanohybrid of graphene oxide (GO) and magnesium oxide (MgO) has been designed by Mohan et al. by using simple precipitation method. The GO is perfectly oriented by MgO in the GOMO nanohybrid structure as characterized by FTIR, XRD, SEM, EDX Raman, BET, and XPS analysis. The experimental data fitted well with the Langmuir isotherm and GOMO showed excellent adsorption having uptake capacity 190 mg.g^{-1} at a low dose of 0.4 g.L^{-1} within 30 min of equilibrium time, 30°C temperature, and 6.5 pH (Mohan et al. 2017).

10.5.5 Nanocomposite Nanomaterials

Nanoparticles regularly show some problems, such as aggregation, and poor separation. In addition, an excessive pressure drop is faced by nanoparticle when these are used in fixed-bed and flow-through systems (Hotze et al. 2010). It has been observed that synthesisation of hybrid nanocomposites is the general technique to resolve these problems (Zhang et al. 2016). Nanocomposites provide high surface area for fast decontamination, improved processability, remarkable stability, cost effectiveness, and selectivity to remove different pollutants in wastewater (Berber 2020). There are nanocomposites based on inorganic and organic polymer supports, together with the magnetic nanocomposites.

Chen et al. have designed bi-functionalized -cyclodextrin (-CD) and polyethyleneimine (PEI) magnetic nanoadsorbents (-CD/PEI- Fe_3O_4) in order to detain methyl orange (MO) and Pb(II) simultaneously from wastewater with spatially separated sorption sites. The efficient adsorption capacity of the composite towards Pb(II) is attributed to the bonding between the edges of -CD with oxygen bearing groups and free amino moieties present on the PEI, acting as active sites for the efficient uptake (Chen et al. 2019). Inorganic materials-based nanocomposites mostly utilized for waste water treatment mainly contain activated carbon (AC), CNTs, bentonite, montmorillonite, zeolite, and so on. The AC-supported nanocomposites are efficient for heavy metal ions removal including Pb(II) as reported by many researchers (Kang et al. 2015, Jayaweera et al. 2018, Fernando et al. 2015).

A multi-walled CNTs/chitosan nanocomposite has been prepared by using sonication technique with the chitosan and CNTs suspension and then crosslinking them with glutaraldehyde. The composite has exhibited high adsorption capacities towards Zn(II), Cu(II), Ni(II), and Cd(II) from waste water (Salam et al. 2011). One batch adsorption experiment has been carried out with fabricated hydroxyapatite nanorods (nHAp) and hydroxyapatite/chitosan nanocomposite (nHApCs) as efficient sorbents for the removal of lead ions from aqueous solutions. Freundlich and Langmuir isotherms are fitted with experimental data. 200 mL solution (pH = 5.6) containing 100 ppm lead ions, and a sorbent dosage of 0.4 g nHAp has 100% removal for lead ions. The adsorption capacities of nHAp and nHApCs to lead ions are reported as 180 and 190 $mg.g^{-1}$, respectively (Mohammad et al. 2017).

Seema and co-workers have synthesized bionanocomposites by titanium dioxide incorporated in a polymer blend (CD-PCL) via solution blending method. The produced bionanocomposites have been characterized using Scanning Electron Microscopy (SEM), transmission electron microscope (TEM) and Brunauer Emmett and Teller (BET). Removal efficiency of the bionanocomposite has been studied under the variation of different parameters such as pH, concentration and adsorbent dose to test effect on adsorption capacity. Maximum adsorption of lead has been obtained as 98% at pH 9.7, 10 ppm with 0.005 g dosage (Seema et al. 2017). Zeolite is another sort of interesting support and additive for nanoparticles due to its attractive ion exchange ability, high surface area, and hydrophilic characteristic. Hydroxyapatite/zeolite nanocomposite (HAp/NaP) has been synthesized to capture Pb(II) and Cd(II) ion from waste water by using batch adsorption method and the maximum adsorption capacities of Pb(II) and Cd(II) have been obtained as 55.55 and 40.16 $mg.g^{-1}$, respectively (Alswata et al.2017, Zendehdel et al. 2016).

Magnetic nanocomposite is prepared with polypyrrole-polyaniline/Fe_3O_4 and it has been investigated for removal of Pb(II) in an aqueous solution. The synthesized nanocomposite has 100% of Pb(II) removal efficiency at 8–10 pH and 20 $mg.L^{-1}$ concentration of Pb(II) (Afshar et al. 2016). Suman et al. have designed another organic based nanocomposite by embedded nanocellulose (NC)-Ag nanoparticles

(AgNPs). It is a pebbles-based composite material which has been applied to remove dyes, heavy metals and microbes in water by column adsorption method. The removal efficiencies have been reported as 99.48% of Pb(II) and 98.30% of Cr(III) from water along with a 99% decontamination efficiency for microbial load (Abhishek et al. 2015). Chitosan is one of the interesting biodegradable adsorbents which has potential of elimination of heavy metals from waste water. ZnO/chitosan nanocomposite is kind of core-shell nanocomposite (ZOCS) which is low cost and eco-friendly. It has been demonstrated that the maximum adsorption capacities for Pb (II), Cd (II), and Cu (II) are 476.1, 135.1, and 117.6 $mg.g^{-1}$, respectively, via batch adsorption (Saad et al. 2018). Carboxymethyl--cyclodextrin (CM--CD) polymer modified Fe_3O_4 magnetic nanocomposite (CDpoly-MNPs) is synthesized for selective removal of Pb(II), Cd(II), and Ni(II) ions from waste water. The adsorption of CDpoly-MNPs has been found dependent on pH, ionic strength, and temperature. In batch adsorption, the maximum adsorption capacities for Pb(II), Cd(II) and Ni(II) are 64.5, 27.7 and 13.2 $mg.g^{-1}$, respectively, at 25°C. The polymer attached on MNPs enhances the adsorption capacity due to the complexing abilities of the multiple hydroxyl and carboxyl groups in polymer backbone with metal ions. It is explained by hard and soft acids and bases (HASB) theory that CDpoly-MNPs can preferably adsorb Pb(II) ions (Badruddoza et al. 2013). Ge et al. have fabricated a Fe@MgO nanocomposite having strong magnetism of nZVI's and MgO's high adsorption capacity. In batch adsorption experiments, the maximum removal capacities of this nanocomposite towards Pb (II) and methyl orange have been reported as 1476.4 and 6947.9 $mg.g^{-1}$, respectively, which represent a great superiority for water treatment (Ge et al. 2018). Hydroxyapatite nanostructures are synthesized by microwave irradiation from egg shells for effective removal of lead ions from wastewater samples (Safatian et al. 2019).

10.6 Future technologies: Localized surface plasmon resonance (LSPR) based optical fibre sensor approach for lead ions removal

Nanotechnology is one of the brilliant way out for removal of heavy metal ions from waste water. In practice, detection and elimination of heavy metal ions is equally important to achieve a healthy eco-system. Thus, remote, fast, real-time detection and removal of heavy metal has been the most discussed matter in research field. Localized surface plasmon resonance (LSPR) is one of the promising technology for optical sensing of heavy metals by using AuNPs.

Au NPs coated optical fibre (OF) design is a simple-to-fabricate, yet effective sensor probe which offers several advantages, including ease of remote sensing capabilities, a potentially high signal-to-noise ratio and resistance to electromagnetic interference. Therefore, several Au NPs coated optical fibre LSPR-based metal ion sensor designs have been explored in the previous literature (Klimant and Otto 1992, Lin and Chung 2008, Bharadwaj and Mukherji 2014, Lin and Chung 2009, Ho et al. 2012, Bhavsar et al. 2017, Raghunandhan 2016). A suitable approach of Pb(II) sensing has been demonstrated by Dhara et al. using the MUA-functionalized LSPR

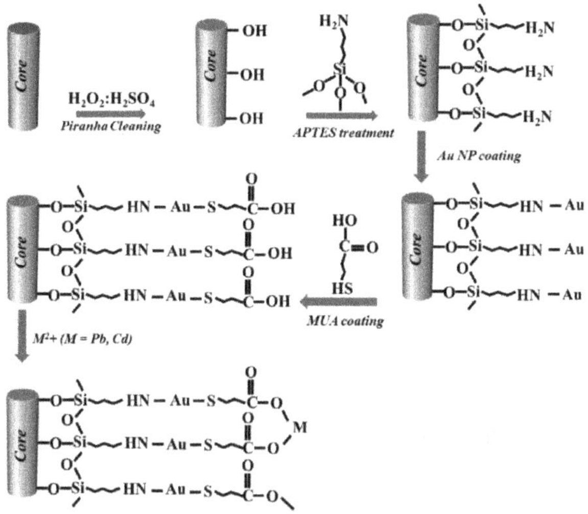

Fig. 3: Schematic of the process involved in optical fibre based sensor preparation. Reproduced with permission of (Dhara et al. 2019).

optical fibre probes (Dhara et al. 2019). The fabricated probe has been dipped in different, known concentrations of aqueous solutions of lead nitrate. Due to the binding of the Pb^{2+}-COOH (carboxyl) group, a change in amplitude as well as the surface plasmon resonance frequency is seen and it is an indicator of the change in the concentration of the Pb(II) or Cd(II) ions studied. The reversibility of the probe has been verified by removing the attached heavy metal ions from the OF, by dipping it into the Ethylenedinitrilotetraacetic acid (EDTA) chelator. The Au NP immobilized optical fibre area is made selective to Pb(II) ions by using MUA as a cross-linker. MUA has two terminal functional groups, one being the thiol group and another being the carboxyl group. The thiol group (one of the two terminal functional groups) in the MUA becomes attached to the Au NPs via a strong Au-S covalent bond, whereas its other terminal group stays unattached ('dangling'). This dangling carboxyl group, chemically bound with Pb(II) ions, thus makes the fibre selective to the heavy metal ions present in the aqueous solution. The process involved in the optical fiber based sensor fabrication has been represented in the Fig. 3.

The experimental set up that is arranged is shown in Fig. 4. The broadband tungsten light source has been used and it is connected to one end of 1×2 fibre coupler. The other two ends of the fibre coupler are connected to the spectrometer and the Au NP-coated fibre. Optical fibre end is reflection based due to the presence of the silver coating at the tip of the Au NP-coated fibre and the spectrometer can collect the reflected light. The real time absorbance spectrum is observed over the wavelength range from 400 to 900 nm. The heavy metal detection ability of the Au NP-only coated fibre has been determined by measuring its response in aqueous Lead Nitrate (LN) solutions of different concentrations (ranging from 0.67 mM to 30 mM). In calibrations carried out, the Pb(II) ion sensitivity is recorded as 0.28 nm/mM, having a very similar response for the Cd(II) ion. This study represents initial research carried out to establish the principle of creating a heavy metal sensor by using such AuNPs coated optical fibre design.

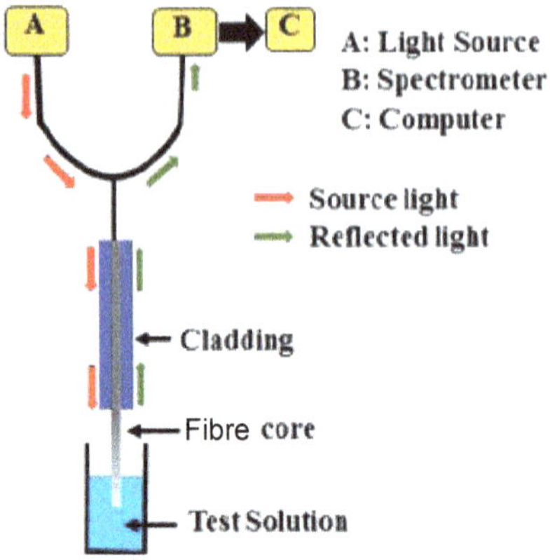

Fig. 4: Schematic of the experimental setup used for the characterization of the Au NP coated fibre. Reproduced with permission of (Dhara et al. 2019).

The world environment, and all the living elements around globe demand for clean and safe water. Pollutants are increasing simultaneously with the rapid increase in population, revolutionized industrialization, urbanization and extensive agriculture practices. There is a requirement for nanomaterial-based wastewater treatment to reuse the water. Optical fibre based sensor technology can be eco-friendly, cost intensive solution to detect the pollutants in aqueous environment and it should be treated urgently to save life.

10.7 Summary

In this chapter, an overview of recent advances in nanotechnology for removing lead from wastewater has been discussed. Across the world, different techniques are presently being used for the decontamination and purification of water but these techniques are not much effective due to its chemical toxicity, energetically and operationally intensiveness. It is established that nanostructure materials have unique properties, such as high surface to volume ratios, high sensitivity and reactivity, high adsorption capacity and ease of functionalization. These properties are helping nanomaterials to overcome the problems associated with traditional methods. Here, chemical precipitation, flotation, coagulation and flocculation, membrane filtration, cementation, adsorption, ion exchange, and electrochemical deposition are briefly described as heavy metal remediation techniques. Among these, adsorption is being focussed on because of its remarkable advantages, such as relatively low-cost process, less volume of chemical and biological sludge, low cost, high efficiency, regeneration of adsorbents and possibility of metal recovery. Mostly, cited nano-absorbents, such as carbon-based nanomaterials, silica-based nanomaterials, zero-valent metal-based nanomaterials, metal oxide-based nanomaterials, and nanocomposite nanomaterials have been discussed elaborately in this scope for lead ions' separation from aqueous environment and some of the results have been tabulated in Table 1. Finally, optical fibre based localized surface plasmon resonance (LSPR) method using gold nanoparticles is deliberated as a promising technology having remote, fast, real-time detection and removal of heavy metals from waste water.

Table 1: Lead metal removal using nano-adsorbents.

Adsorbent	**Adsorption capacity**	**Removal efficiency**	**References**
Activated carbon from grape	1.93 mmol.g^{-1}	-	Sardella et al. 2015
Moso and ma bamboo activated carbons	-	99.9%	Lo et al. 2012
SWCNTs-polysulfone nanocomposite	-	94.2 %	Gupta 2015
Ethylenediamine-grafted multi-walled carbon nanotubes	157.19 mg.g^{-1}	-	Hu et al. 2012
EDTA-RGO	479±46 *mg.g*$^{-1}$		Madadrang 2012
Calcium alginate (CA) beads embedded GO	602 *mg.g*$^{-1}$		Arshad et al. 2018
γ-Fe_2O_3–NH_4OH@SiO_2(APTMS)	96.79 mg. g^{-1}		Kegl et al. 2019
Non-metallic nano silica spheres/CaCO3	266.89 *mg.g*$^{-1}$		Manyangadze et al. 2020
Sil-Phy-NPANI and Sil-Phy-CrossNPANI	900 *mol.g*$^{-1}$ and 1450 *mol.g*$^{-1}$		Mahmoud et al.2016
nZVI		80%–100%	Moazeni et al. 2017
Gel-based membrane of BSA-capped AuNPs		95–106%	Lee et al. 2012
Organodisulfide polymer (PTMT) onto the amino-functionalized magnetite nanoparticles	533.13 mg.g^{-1}		Huang et al. 2018
Dumbbell-like manganese dioxide/gelatin	318.7 mg.g^{-1}		Wang et al. 2018
Rod-like shape ZnO		85%	Le et al. 2019
ZnO/MMT nanocomposite	88.50 mg.g^{-1}	97.2%	Sani et al. 2017
MgO	2614 mg.g^{-1}		Xiong et al. 2015

References

Abbas, M.N. and Alalwan, H.A. 2019. Catalytic oxidative and adsorptive desulfurization of heavy naphtha fraction. Korean Chem. Eng. Res. 57: 283–288.

Abhishek, K., Meeta, G. and Jain, V.K. 2015. A novel reusable nanocomposite for complete removal of dyes heavy metals and microbial load from water based on nanocellulose and silver nano-embedded pebbles. Environ. Technol. 36:706–714.

Abou-Shanab, R.A.I., Ji, M.K., Kim, H.C., Paeng, K.J. and Jeon, B.H. 2013. Microalgal species growing on piggery wastewater as a valuable candidate for nutrient removal and biodiesel production. J. Environ. Manage. 115: 257–264.

Adeleye, A.S., Conway, J.R., Garner, K., Huang, Y., Su, Y. and Keller, A.A. 2016. Engineered nanomaterials for water treatment and remediation: costs, benefits, and applicability. Chem. Eng. J. 286: 640–662.

Afshar, A., Sadjadi, S.A.S., Mollahosseini, A. and Eskandarian, M.R. 2016. Polypyrrole-polyaniline/Fe3O4 magnetic nanocomposite for the removal of Pb(II) from aqueous solution. Korean J. Chem. Eng. 33: 1–9.

Afzal, M.S., Ashraf, A. and Nabeel, M. 2018. Characterization of industrial effluents and groundwater of Hattar industrial estate, Haripur. Adv. Agric. Environ. Sci. 1: 70–77.

Agclidis, T., Fytiano, K., Vasilikiotis, G. and Jannakoudakis, D. 1988. Lead removal from wastewater by Cementation utilising a fixed Bed of Iron Spheres. Environ. Pollut. 50: 243–251.

Agrawal, A., Sahu, K.K. and Pandey, B.D. 2005. Systematic studies on adsorption of lead on sea nodule residues. J. Colloid Interface Sci. 281: 291–298.

Ahluwalia, S.S. and Goyal, D. 2007. Microbial and plant derived biomass for removal of heavy metals from wastewater. Bioresour Technol. 98: 2243–57.
Akbal, F. andCamcı, S. 2010. Comparison of electrocoagulation and chemical coagulation for heavy metal removal. Chem. Eng. Technol. 33: 1655–1664.
Alalwan, H.A., Abbas, M.N., Abudi, Z.N. and Alminshid, A.H. 2018. Adsorption of thallium ion (Tl^{+3}) from aqueous solutions by rice husk in a fixed-bed column: experiment and prediction of breakthrough curves. Environ. Technol. Innov. 12: 1–13.
Alalwan, H.A., Kadhom, M.A. and Alminshid, A.H. 2020a. Removal of heavy metals from wastewater using agricultural byproducts. J. Water Supply Res. T. 69: 99–112.
Alalwan, H.A., Abbas, M.N. and Alminshid, A.H. 2020b. Uptake of cyanide compounds from aqueous solutions by lemon peel with utilising the residue absorbent as rodenticide. Indian Chem. Eng. 62: 40–51.
Ali, H. and Khan, E. 2018.Trophic transfer, bioaccumulation and biomagnification of non-essential hazardous heavy metals and metalloids in food chains/webs: concepts and implications for wildlife and human health. Hum. Ecol. Risk Assess. 25: 1353–1376.
Ali, H., Khan, E. and Ilahi, I. 2019. Environmental chemistry and ecotoxicology of hazardous heavy metals: environmental persistence, toxicity and bioaccumulation. J. Chem. 2019: 6730305.
Alinnor, J. 2007. Adsorption of heavy metal ions from aqueous solution by fly ash. Fuel. 86: 853–857.
Alkesh, I. Shah. 2017. Heavy metal impact on aquatic life and human health – An overview. Proc. International Association for Impact Assessment. Canada.
Alswata, A.A., Ahmad, M.B., Al-Hada, N.M., Kamari, H.M., Hussein, M.Z.B. and Ibrahim, N.A. 2017. Preparation of Zeolite/Zinc Oxide Nanocomposites for toxic metals removal from water. Results Phys.7: 723–731.
Amin, M.T., Alazba, A.A. and Manzoor, U. 2014. A review of removal of pollutants from water/wastewater using different types of nanomaterials. Adv. Mater. Sci. Eng. 825910.
Anirudhan, T.S. and Sreekumari, S.S. 2011.Adsorptive removal of heavy metal ions from industrial effluents using activated carbon derived from waste coconut button. J. Environ. Sci. 23: 1989–1998.
Anitha. K., Namsani, S. and Singh, J.K. 2015. Removal of heavy metal ions using a functionalized single-walled carbon nanotube: a molecular dynamics study. J. Phys. Chem. A. 119: 8349–8358.
Anjum, M., Miandad, R., Waqas, M., Gehany, F. and Barakat, M.A. 2019. Remediation of wastewater using various nanomaterials. Arab. J. Chem. 12: 4897–4919.
Annadhasan, M., Muthukumarasamyvel, T., Babu, V.R.S. and Rajendiran, N. 2014.Green Synthesized Silver and Gold Nanoparticles for Colorimetric Detection of Hg^{2+}, Pb^{2+}, and Mn^{2+} in Aqueous Medium. ACS Sustain. Chem. Eng. 2: 887– 896.
Arshad, F., Selvaraj, M., Zain, J., Banat, F. and Haija, M.A. 2018. Polyethylenimine modified graphene oxide hydrogel composite as an efficient adsorbent for heavy metal ions. Sep. Purif. Technol. 209: 870–880.
Atieh, M.A., Bakather, O.Y., Al-Tawbini, B., Bukhari, A.A., Abuilaiwi, F.A. and Fettouhi, M.B. 2010. Effect of carboxylic functional group functionalized on carbon nanotubes surface on the removal of lead from water. Bioinorg. Chem. Appl. 1: 1–9.
ATSDR. 2020. Toxicological Profile for Lead USA.
Attatsi., I.K. and Nsiah, F. 2020. Application of silver nanoparticles toward Co(II) and Pb(II) ions contaminant removal in groundwater. Appl. Water Sci. 10: 152.
Azizi, S.M., Shahri, M. and Mohamad, R. 2017.Green synthesis of zinc oxide nanoparticles for enhanced adsorption of lead ions from aqueous solutions: equilibrium, kinetic and thermodynamic studies. Molecules. 22: 831.
Baby, R., Saifullah, B. and Hussein, M.Z. 2019. Carbon nanomaterials for the treatment of heavy metal-contaminated water and environmental remediation. Nanoscale Res. Lett. 14: 341.
Badawy, N.A., El-Bayaa, A.A., Abdel-Aal, A.Y. and Garamon, S.E. 2009. Chromatographic separations and recovery of lead ions from a synthetic binary mixture of some heavy metal using cation exchange resin. J. Hazard. Mater. 166: 1266–1271.
Badruddozaa, A.Z.M., Zakir Shawona, Z.B., Jin Daniela, T.W., Hidajata, K. and Uddinb, M.S. 2013. Fe_3O_4/cyclodextrin polymer nanocomposites for selective heavy metals removal from industrial wastewater, Carbohydr. Polym. 91: 322– 332.
Bankole, M.T., Abdulkareem, A.S., Mohammed, I.A., Ochigbo, S.S., Tijani, J.O., Abubakre, O.K. et al. 2019. Selected heavy metals removal from Electroplating wastewater by purified and Polyhydroxylbutyrate functionalized Carbon Nanotubes adsorbents. Sci. Rep. 9: 4475.

Bao, J., Fu, Y. and Bao, Z. 2013. Thiol-functionalized magnetite/graphene oxide hybrid as a reusable adsorbent for Hg^{2+} removal. Nanoscale Res. Lett. 8: 486.

Barakat, M.A. 2011. New trends in removing heavy metals from industrial wastewater. Arab. J. Chem. 4: 361–377.

Belkacem, M., Khodir, M. and Abdelkrim, S. 2008.Treatment characteristics of textile wastewater and removal of heavy metals using the electroflotation technique. Desalination. 228: 245–254.

Berber, M.R. 2020.Current advances of polymer composites for water treatment and desalination. J. Chem. 2020: 1–19.

Beyersmann, D. and A. Hartwig. 2008. Carcinogenic metal compounds: recent insight into molecular and cellular mechanisms. Arch. Toxicol. 82: 493–512.

Bharadwaj, R. and Mukherji, S. 2014. Gold nanoparticle coated U-bend fibre optic probe for localized surface plasmon resonance-based detection of explosive vapours. Sens. Actuators B Chem. 192: 804–811.

Bhavsar, K., Hurston, E., Prabhu, R. and Joseph, G.P. 2017. Fibre optic sensor to detect heavy metal pollutants in water environments. OCEANS 2017 IEEE: 1–4.

Bina, B., M. Amin, A. Rashidi and H. Pourzamani. 2012. Benzene and toluene removal by carbon nanotubes from aqueous solution. Arch. Environ. Prot. 38: 3–25.

Biney,C., A. Amuzu, D. Calamari, N. Kaba and I. Mbome. 1994. Review of heavy metals in the African aquatic environment. Ecotoxicol. Environ. Saf. 28: 134–159.

BIS. 1999. Tolerance Limits for Industrial Effluents Discharged into Inland Surface Waters. Technical Report IS 2490 (Parts 1), Bureau of Indian Standards, Manak Bhawan, New Delhi.

Bohli, T., Ouederni, A., Fiol, N. and Villaescusa, I. 2015. Evaluation of an activated carbon from olive stones used as an adsorbent for heavy metal removal from aqueous phases. C.R. Chimie.18: 88–99

Cai, L.M., Xu, Z.C., Qi, J.Y., Feng, Z.Z. and Xiang, T.S. 2015. Assessment of exposure to heavy metals and health risks among residents near Tonglushan mine in Hubei, China. Chemosphere. 127: 127–135.

Cai, Y., Li, C., Dan, W., Wei, W., Tan, F., Wang, X. et al. 2016. Highly active MgO nanoparticles for simultaneous bacterial inactivation and heavy metal removal from aqueous solution. Chem. Eng. J. 312: 158–166.

ÇAvu¸s, S. and GuRdag, G.L. 2016. Noncompetitive removal of heavy metal ions from aqueous solutions by Poly[2-(acrylamido)-2-methyl-1-propanesulfonic acid-co-itaconic acid Hydrogel. Ind. Eng. Chem. Res. 48: 2652–2658.

Chang, Q., Zhang, M. andWang, J. 2009. Removal of Cu^{2+} and turbidity from wastewater by mercaptoacetyl chitosan. J. Hazard. Mater. 169: 621–625.

Chang, Q. and Wang, G. 2007. Study on the macromolecular coagulant PEX which traps heavy metals. Chem. Eng. Sci. 62: 4636–4643.

Chatterjee, S., Lee, M.W. and Woo, S.H. 2010. Adsorption of congo red by chitosan hydrogel beads impregnated with carbon nanotubes. Bioresour. Technol. 101: 1800–1806.

Chatterjee, S., Chatterjee, T., Lim, S.R. and Woo, S.H. 2011. Effect of the addition mode of carbon nanotubes for the production of chitosan hydrogel core–shell beads on adsorption of Congo red from aqueous solution. Bioresour. Technol. 102: 4402–4409.

Chaturvedi, S., Dave, P.N. and Shah, N.K. 2012. Applications of nanocatalyst in new era. J. Saudi Chem. Soc. 16: 307–325.

Chen, B., Chen, S., Zhao, H., Liu, Y., Long, F. and Pan, X. 2019. A versatile -cyclodextrin and polyethyleneimine bi-functionalized magnetic nanoadsorbent for simultaneous capture of methyl orange and Pb(II) from complex wastewater. Chemosphere. 216: 605–616.

Chen, Q., Luo, Z., Hills, C., Xue, G. and Tyrer, M. 2009. Precipitation of heavy metals from wastewater using simulated flue gas: sequent additions of fly ash, lime and carbon dioxide. Water Res. 43: 2605–2614.

Cheng, Z., KuanTan, A.L., Tao, Y., Shan, D., Ting, K.E. and Yin, X.J. 2012.Synthesis and characterization of iron Oxide nanoparticles and applications in the removal of heavy Metals from industrial wastewater. Int. J. Photoenergy, 2012: 1–5.

Chu, H., Wei, L., Cui, R., Wang, J. and Li, Y. 2010.Carbon nanotubes combined with inorganic nanomaterials: preparations and applications. Coord. Chem. Rev. 254: 1117–1120.

Ciotta, E., Prosposito, P., Moscone, D., Colozza, N. and Pizzoferrato, R. 2019. Detection and removal of heavy-metal ions in water by unfolded-fullerene nanoparticles, AIP Conference Proceedings. 2145: 020008.

Crini, G. 2005. Recent developments in polysaccharide-based materials used as adsorbents in wastewater treatment. Prog. Polym. Sci. 30: 38–70.

Dave, P.N. and Lakhan, V.C. 2014.Application of Iron Oxide nanomaterials for the removal of heavy metals. J. Nanotechnol. 2014: 398569.

Dhara, P., Kumar, R., Binetti, L., Nguyen, H.T., Alwis, L.S., Sun, T. et al. 2019. Optical fiber-based heavy metal detection using the Localized Surface Plasmon Resonance technique. IEEE Sens. 19: 8720–8723.

Dialynas, E. and Diamadopoulos, E. 2009. Integration of a membrane bioreactor coupled with reverse osmosis for advanced treatment of municipal wastewater. Desalination. 238: 302–311.

Dorfner, K. 1991. Ion Exchangers. Walter de Gruyter.

Downtoearth. 2020. Heavy metals emerging as potential threat to public health on urban beaches study 73863. India Science Wire.

El-Sheikh, A.H., Al-Degs, Y.S., Al-As'Ad, R.M. and Sweileh, J.A. 2011. Effect of oxidation and geometrical dimensions of carbon nanotubes on Hg(II) sorption and preconcentration from real waters. Desalination. 270: 214–220.

Ercal, N., Gurer-Orhan, H. and Aykin-Burns, N. 2001.Toxic metals and oxidative stress Part I: mechanisms involved in metal-induced oxidative damage. Curr. Top. Med. Chem. 1: 529–539.

Fan, L., Luo, C., Li, X., Lu, F., Qiu, H. and Sun, M. 2012. Fabrication of novel magnetic chitosan grafted with graphene oxide to enhance adsorption properties for methyl blue. J. Hazard. Mater. 15: 272–279.

Fan, Y., Liu, Z., Wang, L.E. and Zhan, J. 2009. Synthesis of Starch-Stabilized Ag Nanoparticles and Hg^{2+}recognition in aqueous media. Nano Scale Res. Lett. 4: 1230–1235.

FAO. 1992. Committee for Inland Fisheries of Africa Report of the Third Session of the Working Party on Pollution and Fisheries. Accra, Ghana.

Farhan, A.M., Salem, N.M., Al-Dujaili, A.H. and Awwad, A.M. 2012. Biosorption Studies of Cr(VI) Ions from electroplating wastewater by walnut shell powder. Am. J. Environ. Eng. 2: 188–195.

Feng, D., Van Deventer, J.S.J. and Aldrich, C. 2004. Removal of pollutants from acid mine wastewater using metallurgical byproduct slags. Sep. Purif. Technol. 40: 61–67.

Fernando, M.S., Silva, R.M.D. and Silva, K.M.N.D. 2015. Synthesis, characterization, and application of nano hydroxyapatite and nanocomposite of hydroxyapatite with granular activated carbon for the removal of Pb^{2+} from aqueous solutions. Appl. Surf. Sci. 351: 95–103.

Fifield, F.W. and Haines, P.J. 2000. Environmental analytical chemistry. Wiley-Blackwell, USA.

Flores-Cano, J.V., Leyva-Ramos, R., Padilla-Ortega, E. and Mendoza-Barron, J. 2013.Adsorption of heavy metals on diatomite: mechanism and effect of operating variables. Adsorpt. Sci. Technol. 31: 275–291.

Freundlich, H. 1906. Über die adsorption in lösungen. Z. Physik. Chemie (Leipzig). 57A: 385–470.

Fu, F. and Wang. 2011. Removal of heavy metal ions from wastewaters: a review. Journal of Environmental Management. 92407–418.

Ge, F., Li, M.M., Ye, H. and Zhao, B.X. 2012. Effective removal of heavy metal ions Cd^{+2}, Zn^{+2}, Pb^{+2}, Cu^{+2} from aqueous solution by polymer-modified magnetic nanoparticles. J. Hazard. Mater. 211: 366–372.

Ge, L., Wang, W., Peng, Z., Tan, F., Wang, X., Chen, J. et al. 2018. Facile fabrication of Fe@MgO magnetic nanocomposites for efficient removal of heavy metal ion and dye from water. Powder Technol. 326: 393–401.

Gholami, A., Moghadassi, A.R., Hosseini, S.M., Shabani, S. and Gholami, F. 2014.Preparation and characterization of polyvinyl chloride-based nanocomposite nanofiltration-membrane modified by iron oxide nanoparticles for lead removal from water. J. Ind. Eng. Chem. 20: 1517–1522.

Ghorbani, M., Seyedin, O. and Aghamohammadhassan, M. 2020. Adsorptive removal of lead (II) ion from water and wastewater media using carbon-based nanomaterials as unique sorbents: A review. J. Environ. Manage. 254: 109814.

Giraldo, L., Erto, A. and Moreno-Piraján, J.C. 2013. Magnetite nanoparticles for removal of heavy metals from aqueous solutions: Synthesis and characterization. Adsorption. 19: 465–474.

Gode, F. and Pehlivan, E. 2006. Removal of chromium (III) from aqueous solutions using Lewatit S100: the effect of pH, time, metal concentration and temperature. J. Hazard. Mater. 136: 330–337.

Goel, J., Kadirvelu, K., Rajagopal, C. and Kumar Garg, V. 2005. Removal of lead (II) by adsorption using treated granular activated carbon: batch and column studies. J. Hazard Mater. 125: 211–20.

Gordon-Nuñez, F., Vaca-Escobar, K., Villacís-García, M., Fernández, L., Debut, A. et al. 2019. Applicability of goethite/reduced graphene oxide nanocomposites to remove lead from wastewater. Nanomaterials. 9: 1580.

Gonzalez-Serrano, E., Cordero, T., Rodriguez-Mirasol, J., Cotoruelo, L. and Rodriguez, J.J. 2004. Removal of water pollutants with activated carbons prepared from H3PO4 activation of lignin from kraft black liquors. Water Res. 38: 3043–3050.

Gupta, S. DBCNM. 2015. Metal removal studies by composite membrane of polysulfone and functionalized single-walled carbon nanotubes. Sep. Sci. Technol. 50: 9.

Gupta, V.K. and Rastogi, A. 2008. Biosorption of lead(II) from aqueous solutions by non-living algal biomass *Oedogonium* sp. and *Nostoc* sp.: a comparative study. Colloids Surf. B: Biointerfaces. 64: 170– 178.

Gupta, V.K., Agarwal, S. and Saleh, T.A. 2011. Synthesis and characterization of alumina-coated carbon nanotubes and their application for lead removal. J. Hazard. Mater. 185: 17–23.

Gupta, V.K., Tyagi, I., Sadegh, H., Shahryari-Ghoshekand, R., Makhlouf, A.S.H. and Maazinejad, B. 2015. Nanoparticles as adsorbent;a positive approach for removal of noxious metal ions: a review. Sci. Technol. Dev. 34: 195.

Gupta, V.K., Tyagi, I., Sadegh, H., Shahryari-Ghoshekand, R., Makhlouf, A.S.H. and Maazinejad, B. 2015. Nanoparticles as adsorbent; a positive approach for removal of noxious metal ions: a review. Sci. Technol. Dev. 34: 195.

Hamza, I.A.A., Martincigh, B.S., Ngila, J.C. and Nyamori, V.O. 2013. Adsorption studies of aqueous Pb(II) onto sugarcane bagasse/multi-walled carbon nanotube composite. Phys. Chem. Earth. 66: 157–166.

Ho, T.C., Jafri, M.Z.M., San, L.H. and Chong, M.Y. 2012. Detection of heavy metal ions in aqueous solution using fiber optic sensor. International Conference on Computer and Communication Engineering (ICCCE), IEEE: 267–270.

Ho, Y.S., Porter, J.F. and Mckay, G. 2002. Equilibrium Isotherm Studies for the Sorption of Divalent Metal Ions onto Peat: Copper, Nickel and Lead Single Component Systems. Water Air Soil Pollut. 141: 1–33.

Holmberg, J.P. 2006.Competitive adsorption and displacement behavior of heavy metals on peat. Dissertation Chalmes University of Technology, Gothenburg, Sweden.

Hosseini, S.J., Aghaie, H., Ghaedi, M., Dashtian, K. and Purkaitc, M.K. 2016. Lead (II) Adsorption from Aqueous Solutions onto Modified Ag Nanoparticles: Modeling and Optimization. Environ. Prog. Sustain. Energy. 35: 3.

Hotze, E.M., Phenrat, T. and Lowry, G.V. 2010. Nanoparticle aggregation: Challenges to understanding transport and reactivity in the environment. J. Environ. Qual. 39:1909–1924.

Hu, Z.J., Cui, Y., Liu, S., Yuan, Y. and Gao, H.W. 2012. Optimization of ethylenediamine-grafted multi-walled carbon nanotubes for solid-phase extraction of lead cations. Environ. Sci. Pollut. Res. 19: 1237–1244.

Huang, D.L., Chen, G.M., Zeng, G.M., Xu, P., Yan, M., Lai, C. et al. 2015, Synthesis and application of modified zero-valent iron nanoparticles for removal of hexavalent chromium from wastewater. Water Air Soil Pollut. 226: 375.

Huang, P., Ye, Z., Xie, W., Chen, Q., Li, J., Xu, Z. et al. 2013. Rapid magnetic removal of aqueous heavy metals and their relevant mechanisms using nanoscale zero valent iron (nZVI) particles. Water Res. 47: 4050–4058.

Huang, R., Wu, M., Zhang, T., Li, D., Tang, P. and Feng, Y. 2017. Template-free synthesis of large-pore-size porous magnesium silicate hierarchical nanostructures for high efficiency removal of heavy metal ions. ACS Sustain. Chem. Eng. 5: 2774–2780.

Huang, X., Yang, J., Wang, J., Bi, J., Xie, C. and Hao, H. 2018. Design and synthesis of core–shell Fe_3O_4 @PTMT composite magnetic microspheres for adsorption of heavy metals from high salinity wastewater. Chemosphere. 206: 513.

Hube, S., Eskafi, M., Hrafnkelsdóttir, K.F., Bjarnadóttir, B., Bjarnadóttir, M.A., Axelsdóttir, S. and Wu, B. 2020. Direct membrane filtration for wastewater treatment and resource recovery: A review. Sci. Total Environ. 710: 136375.

Hussain, M., Syed, Q., Bashir, R. and Adnan, A. 2021. Electrochemical process for simultaneous removal of chemical and biological contaminants from drinking water. Environ. Sci. Pollut. Res.

Igwe, J.C., Ogunewe, D.N. and Abia, A.A. 2005. Competitive adsorption of Zn(II), Cd(II) and Pb(II) ions from aqueous and non-aqueous solution by maize cob and husk. Afr. J. Biotechnol. 4: 1113–1116.
Jabbari, V., Veleta, J., Zarei-Chaleshtori, M., Gardea-Torresdey, J. and Villagrán, D. 2016. Green synthesis of magnetic MOF@ GO and MOF@ CNT hybrid nanocomposites with high adsorption capacity towards organic pollutants. Chem. Eng. J. 304: 774–783.
Jayaweera, H.D.A.C., Siriwardane, I., Silva, K.M.N.D. and Silva, R.M.D. 2018. Synthesis of multifunctional activated carbon nanocomposite comprising biocompatible flake nano hydroxyapatite and natural turmeric extract for the removal of bacteria and lead ions from aqueous solution. Chem. Central J. 2018: 12–18.
Kabbashi, N.A., Atieh, M.A., Al-Mamun, A., Mirghami, M.E.S., Alam, M.D.Z. and Yahy, N. 2009.Kinetic adsorption of application of carbon nanotubes for Pb(II) removal from aqueous solution. J. Environ. Sci. 21: 539–544.
Kang, A.J., Baghdadi, M. andPardakhti, A. 2015. Removal of cadmium and lead from aqueous solutions by magnetic acid-treated activated carbon nanocomposite. Desalinat. Water Treat. 3994: 1–17.
Kasprzak, K.S. 2002. Oxidative DNA and protein damage in metal-induced toxicity and carcinogenesis. Free Radic. Biol. Med. 32: 958–967.
Kefeni, K.K., Msagati, T.A.M., Nkambule, T.T.I. and Mamba, B.B. 2018.Synthesis and application of hematite nanoparticles for acid mine drainage treatment. J. Environ. Chem. Eng. 6:1865–1874.
Kegl, T., Ban, I., Lobnik, A. and Kosak, A. 2019. Synthesis and characterization of novel γ-Fe2O3-NH4OH@SiO2(APTMS) nanoparticles for dysprosium adsorption. J. Hazard Mater. 378.
Kegl, T., Kosak, A., Lobnik, A., Novaka, Z., Kova Kralj, A. and Ban, I. 2020. Adsorption of rare earth metals from wastewater by nanomaterials: a review. J. Hazard Mater. 386.
Khanna, M., Mathur, A., Dubey, A.K., McLaughlin, J., Moirangthem, I., Wadhwa, S. et al. 2020. Rapid removal of lead(II) ions from water using iron oxide–tea waste nanocomposite –a kinetic study. IET Nanobiotechnol.14: 275–280.
Kim, Y., Johnson, R.C. and Hupp, J.T. 2001.Gold nanoparticle-based sensing of "spectroscopically silent" heavy metal ions. Nano Lett. 1: 165–167.
Klimant, I. and Otto, M. 1992. A fiber optical sensor for heavy metal ions based on immobilized xylenol orange. Microchimica Acta. 108: 11–17.
Kong, N., Huang, X., Cui, L. and Liu, J. 2013. Surface modified graphene for heavy metal ions adsorption. Sci. Adv. Mater. 5: 1083–1089.
Kosa, S.A., Al-Zhrani, G. and Salam, M.A. 2012. Removal of heavy metals from aqueous solutions by multi-walled carbon nanotubes modified with 8-hydroxyquinoline. Chem. Eng. J. 181: 159–168.
Kroto, H.W., Heath, J.R., O'Brien, S.C., Curl, R.F. and Smalley, R.E. 1985. C60: Buckminsterfullerene. Nature. 318: 162.
Kumar, K.Y., Muralidhara, H.B., Nayaka, Y.A., Balasubramanyam, J. and Hanumanthappa, H. 2013a. Low-cost synthesis of metal oxide nanoparticles and their application in adsorption of commercial dye and heavy metal ion in aqueous solution. Powder Technol. 246: 125–136.
Kumar, K.Y., Muralidhara, H.B., Nayaka, Y.A., Balasubramanyam, J. andHanumanthappa, H. 2013b. Hierarchically assembled mesoporous ZnO nanorods for the removal of lead and cadmium by using differential pulse anodic stripping voltammetric method. Powder Technol. 239: 208–216.
Kumar, M., Chung, J.S. and Hur, S.H. 2019. Graphene composites for lead ions removal from aqueous solutions. Appl. Sci. 9: 2925.
Kumar, R., Khan, M.A. and Haq, N. 2014. Application of carbon nanotubes in heavy metals remediation. Crit. Rev. Environ. Sci. Technol. 44: 1000–1035.
Kyzas, G.Z. and Matis, K.A. 2015. Nanoadsorbents for pollutants removal: a review. J. Mol. Liq. 203: 159–168.
Langmuir, I. 1916. The constitution and fundamental properties of solids and liquids. Part I. Solids. J. Am. Chem. Soc. 38: 2221–2295.
Lalmi, A. and Bouhidel, K.E. 2018. Removal of lead from polluted waters using ion exchange resin with Ca(NO3)2 for elution. Hydrometallurgy. 178: 287–293.
Le, A.T., Pung, S.Y., Sreekantan, S., Matsuda, A. and Huynh, D.P. 2019. Mechanisms of removal of heavy metal ions by ZnO particles. Heliyon. 5: e01440.
Lee, L.Z., Zaini, M.A.A. and Tang, S.H. 2019. Porous nanomaterials for heavy metal removal. *In*: Martínez, L., Kharissova, O. and Kharisov, B. (Eds.). Handbook of Ecomaterials. Springer, Cham.

Lee, Y.F., Nan, F.H., Chen, M.J., Wu, H.Y., Ho, C.W., Chen, Y.Y. et al. 2012. Detection and removal of mercury and lead ions by using gold nanoparticle-based gel membrane. Anal. Methods. 4: 1709–1717.

Li, Y.H., Wang, S., Luan, Z., Ding, J., Xu, C. and Wu, D. 2003. Adsorption of cadmium (II) from aqueous solution by surface oxidized carbon nanotubes. Carbon. 41: 1057–1062.

Liang, B., Lehmann, J., Solomon, D., Kinyangi, J., Grossman, J., O'Neill, B. et al. 2006. Black carbon increases cation exchange capacity in soils. Soil Sci. Soc. Am. J. 70: 1719–1730.

Lin, S., Lai, S. and Leu, H. 2000. Removal of heavy metals from aqueous solution by chelating resin in a multistage adsorption process. J. Hazard. Mater. 76:139–153.

Lin, T.J. and Chung, M.F. 2008. Using monoclonal antibody to determine lead ions with a localized surface plasmon resonance fiber-optic biosensor. Sensors. 8: 582–593.

Lin, T.J. and Chung, M.F. 2009. Detection of cadmium by a fiber-optic biosensor based on localized surface plasmon resonance. Biosens. Bioelectron. 24: 1213–1218.

Lin, Y., Huang, C. and Chang, H. 2011. Gold nanoparticle probes for the detection of mercury, lead and copper ions. Analyst. 136: 863–871.

Liu, C., Wu, T., Hsu, P.C., Xie, J., Zhao, J., Liu, K. et al. 2019. Direct/alternating current electrochemical method for removing and recovering heavy metal from water using graphene oxide electrode. ACS Nano. 13: 6431–6437.

Lo, S.F., Wang, S.Y., Tsai, M.J. and Lin, L.D. 2012. Adsorption capacity and removal efficiency of heavy metal ions by Moso and Ma bamboo activated carbons. Chem. Eng. Res. Des. 90: 1397–1406.

Lu. H, Zhang, W., Yang, Y., Huang, X. and Qiu, R. 2011. Relative distribution of Pb^{2+} sorption mechanisms by sludge-derived biochar. Water Res. 46 :854–862.

Lubick, N. and Betts, K. 2008. Silver socks have cloudy lining| Court bans widely used flame retardant. Environ. Sci. Technol. 42: 3910–3910.

Lucenab, R., Simonets, M. andValcárcel, C.M. 2011. Potential of nanoparticles in sample preparation. J Chromatogr A. 28: 620–637.

Luo, T., Cui, J., Hu, S., Huang, Y. and Jing, C. 2010. Arsenic removal and recovery from copper smelting wastewater using TiO2.Environ. Sci. Technol. 44: 9094–9098.

Madadrang, C.J., Kim, H.Y., Gao, G., Wang, N., Zhu, J., Feng, H. et al. 2012. Adsorption behavior of EDTA-graphene oxide for Pb(II) removal. ACS Appl. Mater. Interfaces. 4: 1186–1193.

Madrakian, T., Afkhami, A., Zadpour, B. and Ahmadi, M. 2015. New synthetic mercaptoethylamino homopolymer-modified maghemite nanoparticles for effective removal of some heavy metal ions from aqueous solution. J. Ind. Eng. Chem. 21: 1160–1166.

Mahdavi, S., Jalali, M. and Afkhami, A. 2013. Heavy metals removal from aqueous solutions using TiO_2, MgO, and Al_2O_3 nanoparticles. Chem. Eng. Commun. 200: 448–470.

Mahmood, Q., Wang, J., Pervez, A., Meryem, S.S., Waseem, M. and Ullah, Z. 2015. Health risk assessment and oxidative stress in workers exposed to welding fumes. Toxicol. Environ. Chem. 97: 634–639.

Mahmoud, M.E., Fekry, N.A. and El-Latif, M.M.A. 2016. Nanocomposites of nanosilica-immobilized-nanopolyaniline and crosslinked nanopolyaniline for removal of heavy metals. Chem. Eng. J. 304: 679–691.

Makhloufi, L., Saidani, B. and Hammache, H. 2000. Removal of lead ions from acidic aqueous Solutions by cementation on iron. Wat. Res. 34: 2517–2524.

Manawia,Y., Gordon, M., Ismailc, N., Farda, A.K., Kochkodana, V. and Atieha, M.A. 2018. Enhancing lead removal from water by complex-assisted filtration with acacia gum. Chem. Eng. J. 352: 828–836.

Manyangadze, M., Nyaradzai, M.H.C., Narsaiah, T.B., Chakra, C.S., Charis, G., Danha, G. et al. 2020. Adsorption of lead ions from wastewater using nano silica spheres synthesized on calcium carbonate templates. Heliyon. 6: e05309.

Matouq, M., Jildeh, N., Qtaishat, M., Hindiyeh, M. and Syouf, M.Q.A. 2015. The adsorption kinetics and modeling for heavy metals removal from wastewater by Moringa pods. J. Environ. Chem. Eng. 3: 775–784.

Mehdipour, S., Vatanpour, V. and Kariminia, H.R. 2015. Influence of ion interaction on lead removal by a polyamide nanofiltration membrane. Desalination. 362: 84–92.

Meng, H., Xue, M., Xia, T., Zhao, Y.L., Tamanoi, F., Stoddart, J.F. et al. 2010. Autonomous *in vitro* anticancer drug release from meso-porous silica nanoparticles by pH-sensitive nanovalves. J. Am. Chem. Soc.132: 12690–12697.

Moazeni, M., Ebrahimi, A., Rafiei, N. and Pourzamani, H.R. 2017. Removal of lead ions from aqueous solution by Nano Zero-Valent Iron (nZVI). Health Scope. 6: e40240.

Mohammad, A.M., Eldin, T.A.S., Hassan, M.A. and El-Anadouli, B.E. 2017. Efficient treatment of lead-containing wastewater by hydroxyapatite/chitosan nanostructures. Arab. J. Chem. 10: 683–690.

Mohan, S., Kumar, V., Singh, D.K. and Hasan, S.H. 2017. Effective removal of lead ions using graphene oxide-MgO nanohybrid from aqueous solution: Isotherm, kinetic and thermodynamic modeling of adsorption. J. Environ. Chem. Eng. 5: 2259–2273.

Mohan, V.B., Lau, K., Hui, D. and Bhattacharyya, D. 2018.Graphene-based materials and their composites: a review on production, applications and product limitations. Compos Part B. 142: 200–220.

Morris, T., Copeland, H., Mclinden, E., Wilson, S. and Szulczewski, G. 2002. The effects of mercury adsorption on the optical response of size-selected gold and silver nanoparticles. Langmuir. 18: 7261–7264.

Nassar, N.N. 2010. Rapid removal and recovery of Pb (II) from wastewater by magnetic nanoadsorbents. J. Hazard. Mater. 184: 538–546.

Nica, D.V., Bura, M., Gergen, I., Harmanescu, M. and D.M. Bordean. 2012. Bioaccumulative and conchological assessmentof heavy metal transfer in a soil-plant-snail food chain. Chem. Cent. J. 6: 55.

Nitti, F. 2014. Synthesis of gold nanoparticles and their application for detection and removal of water contaminants. Media Sains. 13, No. 2 Edisi Desember: ISSN 1829-751X.

Noh, K.C., Nam, Y.S., Lee, H.J. and Lee, K.B. 2015. A colorimetric probe to determine Pb^{2+} using functionalized silver nanoparticles. Analyst. 140: 8209–8216.

Nosier, S.A. and Sallam, S.A. 2000. Removal of lead ions from wastewater by cementation on a gas-sparged zinc cylinder. Sep. Purif. Technol. 18: 93–101.

O'Carroll, D., Sleep, B., Krol, M., Boparai, H. and Kocur, C. 2013. Nanoscale zero valent iron and bimetallic particles for contaminated site remediation. Adv. Water Resour. 51: 104–122

Omar, H.E.D.M. 2013. Seasonal variation of heavy metals accumulation in muscles of the African Catfish Clarias gariepinus and in River Nile water and sediments at Assiut Governorate, Egypt. Journal of Biology and Earth Sciences. 3: 236–248.

Özacar, M., Şengil, İA. and Türkmenler, H. 2008. Equilibrium and kinetic data, and adsorption mechanism for adsorption of lead onto valonia tannin resin. Chem. Eng. J. 143: 32–42.

Pan, B.C., Zhang, Q.R., Zhang, W.M., Pana, B.J., Dua, W., Lvb, L. et al. 2007. Highly effective removal of heavy metals by polymer-based zirconium phosphate: a case study of lead ion. J. Colloid Interface Sci. 310: 99–105.

Pang, F.M., Kumar, P., Teng, T.T., Omar, A.M. and Wasewar, K.L. 2011. Removal of lead, zinc and iron by coagulation-flocculation. J. Taiwan Inst. Chem. Eng. 42: 809–815.

Peng, W., Li, H., Liu, Y. and Song, S. 2017. A review on heavy metal ions adsorption from water by graphene oxide and its composites. J. Mol. Liq. 230: 496–504.

Peng, W., Han, G., Cao, Y., Sun, K. and Song, S. 2018. Efficiently Removing Pb(II) from wastewater by graphene oxide using foam flotation. colloids and surfaces a: physicochem. Eng. Aspects. 556: 266–272.

Peng, X., Luan, Z., Di, Z., Zhang, Z. and Zhu, C. 2005. Carbon nanotubes-iron oxides magnetic composites as adsorbent for removal of Pb(II) and Cu(II) from water. Carbon. 43: 880–883.

Prosposito, P., Burratti, L. and Venditti, I. 2020. Silver Nanoparticles as Colorimetric Sensors for Water Pollutants.Chemosensors. 8: 26.

Qian, H., Pretzer, L.A., Velazquez, J.C., Zhao, Z. and Wong, M.S. 2013. Gold nanoparticles for cleaning contaminated water. J. Chem. Technol. Biotechnol. 88: 735–741.

Raghunandhan, R., Chen, L.H., Long, H.Y., Leam, L.L., So, P.L., Ning, X. et al .2016. Chitosan/PAA based fiber-optic interferometric sensor for heavy metal ions detection. Sens. Actuators B Chem. 233: 31–38.

Rahman, M.S., Molla, A.H., Saha, N. and Rahman, A. 2012. Study on heavy metals levels and its risk assessment in some edible fishes from Bangshi River, Savar, Dhaka, Bangladesh. Food Chem. 134: 1847–1854.

Rajeswari, T.R. and Namburu, S. 2014. Impact of heavy metals on environmental pollution. J. Chem. Pharm. Sci. ISSN: 0974-2115. 3: 175.

Rajput, S., Singh, L.P. and Jr., D.Mohan, C.U.P. 2017. Lead (Pb^{2+}) and copper (Cu^{2+}) remediation from water using Superparamagnetic maghemite (Fe_2O_3) nanoparticles synthesized by Flame Spray Pyrolysis (FSP). J. Colloid Interface Sci. 492: 176–190.

Repo, E., Warchol, J.K., Kurniawan, T.A. and Sillanpää, M.E.T. 2010. Adsorption of Co(II) and Ni(II) by EDTA- and/or DTPA-modified chitosan: Kinetic and equilibrium modelling. Chem. Eng. J. 161: 73–82.

Repo, E., Warchoł, J.K., Bhatnagar, A. andSillanpää, M. 2011. Heavy metals adsorption by novel EDTA modified chitosan–silica hybrid materials. J. Colloid Interface Sci. 358: 261–267.

Rodríguez, A., Ovejero, G., Sotelo, J.L., Mestanza, M. and Garcı´a, J. 2010. Adsorption of dyes on carbon nanomaterials from aqueous solutions. J. Environ. Sci. Health Part A. 45: 1642–1653.

Roy, A. and Bhattacharya, J. 2012. Removal of Cu (II), Zn (II) and Pb (II) from water using microwave-assisted synthesized maghemite nanotubes. Chem. Eng. J. 211-212: 493–500.

Roy, Dr. A. and Prof. J. Bhattacharya. 2015. Nanotechnology in Industrial Wastewater Treatment. IWA Publishing London SW1H 0QS, UK.

Saad, A.H.A., Azzam, A.M., El-Wakeel, S.T., Mostafa, B.B. and El-Latif, M.B.A. 2018. Removal of toxic metal ions from wastewater using ZnO@Chitosan core-shell nanocomposite. Environ. Nanotechnol. Monitor. Manag. 9:67–75.

Sadak, O., Hackney, R., Sundramoorthy, A.K., Yilmaz, G. and Gunasekaran, S. 2020. Azo dye-functionalized magnetic Fe_3O_4/polyacrylicacid nanoadsorbent for removal of lead (II) ions. Environ. Nanotechnol. Monit. Manag. 14: 100380.

Safatian, F., Doago, Z., Torabbeigi, M., Shams, H.R. and Ahadi, N. 2019. Lead ion removal from water by hydroxyapatite nanostructures synthesized from egg sells with microwave irradiation. Applied Water Science. 9: 108.

Salam, M.A., Makki, M.S.I. and Abdelaal, M.Y.A. 2011. Preparation and characterization of multi-walled carbon nanotubes/chitosan nanocomposite and its application for the removal of heavy metals from aqueous solution. J. Alloys Compd. 509: 2582–2587.

Samper, E., Rodríguez, M., De la Rubia, M. and Prats, D. 2009. Removal of metal ions at low concentration by micellar-enhanced ultrafiltration (MEUF) using sodium dodecyl sulfate (SDS) and linear alkylbenzene sulfonate (LAS). Sep. Purif. Technol. 65: 337–342.

Sani, H.A., Ahmad, M.B., Hussein, M.Z., Ibrahim, N.A., Musa, A. and Saleh, T.A. 2017. Nanocomposite of ZnO with montmorillonite for removal of lead and copper ions from aqueous solutions. Process Safety and Environmental Protection. 109: 97–105.

Sardella, F., Gimenez, M., Navas, C., Morandi, C., Deiana, C. and Sapag, K. 2015.Conversion of viticultural industry wastes into activated carbons for removal of lead and cadmium. J. Environ. Chem. Eng. 3: 253–260.

Seema, K.M., Mamba, B.B., Njuguna, J., Bakhtizin, R.Z. and Mishra, A.K. 2017. Removal of lead (ii) from aqeouos waste using (cd-pcl-tio2) bio-nanocomposites. Int. J. Biol. Macromol. PII: S0141-8130(17)33607-3.

Sha, L., Guo, X., Feng, N. and Tian, Q. 2009. Adsorption of Cu and Cd from aqueous solution by mercapto-acetic acid modified orange peel. Colloids Surf. B Biointerfaces. 73: 10–14.

Shakir, I.K. and Husein, B.I. 2009. LEAD removal from industrial wastewater by electrocoagulation process. Iraqi Journal of Chemical and Petroleum Engineering. 10: 35–42.

Shamsizadeh, A.A., Ghaedi, M., Ansari, A., Azizian, S. and Purkait, M.K. 2014. Tin oxide nanoparticle loaded on activated carbon as new adsorbent for efficient removal of malachite green-oxalate: nonlinear kinetics and isotherm study. J. Mol. Liq. 195: 212–218.

Singh, S., Barick, K. and Bahadur, D. 2011. Novel and efficient three-dimensional mesoporous ZnO nanoassemblies for environmental remediation. Int. J. Nanosci. 10: 1001–1005.

Singh, S., Kapoor, D., Khasnabis, S., Singh, J. and Ramamurthy, P.C. 2021. Mechanism and kinetics of adsorption and removal of heavy metals from wastewater using nanomaterials. Environ. Chem. Lett. 19: 2351–2381.

Sitko, R., Turek, E., Zawisza, B., Malicka, E., Talik, E., Heimann, J. et al. 2013.Adsorption of divalent metal ions from aqueous solutions using graphene oxide. Dalton Trans. 42: 5682–5689.

Stoian, O., Covaliu, C.I., Paraschiv, G., Catrina Traistaru, G.A., Niţă-Lazăr, M., Matei, E. et al. 2021. Magnetite oxide nanomaterial used for lead ions removal from industrial wastewater. Materials (Basel, Switzerland). 14: 2831.

Sud, D., Mahajan, G. and Kaur, M. 2008. Agricultural waste material as potential adsorbent for sequestering heavy metal ions from aqueous solutions – A review. Bioresour. Technol. 99: 6017–6027.

Tabish, T.A., Memon, F.A., Gomez, D.E., Horsell, D.W. and Zhang, S. 2018. A facile synthesis of porous graphene for efficient water and wastewater treatment. Sci. Rep. 8: 1817.

Tahoon, A.M., Siddeeg, S.M., Alsaiari, N.S., Mnif, W. and Rebah, F.B. 2020. Effective heavy metals removal from water using nanomaterials: a review. Processes. 2020: 645.

Tassel, F., Rubio, J., Misra, M. and Jena, B. 1997. Removal of mercury from gold cyanide solution by dissolved air flotation. Miner. Eng. 10: 803–811.

Tchounwou, P.B., Yedjou, C.G., Foxx, D., Ishaque, A. and Shen, E. 2004. Lead-induced cytotoxicity and transcriptional activation of stress genes in human liver carcinoma cells (HepG2). Mol. Cell. Biochem. 255: 161–170.

Theodore, L. and Ricci, F. 2011. Mass Transfer Operations for the Practicing Engineer; John Wiley & Sons, Inc.: Hoboken, NJ, USA.

Tratnyek, P.G., Sarathy, V., Nurmi, J., Baer, D.R., Amonette, J.E., Chan, L.C., Penn, R.L. et al. 2008. Aging of Iron Nanoparticles in Water: Effects on Structure and Reactivity. J. Phys. Chem. C. 112: 2286–2293.

Ucun, H., Bayhana, Y.K., Kaya, Y., Cakici, A. and Algur, O.F. 2003. Biosorption of lead (II) from aqueous solution by cone biomass of Pinus sylvestris. Desalin. 154: 233–8.

UNEP. 1999. Guidelines for Municipal Solid Waste Management: Planning in Small Islands Developing States in the Pacific Region. South Pacific Regional Environment Programme (SPREP) Western Samoa, Apia.

USEPA. 1999. Development Document for Effluent Limitations Guidelines and Standards for the Metal Finishing Point Source Category, Technical report, US EPA, Washington, DC.

USEPA. 2020. United States Environmental Protection Agency, Basic Information about Lead in Drinking Water, Technical Report, US EPA, Washington, DC.

Vilar,V.J., Botelho, C.M. and Boaventura, R.A. 2005. Influence of pH, ionic strength and temperature on lead biosorption by Gelidium and agar extraction algal waste. Process Biochem. 40: 3267–75.

Vilela, D., Parmar, J., Zeng, Y., Zhao, Y. and Sánchez, S. 2016. Graphene-based microbots for toxic heavy metal removal and recovery from water. Nano Lett. 16: 2860–2866.

Visa, M., Carcel, R.A., Andronic, L. and Duta, A. 2009.Advanced treatment of wastewater with methyl orange and heavy metals on TiO2, fly ash and their mixtures. Catal. Today. 144: 137–142.

Vukojevi´c, N.M., Peri´, J. and Trgo, M. 2006. Column performance in lead removal from aqueous solutions by fixed bed of natural zeolite–clinoptilolite. Sep. Purif. Technol. 49: 237–244.

Wadhawana, S., Jaina, A., Nayyara, J. and Mehtab, S.K. 2020. Role of nanomaterials as adsorbents in heavy metal ion removal from waste water: A review. J. Water Process Eng. 33: 101038.

Wan, S., Wu, J., Zhou, S., Rui, W., Gao, B. and Feng, H. 2018. Enhanced lead and cadmium removal using biochar-supported hydrated manganese oxide (HMO) nanoparticles: Behavior and mechanism. Sci. Total Environ. 616–617: 1298–1306.

Wang, S.G., Gong, W.X., Liu, X.W., Yao, Y.W., Gao, B.Y. and Yue, Q.Y. 2007. Removal of lead (II) from aqueous solution by adsorption onto manganese oxide-coated carbon nanotubes. Sep. Purif. Technol. 58: 17–23.

Wang, L., Li, J., Jiang, Q. and Zhao, L. 2012.Water-soluble Fe3O4 nanoparticles with high solubility for removal of heavy-metal ions from waste water. Dalton Trans. 41: 4544–4551.

Wang, X., Huang, K., Chen, Y., Liu, J., Chen, S., Cao, J. et al. 2018. Preparation of dumbbell manganese dioxide/gelatin composites and their application in the removal of lead and cadmium ions. J. Hazard. Mater. 350: 46–54.

Wang, Z.D., Lee, J.H. and Lu, Y. 2008. Label-free colorimetric detection of lead ions with a nanomolar detection limit and tunable dynamic range by using gold nanoparticles and DNAzyme. Adv. Mater. 20: 3263–3267.

Warner, C.L., Addleman, R.S., Cinson, A.D., Droubay, T.C., Engelhard, M.H., Nash, M.A. et al. 2010. High-performance, superparamagnetic, nanoparticle-based heavy metal sorbents for removal of contaminants from natural waters. ChemSusChem. 3: 749–757.

Waters, A. 1990. Dissolved air flotation used as primary separation for heavy metal removal. Filtr. Sep. 27:70–73.

WHO. 1993. Guidelines for Drinking Water Quality (2nd ed.), Technical report, World Health Organization, Geneva.

WHO. 2021. Lead poisoning and health. World Health Organization, Geneva.

Xin, X., Wei, Q. and Yang, J. 2012. Highly efficient removal of heavy metal ions by amine-functionalized mesoporous Fe_3O_4 nanoparticles. Chem. Eng. J. 184: 132–140.

Xiong, C., Wang, W., Tan, F., Luo, F., Chen, J. and Qiao, X. 2015. Investigation on the efficiency and mechanism of Cd(II) and Pb(II) removal from aqueous solutions using MgO nanoparticles. J. Hazard. Mater. 299: 664–674.

Xu, D., Tan, X., Chen, C. and Wang, X. 2008. Removal of Pb(II) from aqueous solution by oxidized multiwalled carbon nanotubes. J. Hazard. Mater. 154: 407–416.

Xu, P., Zeng, G.M. and Huang, D.L. 2012. Use of iron oxide nanomaterials in wastewater treatment: a review. Sci. Total Environ. 424: 1–10.

Xu., Z., Gu, S., Rana, D., Matsuura, T. and Lan, C.Q. 2021. Chemical precipitation enabled UF and MF filtration for lead removal. J. Water Process Eng. 41: 101987.

Yan, X.M., Shi, B.Y., Lu, J.J., Feng, C.H., Wang, D.S. and Tang, H.X. 2008. Adsorption and desorption of atrazine on carbon nanotubes. J. Colloid Interface Sci. 321: 30–38.

Yang, H., Tian, Z., Wang, J. and Yang, S. 2012. A magnetic resonance imaging nanosensor for Hg (II) based on thymidine functionalized supermagnetic iron oxide nanoparticles. Sens. Actuators B Chem.161: 429–433.

Yang, J., Hou, B., Wang, J., Tian, B., Bi, J., Wang, N. et al. 2019. Nanomaterials for the removal of heavy metals from wastewater. Nanomaterials. 9: 424.

Yuan, X., Meng, Y., Zeng, G., Fang, Y. and Shi, J. 2008. Evaluation of tea-derived biosurfactant on removing heavy metal ions from dilute wastewater by ion flotation. Colloids Surf. A Physicochem. Eng. Asp. 317: 256–261.

Yusuf, M., Elfghi, F.M., Zaidi, S.A., Abdullaha, E.C. and Khan, M.A. 2015.Applications of graphene and its derivatives as an adsorbent for heavy metal and dye removal: a systematic and comprehensive overview. RSC Adv. 5: 50392–50420.

Youssef, Y.M., Moukhtar, N., Hassan, I. and Abdel-Aziz, M.H. 2021. Recovery of heavy metals from liquid effluent by galvanic cementation. Mining Metall. Explor. 38: 177–186.

Zarime, N.A., Wan, Z.W.Y. and Jamil, H. 2017. Removal of heavy metals using bentonite supported nano-zero valent iron particles. In Proceedings of the Ukm Fst Postgraduate Colloquium,University Kebangsaan Malaysia, Faculty of Science & Technology Postgraduate Colloquium, Univ Kebangsaan Malaysia, Fac Sci & Technol, Selangor, Malaysia.

Zendehdel, M., Shoshtari-Yeganeh, B. and Cruciani, G. 2016. Removal of heavy metals and bacteria from aqueous solution by novel hydroxyapatite/zeolite nanocomposite, preparation, and characterization. J. Iran. Chem. Soc. 13: 1915–1930.

Zewail, T.M. and Yousef, N.S. 2015. Kinetic study of heavy metal ions removal by ion exchange in batch conical air spouted bed. Alex. Eng. J. 54: 83–90.

Zhang, C., Sui, J., Li, J., Tang, Y. and Cai, W. 2012. Efficient removal of heavy metal ions by thiol-functionalized supermagnetic carbon nanotubes. Chem. Eng. J. 210: 45–52.

Zhang, Q., Xu, R., Xu, P., Chen, R., He, Q., Zhong, J. et al. 2014a. Performance study of ZrO2 ceramic micro-filtration membranes used in pretreatment of DMF wastewater. Desalination. 346: 1–8.

Zhang, Wx. 2003. Nanoscale iron particles for environmental remediation: an overview. J. Nanopart. Res. 5: 323–332.

Zhang, X., Lin, S., Chen, Z., Megharaj, M. and Naidu, R. 2011. Kaolinite-supported nanoscale zero-valent iron for removal of Pb^{2+} from aqueous solution: reactivity, characterization and mechanism. Water Res. 45: 3481–3488.

Zhang, Y., Yan, L., Xu, W., Guo, X., Cui, L., Gao, L. et al. 2014b. Adsorption of Pb (II) and Hg (II) from aqueous solution using magnetic $CoFe_2O_4$ reduced graphene oxide. J. Mol. Liq. 191: 177–182.

Zhang, Y., Bing, W., Hui, X., Hui, L., Wang, M., He, Y. et al. 2016. Nanomaterials-enabled water and wastewater treatment. Nanoimpact. 3-4: 22–39.

Zhang, Y. and Duan, X. 2020. Chemical precipitation of heavy metals from wastewater by using the synthetical magnesium hydroxy carbonate. Water Sci. Technol. 81: 1130–1136.

Zhao, G., Ren, X., Gao, X., Tan, X., Li, J., Chen, C. et al. 2011. Removal of Pb(II) ions from aqueous solutions on few-layered graphene oxide nanosheets. Dalton Trans. 40: 10945–10952.

Zhao, W., Liu, D. and Feng, Q. 2020. A visualization method for studying the adsorption of lead species in salicylhydroxamic acid flotation of hemimorphite. Miner. Eng. 154: 106434.

Zhu, W.P., Sun, S.P., Gao, J., Fu, F.J. and Chung, T.S. 2014. Dual-layer polybenzimidazole/polyethersulfone (PBI/PES) nanofiltration (NF) hollow fiber membranes for heavy metals removal from wastewater. J. Membr. Sci. 456: 117–127.

11

Water Management by the Phytoremediation of Toxic Metals Present in Water

A Review

Mamta Sharma,[1] *Himani Pathania*,[1] *Priyanka Chauhan*,[1] *Shriya*,[1] *Sheetal Choudhary*,[1] *Diksha Pathania*,[1] *Anil Kumar*[2] and *Rajesh Maithani*[2,*]

11.1 Introduction

Heavy metal contamination is one of the most hazardous forms of pollution on the planet. There are some strategies for cleaning up the atmosphere from these types of toxins, but the majority of them are expensive and difficult to achieve optimal results. Phytoremediation is currently an efficient and cost-effective technical method for extracting or removing inactive metals and metal contaminants from contaminated soil and water (Tangahu et al. 2011). Since 1991, the word "phytoremediation" has been used to characterize the use of plants to minimize the amount, movability and toxicity of pollutants in soil, groundwater, and other polluted media (Etim 2012). We may define it as a collection of ecological strategies that employ plants *in situ* to aid in the breakdown, immobilisation, and removal of pollutants from the environment (Shrestha et al. 2019). It is a low-cost remediation method for removing contaminants (primarily heavy metals and organics) from polluted soils and waters at the site level with minimal disruption to the environment (Itanna and Coulman 2003). This method is used to transport hazardous waste to a landfill or a recycling facility that is off-site (Salt et al. 1998).

[1] School of Biological and Environmental Sciences, Shoolini University, Solan (H.P), 173229.
[2] Department of Mechanical Engineering, University of Petroleum and Energy Studies, Dehradun (U.K), 248007.
* Corresponding author: rmaithani@ddn.upes.ac.in

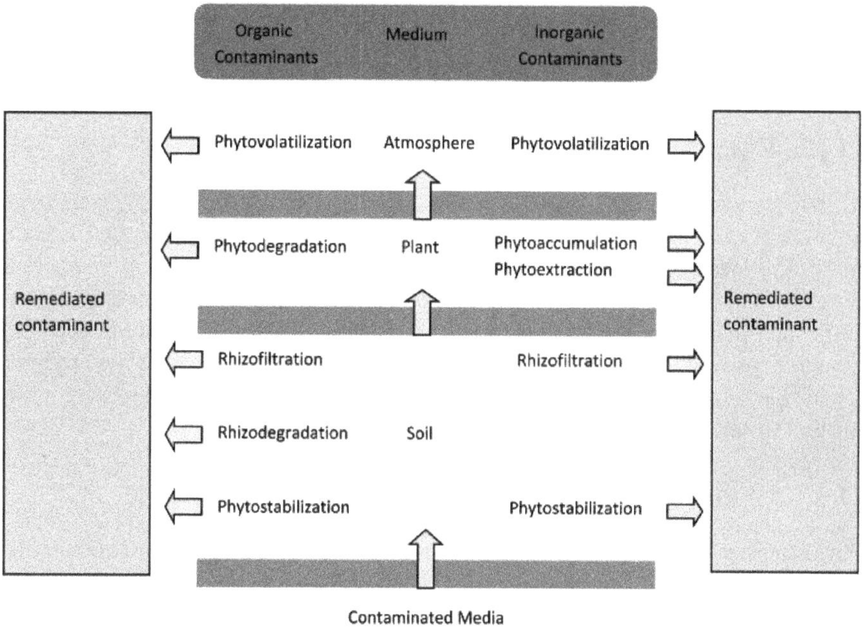

Fig. 1: Phytoremediation technology uptake mechanisms (Tangahu et al. 2011).

Phytoremediation is an environmentally friendly method that involves using living plants to remediate degraded soil, mucks, residue, and ground water *in situ* by foreign material expulsion, corruption, or regulation. Developing new methods and, in some cases, collecting plants on a contaminated site as a remediation strategy is a pleasurable, energy-driven, uninvolved procedure that can be used to clean up locations with shallow, low to moderate levels of contamination. This approach can be used in conjunction with or in place of mechanical cleaning techniques (Dixit et al. 2015). By acting as filters or traps, plants can degrade organic pollutants and absorb and stabilize metal contaminants. Some of the methods that are being tested are described below and shown in Fig. 1.

11.1.1 Phytoextraction

The word "phytoaccumulation" refers to the uptake and transport of metal contaminants in the soil by plant roots into the plants' aboveground portions. In this method, plants are cultivated for several weeks or months in contaminated areas, and then harvested and either incinerated or composted to recycle the metals. This process can be performed as many times as possible to reduce soil contaminant levels to acceptable levels. The ash from incinerated plants must be disposed of in a hazardous waste landfill, but the amount of ash generated would be less than 10% of the volume created if the polluted soil was dug up for treatment. This method is demonstrated by hyper accumulator species in contaminant soil zone and can accumulate 100 times more metal in comparison to other plants (Raksin and Ensley 2000). Hyper accumulators are unusual plants which accumulate particular metals and metalloids in the tissue system of plants (Reeves et al. 2018). By using this method of phytoremediation, the toxic metals can be easily removed permanently from the soil at very low cost. A list of Himalayan region plants used by researchers to accumulate the specific heavy metals is shown in Table 4.

11.1.2 Rhizofiltration

Rhizofiltration is a process that involves the adsorption or precipitation of contaminants in solution onto plant roots in the root region, as well as their absorption into the roots. The cleaning plants are grown in greenhouses with their roots immersed in water instead of soil. The roots are collected when they are fully contaminated with toxins and then burned or composted to recycle the toxins. This process is mainly used to remove Pb, Cd, Cu, Ni, Zn, and Cr from roots, as these metals are primarily retained in roots (Etim 2012). The greater a plants' root system, the greater its ability to absorb radioactive metals from polluted soil.

11.1.3 Phytostabilization

Phytostabilization is a technique that uses specific plant species to immobilize pollutants in the soil and ground water by absorption and accumulation by roots, adsorption onto roots, or precipitation within the root region. This process, also known as in-place-inactivation, decreases contaminant mobility and prevents migration to groundwater or the environment, as well as bioavailability for entry into the food chain. This method can be used effectively in case where rapid immobilization of metal ions is required (Zhai et al. 2018).

11.1.4 Phytodegradation

This method, also known as phytotransformation, involves the breakdown of pollutants taken up by plants by metabolic processes within the plant, as well as the breakdown of pollutants external to the plant by the influence of substances (such as enzymes) produced by the plants. Pollutants are broken down, absorbed by tissue system of plants (Trap et al. 2005), and utilised as nutrients. For sustenance and energy, microorganisms (yeast, fungus, or bacteria) absorb and digest organic molecules. Biodegradation allows some microorganisms to digest dangerous organic compounds such as fuels or solvents and break them down into harmless products. Sugars, alcohols, and acid, which are naturally produced by plant roots, contain organic carbon, which feeds soil microbes and boosts their activity. Plants help biodegradation by loosening the soil and supplying water to the area.

11.1.5 Phytovolatilization

Phytovolatilization is defined as the absorption and transpiration of a pollutant by a plant, followed by the release of the pollutant or a modified version of the contaminant into the atmosphere by the plant. Phytovolatilization occurs when trees and other plants absorb water and organic contaminants. Some of these toxins can pass through the leaves of plants and volatilize into the atmosphere at combustible temperatures.

Several strategies for cleaning up the environment from these types of toxins are currently in use, but the majority of them is expensive and falls short of their potential. Chemical methods produce significant volumes of sludge, which raises expenses (Rakhshaee et al. 2009). Concerns about environmental pollution have prompted the enlargement of devices to determine the existence and motility of metals in soil (Shtangeeva et al. 2004), water, and drainwater. Today, phytoremediation has become an effective and inexpensive technological solution used to extract inactive metals and metal contaminants from contaminated soil. Plants are used to clean up pollutants from soils, sediments, and water in phytoremediation. This process has the potential to be both environmentally benign and cost-effective. Hyper accumulator plants are plants that have a high capacity for storing metals (Cho-Ruk et al. 2006).

Many plant species have shown to be effective at absorbing contaminants such as lead, cadmium, chromium, arsenic, and various radionuclides from soils. One of the phytoremediation categories, phytoextraction, can be utilised to remove heavy metals from soil by utilising its ability to consume ions necessary for plant development (Fe, Mn, Zn, Cu, Mg, Mo and Ni). Also some metals with uncertain biological function (Cd, Cr, Pb, Co, Ag, Se, Hg) can also be absorbed by plants (Cho-Ruk et al. 2006).

Chemical techniques of water filtration have been used for many years, but they are both expensive and dangerous. The problem necessitates a long-term solution, which is bioremediation by the planting of trees, shrubs, and perennial plants in catchment areas and along waterways. Phytoremediation is a gift from nature to humans for the answer to this pressing issue. A variety of phytoremediation plants are available. "The use of green plants to separate, contain, or render harmless environmental pollutants" is how phytoremediation is defined as this whole process itself relates to all biological, microbiological, chemical, and physical processes that are impacted by plants and contribute to the remediation of a polluted site. This technique is also known as green technology, because when used correctly, it's both environmentally benign and visually pleasant to the public (Cunningham and Lee 1995). Phytoremediation does not demand any high-priced tools or highly-specialized personnel; thus, it is relatively easy to implement. It is capable of permanently treating a wide range of contaminants in a wide range of environments. As a result, phytoremediation of a contaminated area provides an environmentally benign, cost-effective, and carbon-neutral method of removing hazardous contaminants from the environment. Phytoremediation is a collection of cutting-edge environmental cleanup solutions that take advantage of plants' inherent extractive and metabolic properties. Organic contaminants are degraded to less hazardous forms by plants, while heavy metals are mineralized and detoxified from contaminated water by plants.

11.1.6 National status review

The Himalayas are a one-of-a-kind biodiversity, ecological, and cultural treasure trove. It is one among the world's largest freshwater resources. It is estimated that more than 1.4 billion people depend on water from the rivers and rivulets that emerges from the Himalayas. The pollution of the surface environment is caused by the presence of a greater quantity of trash from various businesses (Cederlof 2014).

The water of the rivulets passing through cities in hill areas is contaminated with garbage and sewage water, solid waste from household and effluents from industries, affecting the lifeline of rural people. The Hon'ble High Court of Himachal Pradesh, viewing the outbreak of 500 jaundice cases passed an injunction to ban the supply of potable water to the town of Shimla from Ashwani Khud unless and until it is properly treated in 2016. The chemical treatment of water is not the permanent solution. There are many soil remediation methods for withdrawal of toxic heavy metals from soil and by following these techniques, the soil can be made contaminant free (Hasan et al. 2019), and ultimately the water will be purified. Various methods to remove the toxic metals from soil are shown in Fig. 2 in form of flowchart. The holistic treatment lies in educating people not to throw garbage in the rivulets, zero outlets of sewage water, minimal industrial effluents and by planting trees, bushes and herbs along the embankments of the rivulet. The intervention of the High Court in such a dreadful situation is understandable because Shimla, the State capital, is the tourist resort. What about the remaining sick rivulets passing through the cities of different states in the Indian Himalayan Region?

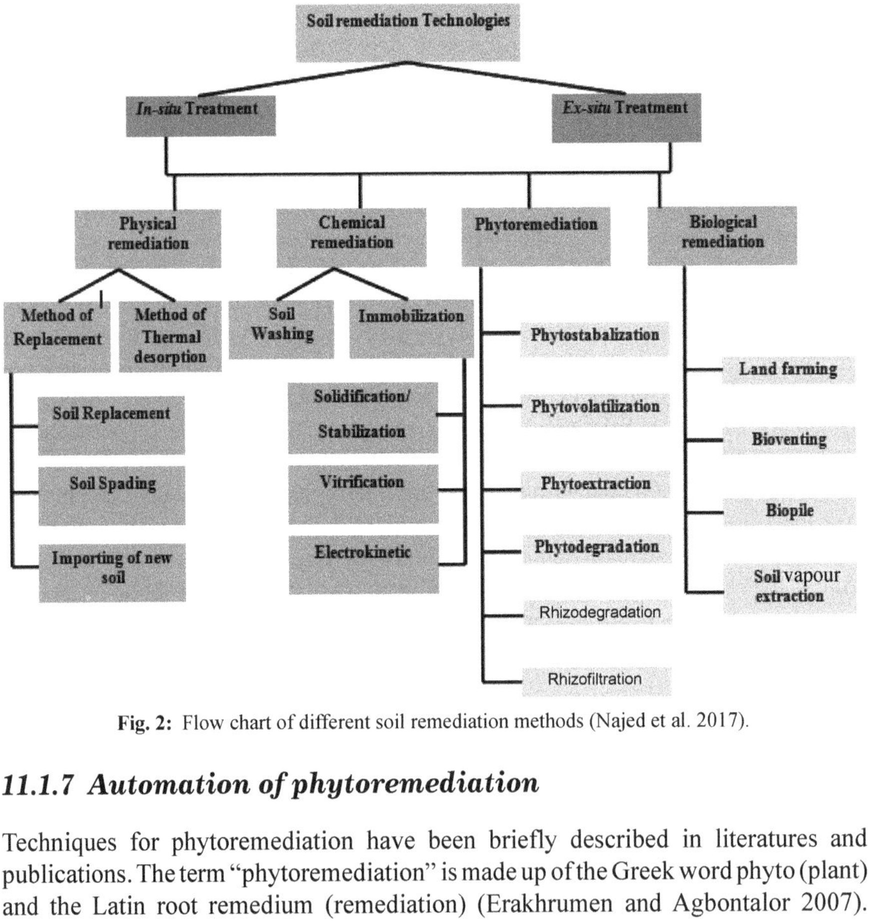

Fig. 2: Flow chart of different soil remediation methods (Najed et al. 2017).

11.1.7 *Automation of phytoremediation*

Techniques for phytoremediation have been briefly described in literatures and publications. The term "phytoremediation" is made up of the Greek word phyto (plant) and the Latin root remedium (remediation) (Erakhrumen and Agbontalor 2007).

The rivulet's water quality is rapidly deteriorating, resulting in a lack of drinking water in the mountainous area (Sargaonkar et al. 2003). The Sorghum plant is being used to examine phytoremediation of Cr metal-damaged soil at Ranipet Tanneries, commonly known as a "hyperaccumulator" for metal polluted site phytoremediation (Revathik et al. 2011). Fluroquinonone is a compound which is not degraded easily and remain in waste water for a long duration of time; thus, the phytoremediation of fluroquinonones is done by some selected wetland plant species like *Taxodium distichium, Canna indica, Chrysopogan zizanioides* without any toxic effect on their growth (Tandon et al. 2013). Including this, the list of plants used for phytoremediation is listed in Table 1. Barog-Kotla rivulet is facing the same problem and is in a dreadful condition. The situation of rivulets gets further aggravated during rains as the speedily flowing muddy water causes havoc to the people downstream. The lift water schemes become non-functional and villagers remain without water supply for days together. This man-made alarming scenario cannot be left unattended. The problem is gigantic and cannot be undertaken in one go but a model rivulet by seeking the cooperation of all stake holders—the rural people, city dwellers and the district administration—needs to be developed. The only solution for this problem is remediation of natural resources by natural means, i.e., phytoremediation. In the Himalayan region, phytoremediation research has only focused on the potential of Phragmites australis in Hokersar Wetland, a Ramsar Site in Kashmir Himalaya (Ahmed et al. 2014). However, no further research and project working on phytoremediation has been listed.

Air, water, and soil contamination and pollution are caused by urbanisation, excessive use of natural resources, and population increase in the country. For the

Table 1: List of some plants showing phytoremediation in Himalayan region.

Sr. No.	Plant species	Contaminant Source	References
1.	*Salix* sp.	Cu	Dickinson et al. 2000
2.	*Zea mays*	Cd, Pb	Mojiri 2011
3.	*Brassica juncea*	Cd, Pb	Gurajala et al. 2019
4.	*Helianthus annus*	As, Cd, Cu, Pb	Madejon et al. 2003
5.	*Quercus* sp.	As, Cd, Cu, Pb	Dominguez et al. 2010
6.	*Nicotiana tobacum*	Cd, Zn	Vangronsveld et al. 2009
7.	*Pelargonium* sp.	Sewage sludge, Cd, Zn	Zehra et al. 2009
8.	*Mentha piperita*	Heavy trace elements of soil	Zheljazkov and Neilsen 1996
9.	*Lavandula angustifolia*	Cd, Zn, Cu	
10.	*Dalbergia sisso*	Heavy metals	Farooq et al. 2006
11.	*Nelumbo nucifera*	Heavy metals	
12.	*Pteris* sp.	Cd, Zn	Vasavi et al. 2010
13.	*Withania Somnifera*	Contaminated water	Rathaur et al. 2012
14.	*Eucalyptus* sp.	Cd, Zn, Cu	Mishra et al. 2012
15.	*Nerium oleander*	Sewage sludge	Mahimairaja et al. 2011
16.	*Jasminum sambac*	Sewage sludge and waste	Mandal et al. 2014

elimination of heavy metals from polluted water and soil, phytoremediation is a cost-efficient and long-term solution that outperforms other remediation techniques. Water pollution is a major concern around the world, and it can be split into three categories: contamination by organic compounds, inorganic chemicals (such as heavy metals), and microorganisms (Raghav and Shrivastava 2016).

Chemical poisoning of water has harmed 1,96,813 of India's 1.42 million communities. Every year, almost a lakh people die as a result of water-borne diseases. The quantity of fluoride, iron, salt, and arsenic in groundwater in one-third of India's 600 districts exceeds the tolerance thresholds, according to reports (Singhal and Gupta 2010). The majority of the poor in these countries do not have access to clean drinking water. Around 80 per cent of diseases in the developing countries are attributed to poor quality of water supply. The World Health Organisation reported that of the 10 million annual deaths in India, 7.8 lakh are due to lack of basic health care amenities like effective sewage system, safe drinking water supply, elementary sanitary facilities and hygienic conditions. Almost 90 per cent of diarrhoea cases are due to contaminated water (Singhal and Gupta 2010). Certain plants have been found in the wild that meet the potential to absorb heavy metals and organic materials. Plants that hyperaccumulate have been documented in at least 45 families; those families are *Brassicaceae, Fabaceae, Euphorbiaceae, Asteraceae, Lamiaceae*, and *Scrophulariaceae* (Anjum et al. 2012a). To determine the level of heavy metal contamination, a phytoremediation study was conducted at Pariyej reservoir, a globally significant wetland included in the Asian Directory of Wetlands and recognised as a "Wetland of International Importance" and a planned community reserve in Gujarat State, India (Kumar et al. 2008a). The research compared the accumulation of heavy metals in various aquatic macrophytes employed as biomonitors to water and sediments (abiotic monitors) for phytoremediation. Roots, stems and leaves of native aquatic plants (biomonitors) represented by seven species: *Ipomoea aquatica* Forsk, *Eichhornia crassipes*, (Mart.) Solms, *Typha angustata* Bory & Chaub, *Echinochloa colonum* (L.) Link, *Hydrilla verticillata* (L.f.) Royle, *Nelumbo nucifera* Gaerth. and *Vallisneria spiralis* L. along with surface sediments and water, were analyzed for Cd, Co, Cu, Ni, Pb and Zn contamination (Dominguez et al. 2010).

The greater deposition of heavy metals was observed in *Nelumbo nucifera* and *Echinochloa colonum*. However, Zn and Cu showed the highest accumulation with alarming toxicity levels, which are considered as one of the most hazardous pollutants in Pariyej reservoir. Species like *Typha angustata* and *Ipomoea aquatica* are also proposed as bioremediants, which are the two most useful plant species in phytoremediation studies due to their ability to collect heavy metals in high concentration in the roots (Kumar et al. 2008b). Suitable trees and grass species for phytoremediation are trees (*Eucalyptus tereticornis, Populus deltoids, Salix* sp., *Terminalia arjuna, Acacia auriculiformis, Syzigium cumini, Albizia lebbek, Dalbergia sissoo*, and *Pongamia pinnata*) and grasses (para grass, cord grass, lemon grass, and *Setaria* grass). Pesticide residue contamination of food and other agricultural goods is a major concern in India. With their vast root system and high transpiration rates, trees like poplar (*Populus* spp.) and willow (*Salix* spp.) show considerable potential for pesticide and organic matter phytoremediation (Jansson and Douglas 2007). Besides this research, the herbicide tolerance activity in potato

(*Solanum tuberosum* L.) *Nicotiana tabaccum* L. (Didierjean et al. 2002) and rice (*Oryza sativa* L.) (Kawahigashi et al. 2007) was also reported.

Floating *Eichhornia crassipes* (water hyacinth), *Lemna minor* (duckweed), and *Azolla pinnata* (water velvet) have been studied for rhizofiltration, phytodegradation, and phytoextraction (Rahman 2011). In a study conducted by Lone in 2008, the phytoextraction potential of various plant species, including *Pteris cretica, Ipomoea carnea, Datura innoxia, Phragmytes karka, Brassica juncea*, and *Brassica campestris*, was investigated for the removal of cadmium (Cd), chromium (Cr), and lead (Pb) contaminants from the environment. The research aimed to assess the effectiveness of these plants in accumulating and reducing the concentrations of these heavy metals, providing valuable insights into their remediation capabilities. (Lone 2008). Most of the conventional remedial technologies are expensive and inhibit the soil fertility and water quality; this subsequently causes negative impacts on the ecosystem. Phytoremediation is a low-cost, environmentally beneficial, and aesthetically beautiful method that is particularly well suited to underdeveloped countries. Despite its potential, phytoremediation is yet to gain commercial acceptance in India (Ghosh and Singh 2005).

11.1.8 ***International status review***

The idea of employing plants to clean up contaminated surroundings is not new. Plants were recommended for use in wastewater treatment some 300 years ago. *Thlaspi caerulescens* and *Viola calaminaria* were the first plant species documented to collect highest metals in leaves (Lasat 2000). Byers discovered that *Astragalus* plants may accumulate up to 0.6 percent selenium in their dry shoot biomass in 1935. Utsunamyia and Chaney reintroduced and developed the concept of utilising plants to absorb metals from contaminated soil, and the first field trial on Zn and Cd phytoextraction was done in 1991. Extensive research on the biology of metal phytoextraction has been undertaken in the recent decade. The natural presence of plant species capable of storing extremely high metal levels makes this process particularly intriguing to study (Lasat 1999). The phytoremediation potential of *Eichhornia crassipes* (water hyacinth) investigated against some heavy metals like K, Na, Zn, Fe, Pb, Cd, Mg, Cu. Removal of those specific toxic heavy metals by specific plant species is listed in Table 3. The amount of metal uptake by water hyacinth according to its per dry weight was highest for potassium and lowest for lead, i.e., 13.52 ppm and 0.01 ppm, respectively (Oluwalinisol and Ogunkanmi 2010). Plant Family has a worldwide distribution. From temperate to tropical climates, more than 400 plant species belonging to 45 plant families have been found and documented as having the ability to resist and excessively collect heavy metals (Anjum et al. 2012b).

Heavy metals and organic matter pollution have affected all countries worldwide; however, the extent and severity of heavy metal pollution varies greatly. As a result, there has been a growing worry about the buildup of pollutants in the environment, as they constitute a significant danger to both human health and the nature. It is critical to carry out hazardous heavy metals from industrial wastewater (Cearley and Coleman 1973). In order to maintain good quality of soil and water, constant efforts have been made by scientists to develop new technologies that can be easily used, are renewable and commercially feasible. Phytoremediation, or the use of plant species to remediate polluted soils and streams, has attracted growing interest in the recent decade as a potentially less expensive approach.

World is divided into 7 continents and phytoremediation research is going on in every continent as science of phytoremediation has shown promising results in terms of environment as an innovative cleanup technology. However, it is still in a developmental stage and more research is needed to increase the understanding and knowledge of this remediation technology (Sharma et al. 2014).

The phytoremediation market is still emerging in Europe, while in the US revenues are likely to exceed $300 million in 2007. The US market was estimated at less than $50 million in 1999. These figures clearly prove the commercial feasibility of phytoremediation, being in fact one of the innovative technologies promoted by the US Environmental Protection Agency (EPA) after an extensive evaluation of selected field trials. Nearly 150 projects are now being funded by the EPA to remediate a variety of pollutants, including chlorinated solvents, explosives, propellants, pesticides, polycyclic aromatic hydrocarbons (PAHs), petroleum hydrocarbon compounds, radionuclides, and heavy metals. Phytoremediation is gaining popularity as a low-cost alternative to high-energy, high-cost traditional approaches. In the realm of novel cleanup technologies, it is considered a "Green Revolution". Phytoremediation has the potential to clean an estimated 30,000 contaminated waste sites throughout the US according to the EPA's Comprehensive Environmental Response Compensation Liability Information System (CERCLIS). Sites included in this estimate are those that have either been owned or contaminated with solvents, organic matter, coated glass, paints, leather and chemicals, etc. (Alcock and Jones 1993).

In South Africa, industrialization has resulted in an increase in industrial waste of environmental concern. Thus, the "Ecological Engineering and Phytoremediation Research Programme" was initiated in 1995 by AngloGold Ashanti (then Anglo American Gold Division) and the School of Animal, Plant and Environmental Sciences (APES) of the University of the Witwatersrand, Johannesburg (Wits University). AngloGold Ashanti has planted around half a million trees over the last decade for research on using phyto (plant) technologies to prevent and repair environmental impacts from the company's tailings storage facilities (TSFs) in South Africa. Most TSFs emit dust and seepage containing salts and metals into soil and water which contaminate the surrounding environment. Around 80 species of plants have been assessed to date in tailings experiments, and almost 60 species of trees are being assessed in woodlands' trials on seepage from TSFs (Kvesitadze et al. 2009). The final phase of the research (taking place between 2009 and 2012) will look at how phytoremediation plants might be able to produce materials for secondary industries, and thus help local communities. Such materials might include precious woods, fibres, chemicals, essential oils, dyes, gums and recoverable minerals and metals.

In Australia, the botany department at the University of Melbourne, The Centre for Mined Land Rehabilitation (CMLR) the Commonwealth Scientific and Industrial Research Organization (CSIRO) and several universities have phytoremediation programmes. Phytolink Australia is a dedicated and well established phytoremediation company which focuses on the use of plants as biopumps (Robinson et al. 2003).

Solar energy is used by the plant to dewater contaminated locations, limit leaching, and bring organic matter and microbial activity into the rhizosphere. Here, rootzone process, thereby augmenting contaminants' deterioration and decreasing the mobility of heavy metals specifically in that area. Rather than phytoextraction of heavy metals, phytoremediation's most essential job is to limit contaminant mobility and destroy organic contaminants. The western treatment plant, Victoria, Australia is working for removal of Phytoremediation of biosolids (Ney and Schnoor 2002). Phytoremediation technique, being useful and cost effective technique for water treatment, seeks appropriate place worldwide.

Unlike heavy metals, organic pollutants offer the prospect of being metabolised by living organisms. The extraction of organic and inorganic pollutants from wastewater is evaluated via *E. crassipes* (Mishra and Maiti 2017). Plants can immobilise, store, volatilize, turn to various degrees (even mineralize) or a combination of these actions on these compounds, depending on the particular compound, environmental conditions, and plant genotypes involved. As a result, numerous phytotechnologies have been established to aid in the reduction of organic pollution: phytoextraction is the process of removing pollutants from soil and storing them in plant tissues (rhizofiltration is the preferred word when water supplies are treated); phytodegradation is the process of chemically altering pollutants to make them less toxic, accompanied by storage or elimination. Phytovolatilization removes toxins from soil or water and releases them into the environment by plant transpiration, or it converts them into more volatile compounds. By immobilising or attaching contaminants to the substrate matrix, phytostabilization decreases their bioavailability. The bioavailability of pollutants is an obvious determinant of phytoremediation performance (Dzantor 2007). Cu phytoremediation is possible with the aquatic fern *Salvinia calcullata.*

Bafeel (2008) conducted a study on the function of mycorrhizae in phytoremediation of lead-contaminated soils by *Eucalyptus rostrata.* Eucalyptus consumed and accumulated more Pb after being inoculated with mycorrhizal fungi. The findings revealed that inoculating the host plants with AMF protects them from the possible toxicity of increased Pb uptake. In all cases, *E. rostrata* growth and other parameters performed better in the presence of *P. vulgaris* plants. Arbuscular mycorrhizae appear to have potential in phytoremediation of heavy metal polluted soils, particularly when legumes are present. It has been determined that inoculating the host plant with the AM fungus protects it from the potential toxicity of Pb. When inoculated with *G. deserticola*, the obtained results indicate that *E. rostrata* is ideal for growing and rehabilitating heavy-metal contaminated soils. The association of *Eucalyptus* with heavy-metal-resistant legume varieties can also help to increase *Eucalyptus* resistance to heavy metals.

The presence of organic acids in root system of plants have ability to bind the heavy toxic metals present in surrounding soil. Exudation of organic acids from the root system of plants binds and forms complex molecules with the heavy metals (Hale and Grittin 1974). Presence of organic acids like carboxylic acid and amino acids in plants increases the uptake of Cr metal ions (Srivastava et al. 1999). Thus, organic acids play an important role in aggregation of heavy metals. Plant species used to remove contaminants by application and presence of organic acids are listed below in Table 2.

Table 2: Plants having ability for phytoremediation by means of organic acids.

Sr. No.	Plant species	References
1.	*Populus* sp.	Jansson and Douglas 2007
2.	*Salix* sp.	Jansson and Douglas 2007
3.	*Helianthus tuberosus*	Robineau et al. 1998
4.	*Glycine max*	Robineau et al. 1998
5.	*Solanum tuberosum* L.	Antonious et al. 2007
6.	*Oryza sativa* L.	Kawahigashi et al. 2007
7.	*Nicotiana tabaccum* L.	Didierjean et al. 2002
8.	*Datura stramonium*	Shirkhani et al. 2018
9	*Taxodium distichum,* *Canna indica* *Chrysopogon zizanioides*	Tandon et al. 2013
10.	Sorghum plant	Revathi et al. 2011

Table 3: Example of plants widely used and studied in bioremediation (Bhatnagar and Kumari 2013).

Sr. No.	Plants	Toxic chemicals/Elements
1.	*Ambrosia artemisifolia*	Pb
2.	*Apocynum cannabinum*	Pb
3.	*Brassica juncea*	Se, Pb and Cu
4.	*Helianthus annuus*	As and Ur
5.	*Medicago sativa*	Benzopurene, PAEs and PAHs
6.	*Melastoma malabathricum*	Al
7.	*Nephrolepis exaltata*	Hg
8.	*Pteridium esculentum*	As
9.	*Pteris vitata*	As, Hg, Cs and Sr
10.	*Salix viminalais*	Cd, Zn, and Cu
11.	*Raphanus sativa*	Cu
12.	*Silene vulgaris*	Zn and Cd
13.	Thlaspi careulescens	Cd and Zn

Calandulla officinalis, an ornamental plant, was examined in terms of growth responses, photosynthetic activities and antioxidant enzymes such as SOD, CAD and GPX. This plant shows high Cu tolerance from the phytotoxic range, i.e. upto 400 mg/k. Extraction coefficient indicates the uptake capacity of heavy metals from soil by plants whereas the translocation factor describes the movement of heavy metals from the root to the shoots (Goswami and Das 2016).

11.1.9 Mechanism of phytoremediation

According to studies, using plants in phytoremediation technology to extract heavy metals from polluted areas is an alternative approach. The soil is an essential component of our agricultural resource, playing a key role in the green revolution and food security. Agricultural lands have become increasingly polluted by organic, inorganic, and metallic contaminants as a result of rapid urbanisation and industrialization (Lichtfouse and Eglinton 1995, Muthusaravanan et al. 2018).

The whole mechanism of phytoremediation relies on the contaminant type and its presence in soil. Various methods are included in this review which are used in removal of contaminants. There are lots of similarities in every method, but as their nomenclature varies, slight differences are there for all the methods as these six methods are fully explained before. The six methods used for phytoremediation are shown in Fig. 3.

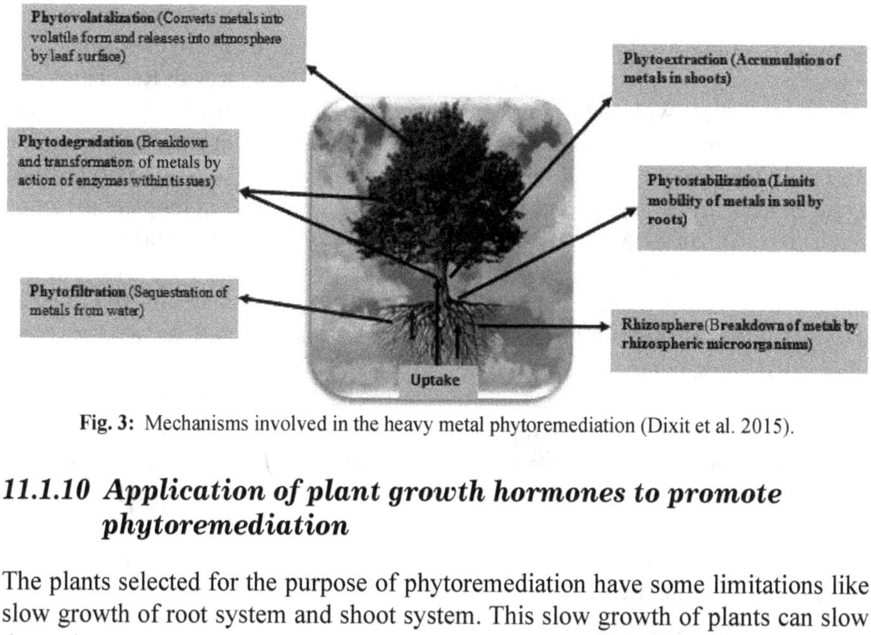

Fig. 3: Mechanisms involved in the heavy metal phytoremediation (Dixit et al. 2015).

11.1.10 Application of plant growth hormones to promote phytoremediation

The plants selected for the purpose of phytoremediation have some limitations like slow growth of root system and shoot system. This slow growth of plants can slow down the process of toxic metal removal process at large scale. Hence, to minimize these limitations, plant growth hormones can be applied on plants to increase their growth as well as to ensure the working capacity of these plants for phytoremediation. Three criteria (time, concentration, and utilisation technique) should be considered while using Plant Growth Regulators. One of the most significant aspects of using these materials is concentration, which has a direct impact on plant growth. In most investigations, Le100 ppm concentrations were employed. Khan et al. examined the effects of different cytokinin concentrations on the Salix tetrasperma plant's features. The results showed that at a concentration of 5 mm, 6-benzylaminopurine (BA) functioned best and boosted the growth while on the other hand, increasing the concentration had negative consequences and stunted plant development. This could have been caused by the hyperhydricity of the shoots (Khan et al. 2012). Israr et al. discovered that combining 10 mm IAA and NAA with EDTA reduced the development of *Sesbania drummondii*. However, increasing the doses up to 100 mM boosted plant growth due to hormonal restorative effects on lead toxicity (Israr and Sahi 2008). Spraying plant growth regulators on plants, water, or soil is another option. The plant growth regulators are sprayed uniformly over the plant tissues and subsequently absorbed by the plant in the plant spraying method. As a result, uniform application of the material to the plant's tissue is crucial in this procedure, and failure to spray particular regions of the plant can have an impact on its morphological traits. Cassina et al. sprayed cytokinin three times on plant leaves at 5-day intervals 34 days after seed germination (Cassina et al. 2011). The indole acetic acid was also utilised three times with 5-day intervals 35 days after sorghum germination (Rostami et al. 2016). Spraying Plant Growth Regulators onto the soil is another option. After irrigation, the chemical will be available to the plant's root. Another strategy involves spraying a large amount of Plant Growth Regulators at a low concentration, with the excess flowing into the soil as runoff. Another method of employing Plant Growth Regulators is to irrigate the plant with these compounds at extremely low quantities (Latimer and Whipker 2013). Auxin was added to distilled water in the study by Liphadzi et al. which was subsequently used for plant irrigation (Liphadzi et al. 2006). After 20 days of development, Cassina et al. sprayed alyssum murale with 10 ml of cytokinin at a concentration of 15 ppm. They also injected 50 ml of cytokinin to the soil at a concentration of 3 ppm (Cassina et al. 2011). One of the most crucial elements is the period of time that Plant Growth Regulators are used. In practice, these compounds should be employed before the plant's morphological and physiological parameters are negatively affected by the environment. To obtain a stronger impact, plant growth regulators should be applied early in the morning or late in the evening when the rate of plant vaporisation is low. In order to maximize absorption, plant growth regulators should have a long contact period with plant tissues.

Phytoremediation is a green and cost-effective way to clean up toxic metals.It can also improve soil quality, reduce erosion, and create jobs. As the technology develops, more economic benefits will be identified as shown in Fig. 4. Phytoremediation is a promising technology with a bright future. It is a sustainable and environmentally friendly way to remediate polluted sites. Root system and shoot system of plants stabilizes the movement of metal ions and from there heavy metals can easily runoff and blow up with wind. As phytoremediation is a totally natural technique, there is no requirement of transportation means and saves the onsite and offsite processing costs.

11.1.11 Role of aromatic plants in phytoremediation

A significant range of aromatic plants have recently been investigated for their phytoremediation capabilities. Families Poaceae, Lamiaceae, Asteraceae, and

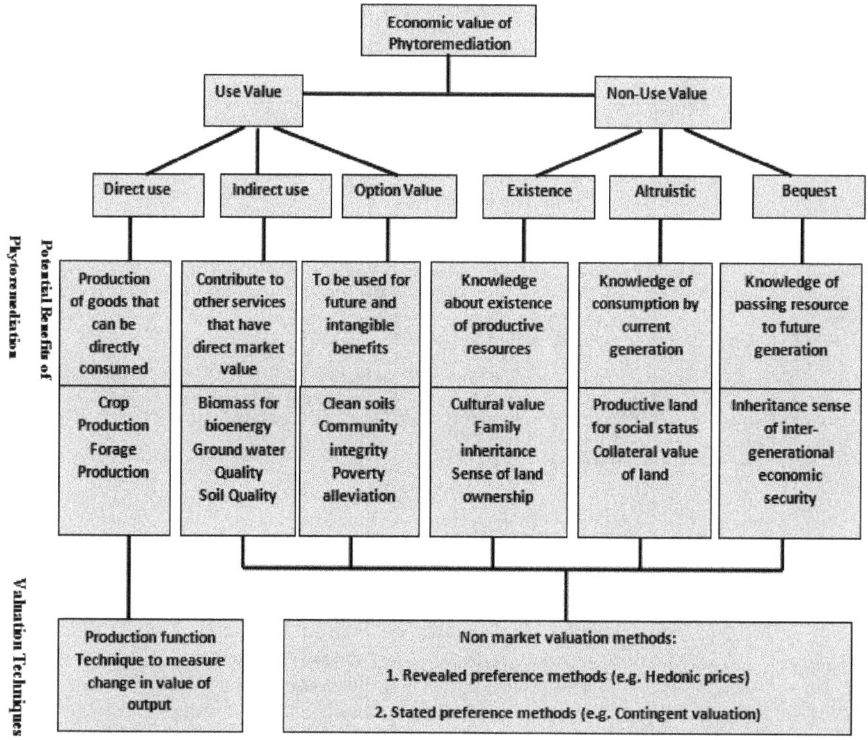

Fig. 4: Flow chart showing the benefits with total economic value of phytoremediation.

Geraniaceae have been recognised as the most promising aromatic plants for phytoremediation of heavy metal polluted environments. Perennial aromatic grasses produce a huge amount of biomass and are commonly used to create high-value essential oils such as Vetiver (*Chrysopogon zizanioides* (L.) Nash), Lemon grass (*Cymbopogon flexuosus* (Nees ex Steud.) Watson), Palmarosa (*Cymbopogon martinii* (Roxb.) Watson) and Citronella (*Cymbopogon winterianus* Jowitt ex Bor.) These aromatic grasses have a lot of potential when it comes to phytoremediation of heavy metal contaminated environments. Plants such as Ocimum, Mentha, Lavender, Salvia, Rosemary (Family – Lamiaceae), Chamomile (Family – Asteraceae), and Geranium (Family – Asteraceae) have been identified for the same (Pandey et al. 2019).

Aromatic plants have a ton of potential when it comes to phytoremediation of heavy metal contaminated environments. They are phytostabilizers, hyper accumulators of certain heavy metals, bio-monitors, and facultative metallophytes. Heavy metal stress has been found to increase the amount of essential oils in certain fragrant crops. As a result, planting these plants in hazardous areas can provide economic benefits.

Table 4: Comparative study of different Himalayan plants showing their role in accumulation of heavy metals.

S. no.	Plant species	Metals absorbed	Result	References
1	*Jatropha curcas*	Cr, Ni, Zn,Cu, Pb (30–70%) Fe (> 3000 mg kg^{-1} plant) Cd, (Completely removed) Hg (Completely removed) Sn (Completely removed)	Fe metal was absorbed in maximum amount, i.e., 15%	Mateos et al. 2019
2	*Jatropha curcas*	Cu (0.27–63.14 mg/kg) Zn (4.82–54.18 mg/kg) Cr (0.48–3.47 mg/kg) Pb (1.75–72.37 mg/kg) Ni (0.05–3.23 mg/kg) in the shoots	*Jatropha curcas, Jatropha gossypifolia, and Jatropha multifida* spp. were able to accumulate heavy metals into their roots and shoots and could therefore be a phytoremediator for studied metals	Awotedu and Ogunbamowo 2019
		Cu (0.25–33.36 mg/kg Zn (5.40–12.48 mg/kg Cr (0.29–1.45 mg/kg Pb(0.58–8.26 mg/kg Ni (0.05–3.23 mg/kg in the roots		
3	*Jatropha curcas*	In Jatropha remediated soil, there was a loss of waste lubricating oil (56.6 percent and 67.3 percent, respectively)	When organic waste (BSG) is added to Jatropha remediation, in oil-contaminated soil, the elimination of waste lubricating oil increases to 89.6 percent and 96.6 percent	Agamuthua et al. 2010
4	*Jatropha curcas*	There were four levels of mercury concentration in the soil in the experimental treatments—T0, T1, T5, and T10 (0, 1, 5, and 10 µg Hg per g soil, respectively). T0 (Not detected) T1 (0.85 ± 0.08) T2 (2.77 ± 0.17) T3 (6.67 ± 1.10)	Hg conc. was found max. in roots followed by leaves and the stem	Negrete et al. 2015
5	*Eucalyptus camaldulenses*	Plants were grown under greenhouse conditions in nutrient solutions with increasing concentrations of Cd (0, 15, 25, 45, 90 µmol m^{-3}) with total accumulation rate of Cd (0.4, 13.6, 15.8, 15.3, 31.8 µg)	There is high biomass production of this species for root part, showing significant potential for use in Cd phytoremediation programmes	Gomes et al. 2012

6	*Eucalyptus camaldulensis*	Bio concentration factor (BCF) Pb in the shoot (0.87) and the root (0.51) Bio concentration factor (BCF) Zn in the shoot (2.02) and root (1.16) Accumulation of Pb (30–300 $mgkg^{-1}$ Pb). Accumulation of Zn (100–400 mg kg-)	The plant species showed a higher BCF shoot value than BCF root, indicating that it could be used as an accumulator and had a high efficacy in phytoextraction of metal-contaminated soils	Ebrahimi 2014
7	*Eucalyptus globulus*	Accumulation of Cd kg^{-1} concentration of four sites (0.43, 0.42, 0.40 and 0.43 mg) Accumulation of Hg kg-1 concentration of four sites (0.59, 0.65, 0.63 and 0.61 mg)	Eucalyptus globulus is a promising phytoremediation species, capable of stabilising contaminants and accumulating metals	Luo et al. 2015
8	*Hibiscus cannabinus*	Roots acquired more than 85% of total plant Pb in Pb-spiked treatments. The immobilisation of Pb in the roots, and hence the restriction of upward mobility (translocation factor < 1) may explain kenaf's ability to tolerate Pb and prevent phytotoxicity	Because of its strong absorption ability, especially in the roots, kenaf could be employed for phytoremediation of Pb-contaminated sand tailings. The immobilisation of Pb in the root and the formation of Pb deposits in cell walls prevented Pb from reaching the aerial portions, showing kneaded tolerance to Pb via the exclusion mechanism.	Mun et al. 2008

11.2 Conclusion

Today, soil and water contamination is a major environmental concern. Metals and other organic and inorganic contaminants are among the most common types of contamination encountered at waste sites, and their removal from soils and sediments is one of the most technically challenging tasks. Due to the high expense of present cleanup technologies, researchers are looking for novel cleanup tactics that are low-cost, low-impact, visually appealing, and environmentally friendly. Various wet land plant species are able to treat the municipal waste water effectively. Plants are used to repair or stabilize contaminated areas in phytoremediation, a new cleanup technique. Phytoremediation is a potential remediation approach for soils that have been contaminated with inorganic contaminants. Because it is a low-cost technology, research connected to it should be fostered, stressed, and extended in developing countries. Phytoremediation is a new contaminant-remediation method that is appealing due to its low cost and adaptability. It isn't a panacea for hazardous waste concerns, but it has a lot of potential in a variety of applications for treating metals and organics at sites with superficial contamination. Phytoremediation is seen as a more environmentally friendly "green" and low-tech alternative to more active and intrusive cleanup processes. Finally, there are genuine dangers involved with phytoremediation, which necessitate a risk assessment and identification of management solutions prior to any field-based operations. Alternatives for management using containment measures like as onsite processing, discing, harvesting before seed set, and volunteer control, pollen and seed movement may be reduced, decreasing possible dangers. If phytoremediation is to gain widespread acceptability, data collection, interpretation, and risk communication must all be examined. This supports a balanced approach to the consideration of phytoremediation's advantages and hazards. We believe that the risks of phytoremediation are minor in comparison to the risks of doing nothing or the financial and engineering risks of a "dig and haul" operation. Through community-based management, phytoremediation is one of the most potent and viable solutions for poor farmers. Phytoremediation will serve as a link between researchers from various institutions and farmers. The use of polluted soils for agricultural purposes will increase as a result of these connections between farmers and researchers. Because this technique is biological, it has a bright future ahead of it. The study of phytoremediation necessitates a thorough understanding of the processes and mechanisms that promote it, as well as plant species' tolerance for saline soils and the markets for agricultural products produced as a result of phytoremediation. To commercialise this approach on a wide scale, additional study and expertise is required, which will assure food security in a sustainable manner while also making the planet Earth a more beautiful place to live.

References

Ahmad, S.S., Reshi, Z.A., Shah, M.A., Rashid, I., Ara, R. and Andrabi, S.M. 2014. Phytoremediation potential of *Phragmites australis* in hokersar wetland-a ramsar site of Kashmir Himalaya. International Journal of Phytoremediation. 16(12): 1183–1191.

Akhtar, M.S., Chali, B. and Azam, T. 2013. Bioremediation of arsenic and lead by plants and microbes from contaminated soil. Research in Plant Sciences. 1(3): 68–73.

Alcock, R.E. and Jones, K.C. 1993. Polychlorinated biphenyls in digested UK sewage sludges. Chemosphere. 26(12): 2199–2207.

Álvarez-Mateos, P., Alés-Álvarez, F.J. and García-Martín, J.F. 2019. Phytoremediation of highly contaminated mining soils by *Jatropha curcas* L. and production of catalytic carbons from the generated biomass. Journal of Environmental Management. 2019 Feb 1; 231: 886–95.

Anjum, N.A., Ahmad, I., Mohmood, I., Pacheco, M., Duarte, A.C., Pereira, E. et al. 2012a. Modulation of glutathione and its related enzymes in plants' responses to toxic metals and metalloids—a review. Environmental and Experimental Botany. 2012; (75): 307–324.

Anjum, N.A., Gill, S.S., Ahmad, I., Pacheco, M., Duarte, A.C., Umar, S. et al. 2012b. The plant family Brassicaceae: An introduction. In The Plant Family Brassicaceae. Springer. 2012(21): 1–33.

Antonious, G.F. and Snyder, J.C. 2007. Accumulation of heavy metals in plants and potential phytoremediation of lead by potato, *Solanum tuberosum* L. Journal of Environmental Science and Health, Part A. 2007 May 7; 42(6): 811–6.

Awotedu, O.L. and Ogunbamowo, P.O. 2019. Comparative heavy metal uptake and phytoremediation potential of three *Jatropha* Species. Environment & Ecosystem Science (EES). 3(2): 26–30.

Bafeel, S.O. 2008. Contribution of mycorrhizae in phytoremediation of lead contaminated soils by *Eucalyptus rostrata* plants. World Appl. Sci. J. 5(4): 490–8.

Bhatnagar, S. and Kumari, R. 2013. Bioremediation: a sustainable tool for environmental management–a review. Annual Review and Research in Biology. 3(4): 974–993.

Cassina, L., Tassi, E., Morelli, E., Giorgetti, L., Remorini, D., Chaney, R.L. et al. 2011. Exogenous cytokinin treatments of an Ni hyper-accumulator, Alyssum murale, grown in a serpentine soil: Implications for phytoextraction. International Journal of Phytoremediation. 2011 Jan 1;13(sup1): 90–101.

Cearley, J.E. and Coleman, R.L. 1973. Cadmium toxicity and accumulation in southern naiad. Bulletin of Environmental Contamination and Toxicology. 9(2): 100–101.

Cederlöf, G. 2014. Monsoon Landscapes: patial Politics and Mercantile Colonial Practice in India. RCC Perspectives. 2014 Jan 1(3): 29–36.

Cho-Ruk, K., Kurukote, J., Supprung, P. and Vetayasuporn, S. 2006. Perennial plants in the phytoremediation of lead-contaminated soils. Biotechnology. 5(1): 1–4.

Cunningham, S.D. and Lee, C.R. 1995. Phytoremediation: Plant-based remediation of contaminated soils and sediments. Bioremediation: Science and Applications. 145–56.

Das, S. and Goswami, S. 2017. Copper phytoextraction by *Salvinia cucullata*biochemical and morphological study. Environmental Science and Pollution Research. 24(2): 1363–1371.

Dickinson, N.M., Mackay J.M., Goodman, A. and Putwain, P. 2000. Planting trees on contaminated soils: issues and guidelines. Land Contamination & Reclamation. 8(2): 87–102.

Didierjean, L., Gondet, L., Perkins, R., Lau, S.M., Schaller, H. et al. 2002. Engineering herbicide metabolism in tobacco and Arabidopsis with CYP76B1, a cytochrome P450 enzyme from Jerusalem artichoke. Plant Physiology. 130(1): 179–189.

Dixit, R., Malaviya, D., Pandiyan, K., Singh, U.B., Sahu, A., Shukla, R. et al. 2015. Bioremediation of heavy metals from soil and aquatic environment: an overview of principles and criteria of fundamental processes. Sustainability. 7(2): 2189–212.

Domínguez, M.T., Madejón, P., Marañón, T. and Murillo, J.M. 2010. Afforestation of a trace-element polluted area in SW Spain: woody plant performance and trace element accumulation. European Journal of Forest Research. 129(1): 47.

Dzantor, E.K. 2007. Phytoremediation: the state of rhizosphere engineering for accelerated rhizodegradation of xenobiotic contaminants. Journal of Chemical Technology & Biotechnology: International Research in Process, Environmental & Clean Technology. 82(3): 228–232.

Ebrahimi, M. 2014. Effect of EDTA and DTPA on phytoremediation of Pb-Zn contaminated soils by *Eucalyptus camaldulensis* Dehnh and Effect on Treatment Time. Desert. 2014 Jan 1; 19(1): 65–73.

Erakhrumen, A.A. and Agbontalor, A. 2007. Phytoremediation: an environmentally sound technology for pollution prevention, control and remediation in developing countries. Educational Research and Review. 2(7): 151–6.

Etim, E.E. 2012. Phytoremediation and its mechanisms: a review. international Journal of Environment and Bioenergy. 2012 Jan 1;2(3): 120–36.

Farooq, H., Siddiqui, M.T., Farooq, M., Qadir, E. and Hussain, Z. 2006. Growth, nutrient homeostatis and heavy metal accumulation in *Azadirachta indica* and *Dalbrgiasissoo* Seedlings Raised from Waste Water. International Journal of Agriculture Biology. 8(4): 504–507.

Friesl, W., Friedl, J., Platzer, K., Horak, O. and Gerzabek, M.H. 2006. Remediation of contaminated agricultural soils near a former Pb/Zn smelter in Austria: batch, pot and field experiments. Environmental Pollution. 144(1): 40–50.

Ghosh, M. and Singh, S.P. 2005. A review on phytoremediation of heavy metals and utilization of it's by products. Asian Journal of Energy and Environment. 6(4): 18.

Gomes, H.I. 2012. Phytoremediation for bioenergy: challenges and opportunities. Environmental Technology Reviews. 2012 Nov 1;1(1): 59–66.

Goswami, S. and Das, S. 2016. Copper phytoremediation potential of *Calandula officinalis* L. and the role of antioxidant enzymes in metal tolerance. Ecotoxicology and Environmental Safety. 126: 211–218.

Gurajala, H.K., Cao, X., Tang, L., Ramesh, T.M., Lu, M. and Yang, X. 2019. Comparative assessment of Indian mustard (*Brassica juncea* L.) genotypes for phytoremediation of Cd and Pb contaminated soils. Environmental Pollution. 2019 Nov 1; 254: 113085.

Hale, M.G. and Griffin, G.J. 1974. Effect of injury on exudation from immature and mature plant fruits. Plants Physiology. 13.

Hasan, M.K., Ahammed, G.J., Sun, S., Li, M., Yin, H. et al. 2019. Melatonin inhibits cadmium translocation and enhances plant tolerance by regulating sulfur uptake and assimilation in *Solanum lycopersicum* L. Journal of Agricultural and Food Chemistry. 5; 67(38): 10563–76.

Israr, M. and Sahi, S.V. 2008. Promising role of plant hormones in translocation of lead in *Sesbania drummondii* shoots. Environmental Pollution. 2008 May 1;153(1): 29–36.

Itanna, F. and Coulman, B. 2003. Phyto-extraction of copper, iron, manganese, and zinc from environmentally contaminated sites in Ethiopia, with three grass species. Communications in Soil Science and Plant Analysis. 1; 34(1-2): 111–24.

Jansson, S. and Douglas, C.J. 2007. Populus a model system for plant biology. Annual Review of Plant Biology. 58: 435–458.

Kawahigashi, H., Hirose, S., Ohkawa, H. and Ohkawa, Y. 2007. Herbicide resistance of transgenic rice plants expressing human CYP1A1. Biotechnology Advances. 25(1): 75–84.

Khan, M.I. and Anis, M. 2012. Modulation of *in vitro* morphogenesis in nodal segments of *Salix tetrasperma* Roxb. through the use of TDZ, different media types and culture regimes. Agroforestry systems. 2012 Sep. 1; 86(1): 95–103.

Kumar, J.N., Soni, H., Kumar, R.N. and Bhatt, I. 2008a. Macrophytes in phytoremediation of heavy metal contaminated water and sediments in Pariyej Community Reserve, Gujarat, India. Turkish Journal of Fisheries and Aquatic Sciences. 8(2).

Kumar, J.N., Soni, H. and Kumar, R.N. 2008b. Evaluation of biomonitoring approach to study lake contamination by accumulation of trace elements in selected aquatic macrophytes: a case study of Kanewal Community Reserve, Gujarat, India. Applied Ecology and Environmental Research. 6(1): 65–76.

Kvesitadze, E., Sadunishvili, T. and Kvesitadze, G. 2009. Mechanisms of organic contaminants uptake and degradation in plants. World Academy of Science, Engineering Technology. 55(6): 458–468.

Lasat, M.M. 1999. Phytoextraction of metals from contaminated soil: a review of plant/soil/metal interaction and assessment of pertinent agronomic issues. Journal of Hazardous Substance Research. 2(1): 5.

Lasat, M.M. 2000. The use of plants for the removal of toxic metals from contaminated soils. US Environmental Protection Agency.

Lichtfouse, E. and Eglinton, T.I. 1995. 13C and 14C evidence of pollution of a soil by fossil fuel and reconstruction of the composition of the pollutant. Organic Geochemistry. 1995 Oct 1; 23(10): 969–73.

Liphadzi, M.S. and Kirkham, M.B. 2006. Availability and plant uptake of heavy metals in EDTA-assisted phytoremediation of soil and composted biosolids. South African Journal of Botany. 2006 Aug 1; 72(3): 391–7.

Lone, M.I., He, Z.L., Stoffella, P.J. and Yang, X.E. 2008. Phytoremediation of heavy metal polluted soils and water: progresses and perspectives. Journal of Zhejiang University Science. 9(3): 210–220.

Luo, J., Qi, S., Peng, L. and Wang, J. 2016. Phytoremediation efficiency of Cd by Eucalyptus globulus transplanted from polluted and unpolluted sites. International Journal of Phytoremediation. 2016 Apr 2;18(4): 308–14.

Madejón, P., Murillo, J.M., Marañón, T., Cabrera, F. and Soriano, M.A. 2003. Trace element and nutrient accumulation in sunflower plants two years after the *Aznalcóllar* mine spill. Science of the Total Environment. 307(1-3): 239–257.

Mahimairaja, S., Shenbagavalli, S. and Naidu, R. 2010. Remediation of Chromium-Contaminated Soil due to Tannery Waste Disposal: Potential for Phyto and Bioremediation (Symposium 3.5. 1 Heavy Metal Contaminated Soils,< Special Issue> International Symposium: Soil Degradation Control, Remediation, and Reclamation, Tokyo Metropolitan University Symposium. 54(3): 175–181.

Mandal, A., Purakayastha, T.J., Ramana, S., Neenu, S., Bhaduri, D. et al. 2014. Status on Phytoremediation of Heavy Metals in India-A Review. International Journal of Bio Resources and Stress Management. 5: 553–560.

Marrugo-Negrete, J., Durango-Hernández, J., Pinedo-Hernández, J., Olivero-Verbel, J. and Díez, S. 2015. Phytoremediation of mercury-contaminated soils by Jatropha curcas. Chemosphere. 2015 May 1; 127: 58–63.

Mishra, S. and Maiti, A. 2017. The efficiency of *Eichhornia crassipes* in the removal of organic and inorganic pollutants from wastewater: a review. Environmental Science and Pollution Research. 24(9): 7921–7937.

Mishra, V., Majumder, C.B. and Agarwal, V.K. 2012. Sorption of Zn (II) ion onto the surface of activated carbon derived from eucalyptus bark saw dust from industrial wastewater: isotherm, kinetics, mechanistic modeling, and thermodynamics. Desalination and Water Treatment. 46(1-3): 332–351.

Mojiri, A. 2011. The potential of corn (*Zea mays*) for phytoremediation of soil contaminated with cadmium and lead. J. Biol. Environ. Sci. 5(13): 17–22.

Moosavi, S.G. and Seghatoleslami, M.J. 2013. Phytoremediation: a review. Advance in Agriculture and Biology. 1(1): 5–11.

Mun, WHO, Hoe, L.A. and Koo, L.D. 2008. Assessment of Pb uptake, translocation and immobilization in kenaf (*Hibiscus cannabinus* L.) for phytoremediation of sand tailings. Journal of Environmental Sciences. 20: 1341–1347.

Muthusaravanan, S., Sivarajasekar, N., Vivek, J.S., Paramasivan, T., Naushad, M., Prakashmaran, J. et al. 2018. Phytoremediation of heavy metals: mechanisms, methods and enhancements. Environmental Chemistry Letters. 2018 Dec. 15; 16(4): 1339–59.

Ney, R.A. and Schnoor, J.L. 2002. Incremental life cycle analysis: using uncertainty analysis to frame greenhouse gas balances from bioenergy systems for emission trading. Biomass and Bioenergy. 22(4): 257–269.

Okunowo, W. and Ogunkanmi, L.A. 2010. Phytoremediation potential of some heavy metals by water hyacinth. International Journal of Biological and Chemical Sciences. 4(2).

Pandey, J., Verma, R.K. and Singh, S. 2019. Suitability of aromatic plants for phytoremediation of heavy metal contaminated areas: a review. International Journal of Phytoremediation. 2019 Apr 16; 21(5): 405–18.

Parikh, J. and Datye, H. 2003. Sustainable management of wetlands: biodiversity and beyond. Sage Publications India.

Raghav, N. and Srivastava, J.N. 2016. Toxic pollution in river water and bacterial remediation: an overview. International Journal of Current Microbiology Applied Science. 5(4): 244–266.

Rahman, M.A. and Hasegawa, H. 2011. Aquatic arsenic: phytoremediation using floating macrophytes. Chemosphere. 83(5): 633–646.

Rakhshaee, R., Giahi, M. and Pourahmad, A. 2009. Studying effect of cell wall's carboxyl–carboxylate ratio change of Lemna minor to remove heavy metals from aqueous solution. Journal of Hazardous Materials. 163(1): 165–73.

Rathaur, P., Ramteke, P.W., Raja, W. and John, S.A. 2012. Isolation and characterization of nickel and cadmium tolerant plant growth promoting rhizobacteria from rhizosphere of *Withania somnifera*. Journal of Biological and Environmental Sciences. 6(18).

Reeves, R.D., Baker, A.J., Jaffré, T., Erskine, P.D., Echevarria, G. and van der Ent, A. 2018. A global database for plants that hyperaccumulate metal and metalloid trace elements. New Phytologist. 218(2): 407–11.

Revathi, K., Haribabu, T.E. and Sudha, P.N. 2011. Phytoremediation of chromium contaminated soil using sorghum plant. International Journal of Environmental Sciences. 2(2): 417.

Robineau, T., Batard, Y., Nedelkina, S., Cabello-Hurtado, F., LeRet, M. et al. 1998 The chemically inducible plant cytochrome P450 CYP76B1 actively metabolizes phenylureas and other xenobiotics. Plant Physiology. 118(3): 1049–1056.

Robinson, B., Green, S., Mills, T., Clothier, B., van der Velde, M. et al. 2003. Phytoremediation: using plants as biopumps to improve degraded environments. Soil Research. 41(3): 599–611.

Robinson, W.O., Edgington, G. and Byers, H.G. 1935. Chemical studies of infertile soils derived from rocks high in magnesium and generally high in chromium and nickel. US Department of Agriculture.

Rostami, A., Firoozfar, A., Adhami, B., Asghari, N. 2016. Impact of soil type used in tunneling on land subsidence and mobility effective time under different earthquake records. Open Journal of Geology. 6(11): 1469.

Salt, D.E., Smith, R.D. and Raskin, I. 1998. Phytoremediation. Annual Review of Plant Biology. 49(1): 643–68.

Sargaonkar, A. and Deshpande, V. 2003. Development of an overall index of pollution for surface water based on a general classification scheme in Indian context. Environmental Monitoring and Assessment. 89(1):43–67.

Sharma, P. and Pandey, S. 2014. Status of phytoremediation in world scenario. International Journal Environmental Bioremediation and Biodegradation. 2: 178–191.

Shirkhani, Z., Rad, A.C., Gholami, M. and Mohsenzadeh, F. 2018. Phytoremediation of Cd-contaminated Soils by *Datura stramonium* L. Toxicology and Environmental Health Sciences. 2018 Sep 1; 10(3): 168–78.

Shrestha, P., Bellitürk, K. and Görres, J.H. 2019. Phytoremediation of heavy metal-contaminated soil by switchgrass: A comparative study utilizing different composts and coir fiber on pollution remediation, plant productivity, and nutrient leaching. International Journal of Environmental Research and Public Health. 16(7): 1261.

Shtangeeva, I., Laiho, J.V., Kahelin, H. and Gobran, G.R. 2004. Phytoremediation of metal-contaminated soils. symposia papers presented before the division of environmental chemistry. InAmerican Chemical Society, Anaheim, Calif, USA, http://ersdprojects. science. doe. gov/workshop% 20pdfs/ california 2004 (Vol. 202004, p. p050).

Shukla, A.C. and Asthana, V. 1995. Ganga, a water marvel. APH Publishing.

Singhal, B.B. and Gupta, R.P. 2010. Groundwater quality. In Applied Hydrogeology of Fractured Rocks 2010 (pp. 205–220). Springer, Dordrecht.

Srivastava, S., Srivastava, S., Prakash, S. and Srivastava, M.M. 1999. Fate of trivalent chromium in presence of organic acids. Chemical Speciation and Bioavailability. 10: 147150.

Susarla, S., Medina, V.F. and McCutcheon, S.C. 2002. Phytoremediation an ecological solution to organic chemical contamination. Ecological Engineering. 18(5): 647–658.

Tandon, S.A., Kumar, R. and Yadav, S.A. 2013. Pytoremediation of fluoroquinolone group of antibiotics from waste water. Natural Science. 5(12): 21.

Tangahu, B.V., Sheikh Abdullah, S.R., Basri, H., Idris, M., Anuar, N. and Mukhlisin, M. 2011. A review on heavy metals (As, Pb, and Hg) uptake by plants through phytoremediation. International Journal of Chemical Engineering. 1: 2011.

Tengö, M. and Hammer, M. 2003. Management practices for building adaptive capacity a case from northern Tanzania. Cambridge University Press, Cambridge, UK.

Trap, S., Kohler, A., Larsen, L.C., Zambrano, K.C. and Karlson, U. 2005. Phytotoxicity of fresh and weathered diesel and gasoline to willow and poplar trees. J. Soil Sediments. 1: 71–76.

Vangronsveld, J., Herzig, R., Weyens, N., Boulet, J., Adriaensen, K. et al. 2009. Phytoremediation of contaminated soils and groundwater: lessons from the field. Environmental Science and Pollution Research. 16(7): 765–794.

Vasavi, A., Usha, R. and Swamy, P.M. 2010. Phytoremediation–an overview review. Journal of Industrial Pollution Control. 26: 83–88.

Zehra, S.S., Arshad, M., Mahmood, T. and Waheed, A. 2009. Assessment of heavy metal accumulation and their translocation in plant species. African Journal of Biotechnology. 8(12).

Zhai, X., Li, Z., Huang, B., Luo, N., Huang, M., Zhang, Q et al. 2018. Remediation of multiple heavy metal-contaminated soil through the combination of soil washing and *in situ* immobilization. Science of the Total Environment. 1; 635: 92–9.

Zheljazkov, V.D. and Nielsen, N.E. 1996. Effect of heavy metals on peppermint and cornmint. Plant and Soil. 178(1): 59–66.

Zheljazkov, V.D. and Nielsen, N.E. 1996. Studies on the effects of heavy metals (Cd, Pb, Cu, Mn, Zn and Fe) upon the growth, productivity and quality of lavender (*Lavandula angustifolia* Mill.) production. Journal of Eessential Oil Research. 8(3): 259.

12

Synthesis of Nanomaterials for Protein Extraction

*CH. Harshitha, Aditi Chaturvedi, Surajit Mondal**
and *K. Krishna Koundinya*

12.1 Introduction

It is vital to ensure effective wastewater treatment in order to prevent disease, protect the environment, and keep life and the economy going. Among the technologies present for waste water, photocatalytic process is one of the promising techniques, especially when it comes to fulfilling the "no wastewater objective" in industrial processes. As UV-B rays penetrate the depleting ozone layer, a sharp rise in photochemical processes has been observed. This process occurs as photocatalytic oxidative degeneration in biological toxins such as pesticides, natural fertilizers, and organic matter in the insolation-receiving surface and sub-surface of the larger hydrosphere. Technologically available photocatalytic methods of water purification using organic pollutants generally practice oxidising molecular oxygen, ozone and hydrogen peroxide stimulated by artificial or actual UV where there is stark occurrence of PCs in dissolved or suspended state.

The retrieval of the catalyst particles post-treatment still poses a challenge to making this technology commercially viable (Yaqoob et al. 2020). Another path breaking technique comes from the domain of microbubble and nanobubble technologies where they have an appreciable surface area for chemical reactions to take place (Bouaifi et al. 2001). This provides for the inhibition of the MNBs (micro nano bubbles), responsible for production of hydroxyl radical leading to instability (Khuntia et al. 2015). Other widely adopted water treatment processes like electro-floatation and froth flotation by dissolving air (Miettinen et al. 2010) need MNBs to function owing to its high bioactivity with contaminants found in domestic as well as industrial waste (Kulkarni and Joshi 2005).

Department of Electrical and Electronics Engineering, University of Petroleum and Energy Studies.
Emails: harshithachandra08@gmail.com; aditichaturvedi1098@gmail.com;
krishnakoundinya007@gmail.com
* Corresponding author: smondal@ddn.upes.ac.in

Purifying proteins is either for commercial applications in larger quantities or for analytical applications such extracting the traces of protein in structural or functional research. (Chen et al. 2017) investigated the viability of the use of the process of ultrafiltration (UF) for protein recovery from poultry wastewater. It was found in this study that employing a poly-sulfone membrane led to retainment of most of crude proteins in wastewater, in addition to reduction of the chemical oxygen demand (COD) in the run-off to less than 200 mg L^{-1}. Optimized values for parameters such pH, volumetric flow rate, and transmembrane pressure were identified with the help of response surface methodology (RSM), to significantly improve upon the average flux to well above 200 L m^{-2} h^{-1} while keeping membrane performance as consistent as possible.

Nasrollahzadeh et al. (2021) reviewed the use of green nanomaterials and nanocomposites for removal of proteins ad microbes from food processing wastewater. The study also evaluated the effect of the same on various parameters and the preparation of nanoparticles from wastewater.

Jiang et al. (2011) is an assessment of nano-catalysts and nanomaterials synthesized in an environmental-friendly way for the purpose of wastewater treatment with respect to concerns such as superior surface area, mechanical properties, significant chemical reactivity, and cost effectiveness with lesser energy consumption. It also evaluates how recovery from wastewater has aided with innovations in nanofabrication and smart nanomaterials.

A 2-stage foam separation technology was developed by (Chollangi and Hossain 2007) for protein recovery from wastewater. The enrichment ratio after first stage was 8.5. Almost 80 percent recovery of protein was achieved at the end of second stage.

Zhang et al. (2016) scrutinized ultrafiltration technique for fractional processing of dairy wastewater into lactose and protein concentrated steam. By optimizing the process variables like operating temperature, trans-membrane pressure, and lactose concentration were used for evaluating performance in pure lactose solution, lactose concentration with protein and ultimately in dairy wastewater. The best membrane for fractionation was concluded to be of 10 kDa.

Baruah et al. (2018) elucidated upon various nano-technology methods for wastewater treatment such as—carbon/polymer/metal/zeolite founded absorbents and membrane centred nano filtration techniques, nanomaterials for antimicrobial control, heterogeneous photo catalysts for wastewater treatment, in addition to challenges and threats of using nanomaterials.

Hülsen et al. (2018) expounded on synthesis and physiochemical properties of assorted free nanomaterials, carbon-based nanomaterial, metal and metal oxides, noble metal oxides, noble metal nanoparticles, environmental nanocomposites, nanocomposite membrane and magnetic nanocomposites—their mechanism and performance on application in wastewater treatment.

The large surface to volume ratio of nanomaterials is the key property that makes it so useful in all its versatile applications. This is also the reason for its suitability for protein extraction. Several nanomaterials, including zero valent metal nanoparticles, silver nanoparticles, iron nanoparticles, zinc nanoparticles, metal oxides nanoparticles, and carbon nanotubes are utilized for wastewater treatment (Abdelbasir and Shalan 2019). This idea is also relevant for the domain of microbubble and nanobubble technology.

The large area allowing numerous chemical reactions is responsible for inhibition of micro nano bubbles (MNB) leading to production of hydroxyl radicals, destabilizing the complex organic chains. Hence, nanoparticles are used for protein recovery from waste water owing to their high bioactivity with contaminants found in domestic and industrial waste water (Kulkarni and Joshi 2005).

12.2 Nano technology for waste water treatment

With nano technology, many processes are able to provide effective performance results which are affordable and suitable for waste water treatment. The presence of large surface area and small volume is suitable for this operation and also nanoscale allows them to change their properties –biological, chemical, electrical, physical and optical.

Processing of waste water treatment with nanotechnology not only to solve major factors and challenges which include inefficient purification processes, complexity in recovering the adsorbents and membranes makes the processes cost ineffective which are mentioned in Fig. 1, but also to deliver an efficient process and operation to enhance the purity of water and increase the supply from emerging water sources. This technology is effective for biological and chemical reactions, and chemical affinity towards target is increased as they are adaptable to various reactor groups.

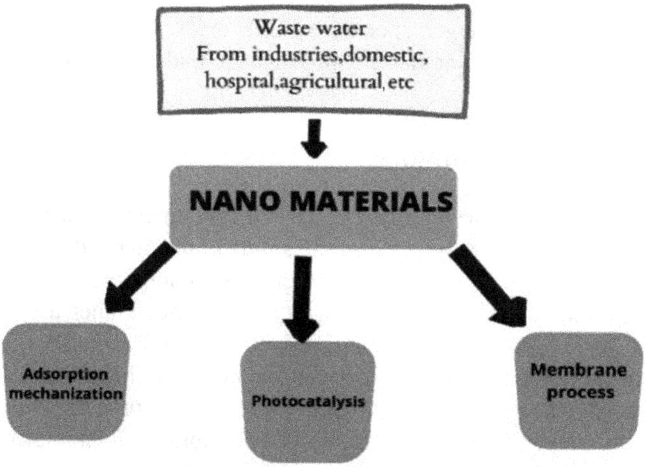

Fig. 1: Wastewater treatment based on nanotechnology.

12.3 Applications of nanotechnology in water treatment

12.3.1 Adsorption

Current limitations of adsorbents are surface area, adsorption kinetics, and active sites, so the efficiency is affected. Nano adsorbents have improved properties like having high specific surface area, associated adsorption sites, pore size, and surface chemistry.

12.3.2 Nano adsorbents for separation of heavy metals

Metals such as arsenic, copper, lead, cadmium, chromium, zinc, nickel, mercury, and cobalt are present in their respective oxidation states in the waste water. These metals can be separated using the process of adsorption with nano-adsorbents. The significance of the utilization of nano-adsorbents is that the adsorption efficiency with these adsorbents will be high and appreciable. The choice of adsorbent depends on the oxidation state of the metal present in the waste water. The separation process for majority of the metals follows the pseudo second order and second order reactions depending on the nature of metals. These nano-adsorbents sometimes follow chemical methods which might leave some toxicity in the water even after removing the heavy metals. These processes are also not economical as the adsorbents used are very expensive.

12.3.3 Carbon based nano-adsorbents

Carbon adsorption for home water filter treatment is a widely used method because of its ability to improve the quality of water by eliminating disagreeable tastes and odours including chlorine, can be effective against microorganisms, and will not affect total dissolved solids during the process (Jain et al. 2021). Adsorption efficiencies for the carbon nanotubes (CNTs) will be more than that of activated carbon. The properties of CNTs like larger specific surface areas, and diverse contaminate interaction properties, and their viability improve their adsorption ability for various chemicals (Nnaji et al. 2018).

12.3.4 Membrane technology

Membrane technology has a significant place in the treatment of waste water as they use membranes that selectively separate the solid contaminants depending on their properties and respective sizes. Several developments to improve the parameters like permeability, durability, viability and strength of the membranes are made which enhanced the applications of membranes (Fane et al. 2011). The membrane separation processes are classified as equilibrium driven and non-equilibrium driven processes. The operational parameters for the equilibrium driven process like membrane distillation are temperature, and pressure. The membrane technologies which are not driven by pressure are forward osmosis, liquid membrane technologies, and electrodialysis which is driven electrically (Jhaveri and Murthy 2016).

Depending on the applications, the choice of material for membrane varies and usually both organic and inorganic materials are used to manufacture membranes. Factors like pore size, structure and material of membranes play a key role in filtration process (Elorm Obotey Ezugbe 2020). The membranes with uniform structure and composition are called isotropic membranes and the membranes with non-uniform structure are called anisotropic membranes which are used in reverse osmosis.

Table 1: List of other nano-membranes of their optimistic characteristics and approaches (Manikandan et al. 2021).

Nano-membranes	**Characteristics**		**Approaches**	**Water treatment Nanomembranes' thickness**
	Optimistic	**Negotiable**		
Nano-composite membranes	Improve hydrophobicity, thermal/mechanical stability, and fouling resistance	Resistant bulk material needed to release potential nanoparticles while using oxidizing nanomaterial	Extremely complex conditions, such as micro-pollutant elimination and reverse osmosis	4.1 to 49.8 μm^2 (Wen et al. 2019)
Nano-filtration membranes	Repulsion dependent on load, low relative pressure, and high selectivity	Blocking of the membrane (polarization of the concentration)	Heavy metals reduction, odour, colour and hardness	Zeolite membranes (thickness $\approx 3\ \mu m$) (Pendergast and Hoek 2011)
Aquaporin based membranes	Permeability and high ionic resistance	Mechanical impairment	Desalination under low pressure	Aquaporin Flat-Sheet Membrane (thickness ~ 110 μm) (Xia et al. 2017)
Nano self-assembly membranes	Nanopores	Accessible in small amounts (laboratory scale)	Ultrafiltration	~ 40 μm thick (Feng et al. 2019)
Nano-fibre membranes	High efficiency of permeability, customized, bactericidal, and porous	Nanofibres may be released, pore-blocking possible	Stand-alone filter, ultrafiltration, cartilage filters, water filtering, pre filtration	PUR10 is mere 6 μm, nylon 6 nanofibre/cellulose 205 μm (Fauzi et al. 2020, Jiříček et al. 2017)

Several development methods are executed to improve the separation efficiency and quality of the process, among which application of nano tubes, nano-fibres and other organic nano membranes have an appreciable potential to fulfil the process requirements.

12.3.5 *Nano fibre membrane*

Nano-fibres have an appreciable significance in various operations such as adsorption, catalysis and separation technologies. They are made of polymers which might be natural or synthesized depending on the application of fibres (Ahn et al. 2006, Lannutti et al. 2007, Vasita and Katti 2006, Reneker and Yarin 2008, SantosMiranda et al. 2006). These nanofibres have many advantages such as higher surface areas, resistance to corrosion, higher strengths, anti-fouling properties and appreciable separation efficiencies. These fibres have several industrial applications like reverse osmosis, and heavy metal separation from industrial solvents and wastewater. In spite of having effective advantages, still these fibres have some disadvantages like regeneration, large scale production, and high manufacturing costs which can be minimized with improvement in the demand.

12.3.6 *Biologically inspired membranes*

Application of engineering technologies using biological methods for the production of bio-inspired membranes which include carbon nano tubes, graphenes, and carbon fibers (Vullev 2011, Song et al. 2018). Both degradable and non-biodegradable contaminants can be removed using these biologically inspired membranes. Wastewater containing various plastic impurities with smaller sizes in microscale can also be treated using the membranes like aquaporin and its composites by consuming minimum energy. But still these membranes have some limitations such as high drops in permeability flux after the operation, efficiency drops, high maintenance requirements and fouling (Singh et al. 2018, Pedersen et al. 2018, Engelhardt et al. 2018, Munshi 2019, Wagh and Escobar 2019). Scaling up the separation processes could be possible by further research on these artificial membranes. They have a good potential to withstand the high pressures which makes them more advantageous with some optimization on the efficiency studies and recovery methods for reuse in the separation methods.

12.3.7 *Photocatlaysis*

Semi-conducting materials such as oxides of titanium and zinc form the brand of photo-catalyzing materials. The mechanism by which they work is that they their function of light absorption and generation of reactive oxygen breaks down pollutants and contaminants through semiconductors that have been doped with metal and adsorption properties. The free hydroxyl radicals culminate into formation of biological molecules that mutilate pollutants in wastewater. The hydroxyl ions result from electron-hole pair generation which in turn is due to charge separation incited by absorption of photons by nanomaterials and the subsequent excitation in the valence and conduction bands. The requirement of a large amount of energy, or simply put, the band gap energy, puts quite a considerable limit on the efficacy of this method. This can be corrected by employing an ultraviolet lamp to fulfill the deficit band gap energy for the excitation to take place. However, it significantly adds to the cost of employing the method of photocatalysis.

Other factors that limit the efficiency of photocatalysis are the rate of recombination, the rate of transfer and mobility of charge carriers. Restructuring and doping can be made with regards to the material to address these by reducing the bandgap and in turn mitigating rapid recombination, in addition to using plasma materials that further reduce energy required for excitation (Baruah et al. 2018). Photo-catalysts have now moved into commercial applications. An example of a commercialized method is use of the titanium oxide as a photocatalyst in purifying water, which is activated by UV. It is widely used in combination with AOP which breaks down several pollutants. Research and development of photocatalytic materials has picked up pace in recent times and is constantly contributing to making this technology affordable, feasible and efficient in mega-scale applications.

12.3.8 Disinfection and microbial control

Waste water can be separated from the microbials and purified by various methods like thermal disinfection, chlorination, activated sludge processes, membrane filtration, reverse osmosis and biological methods for destroying the pathogens. Moreover, many developments have been implemented to remove the microbials and pathogens from the water. Nanomaterials like fullerenes, activated carbon nanotubes, and other polymer membranes have an appreciable potential to separate the microbes with cost effective methods. Numerous studies have been done and development activities are emerging to improve the sustainable technologies for treating the wastewater with economical methods with higher efficiencies by developing various eco-friendly nanomaterials.

12.4 Approaches

Two approaches, namely top-down and bottom-up, can be executed to synthesize the nanomaterials. In the top-down approach, majority of materials are split into nano sized structures. In the production of micron sized particles, this approach adds as an extension. This approach is quite simple—the breaking of bulk into various forms undergoing various fabrication processes to supply the specific requirement of properties. The challenge incudes the imperfection of surface structure. For example, the nanowires are not smooth and have lot of impurities and defects on the surface which can be resolved by lithography. These techniques include high energy wet ball milling, lithography beam, gas-phase condensation, aerosol spray, etc.

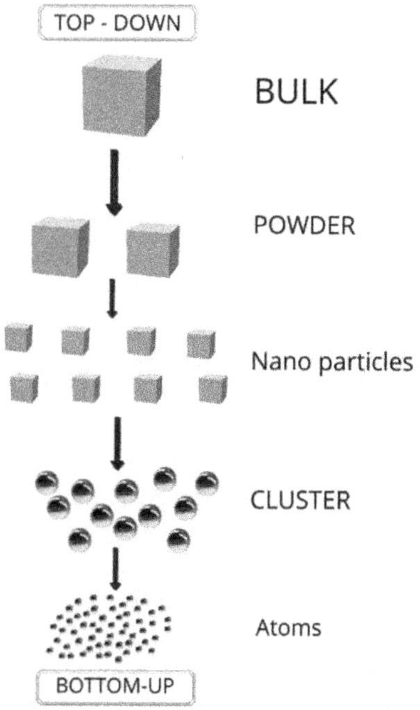

Fig. 2: Synthesis route of nanomaterials.

The bottom down approach is more economical and produce less waste; in this, the materials are made from bottom-to-top : from atoms to molecules, molecules-clusters, clusters to powder and to bulk material. The techniques of this approach are still in their cradle stage to be used in commercial market production of nanotechnology. Some prominent techniques include the reverse micelle route, sol-gel synthesis, colloidal precipitation, etc. (Joshi et al. 2021).

12.4.1 Synthesis techniques of nanomaterials

Nanomaterials can be synthesized either by physical operation or chemical operation depending on the nature of the materials and their applications. A summary describing various methods for synthesizing nano-materials are shown in the Fig. 3.

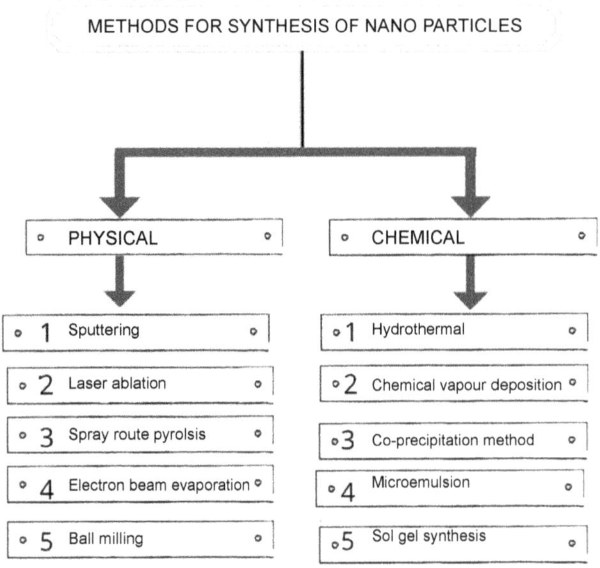

Fig. 3: Summary of methods for synthesizing nano particles.

12.4.2 Chemical operations

12.4.2.1 Hydro thermal

Hydrothermal operation is one of the most common methods used for preparation of nanomaterials, using solution reaction-based approach. The morphology of the nano particles is optimized by the pressure variation. Varieties of nano particles are synthesised using this approach, and the generated nanoparticles are found unstable at aerial temperatures. With high vapour pressures and minimum loss of materials, nanomaterials can be produced by this method. Through liquid and multiphase chemical reactions, nanomaterials' composition can be controlled (Malik et al. 2012).

12.4.2.2 Chemical precipitation technique

This technique is implemented to prepare the nano particles of calcium hydroxide. This synthesis involves a chemical reaction, and it can be effectively manufactured in large scale economically (Roy 2014). In this process, the selection of suitable reactants is a major factor for any kind of chemical synthesis method. In order to select appropriate reactant, knowledge on chemical reactivity of reagents and procedure of reaction is necessary. The composition of the nanomaterial and morphology can be administered if the reaction process and steps are known and can be controlled. The results observed are increase in size of nanoparticles due to the increase in concentration of precursor.

12.4.2.3 Chemical vapour deposition technique

In order to produce high purity and high yield, this method is adopted which involves the chemical reaction between the organometallic compound or halide to be deposited and other gases to produce a non-volatile solid film on the substrate as shown in the Fig. 4.

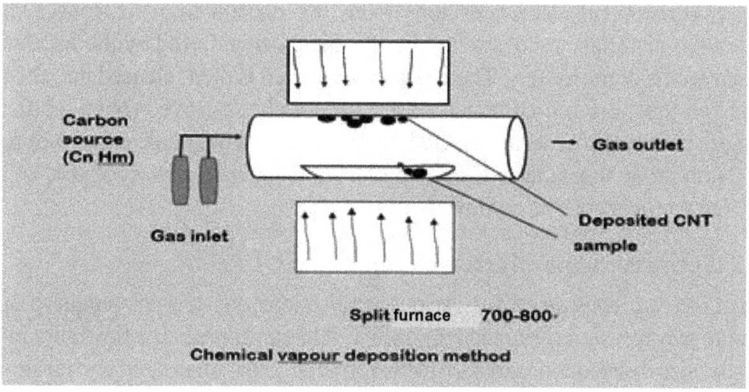

Fig. 4: Chemical vapour deposition technique.

12.4.2.4 Sono chemical technique

Nanoparticles are prepared using ultrasonic probe by irradiating the organic dispersal or liquefaction at room temperature. The particle size depends upon various factors such as sonication time and concentration of the solution. MnO_2 CdSe, and CeO_2 nanoparticles have been synthesised using this technique (Roy 2014).

12.4.2.5 Micro emulsion

The emulsification of suitable reactants is done for the preparation of nanoparticles. Brownian movement of micelles result in inter molecular collisions which are adequately dynamic collisions. Moreover, these collisions are occurred by the combination of micellar contents due to the Brownian movement of micelles. Chemical reaction takes place when there is fusion and fission between the droplets as main component for the mixing of the reactants. After the completion of reaction at the nano droplets, the formation of critical molecules takes place, which further leads to the expansion of nanoparticles after the nuclei formation (Malik et al. 2012).

12.4.3 Physical synthesis

12.4.3.1 Laser ablation technique

An intense laser pulse is used in this method to vaporize the material from the target surface, thus allowing the measurement of the interaction angle of the residual material. As a result, the supersonic jet of the ejected particles is perpendicular to the surface of the source target. The suction tube is almost like a rocket exhaust, distributing away from the target with a transition velocity constant for different particles. Abraded material condenses on the substrate in front of the target. Combustion occurs in a chamber, under a vacuum or in an atmosphere with a background gas. For the oxide film, oxygen is the best known background gas (Joshi et al. 2021).

12.4.3.2 Sputtering technique

A technology for depositing a thin layer of material on a "substrate". It generates gaseous plasma, then accelerates ions from the plasma into the source material target. Source materials are destroyed by the energy transferred by the incident ions, individual atoms or molecules. The energy transferred is then released into the neutral particles and thus, the source materials are ultimately destroyed (Nnaji et al. 2018). When these neutral particles are ejected, they travel on a path unless they approach or connect with other substances. Using sputtering techniques, the synthesis of single-phase TiO2 nanoparticles is achieved.

12.4.3.3 Spray route pyrolysis technique

Nano particles are considered for various applications due to their unique properties. Spray route pyrolysis is a sustained approach as it has an appreciable flexibility in crystal formations, size distribution, and narrow distribution to obtain desired crystal structure.

In this process, a thin coating is deposited by spraying the solution onto a heated surface, where the particles of the solution merge to form a compound. Chemical testing agents are chosen so that undesirable compounds in the final product are stable at the sublimation temperature (Fauzi et al. 2020, Jiříček et al. 2017). This process becomes particularly useful in depositing oxides such as SnOx—it is common to apply a transparent conductive thin film onto the glass.

12.4.3.4 Inert Gas Condensation technique

By evaporation and accumulation in a flowing stream of inert gas using nano particles, metallic iron is prepared using inert gas condensation technique. The most advanced and expensive technique, it can exactly synthesise and deliver the desired size of nano particles. Nano particles, after formation, collide with inert gas under low pressure conditions resulting in small sized particles of nano materials (Nnaji et al. 2018).

12.5 Protein recovery with nanomaterials

Excess domestic sludge has been studied for its ability to separate proteins such as lipase, amylase, protease, dihydrocaniase and glucosidase by dynomil destruction method. Enzymes represent potential applications to improve the hydrolysis site of proteins in domestic wastewater treatment equipment (Veluchamy et al. 2021). Despite the high rate of protease isolation, the economic costs appear to be very high. Some researchers have shown that alkaline protease assembly is usually completed at a concentration of 4.58 IU mL at the same time as the biopesticide, and protease assembly in a scale bioreactor. The pilot will calculate an accurate amount. Molecular phylogenetics using RNA properties has shown that up to 20% of microorganisms present in sludge can be cultured using standard culture techniques. Due to various constraints, commercialization of this product has not yet been achieved and multi production of enzymes and biopesticide can lead to cost savings. Producing proteins and enzymes is a unique business with less investment and availability (Yadav et al. 2020). Common sludge-derived substances and fermentation products from domestic wastewater treatment plants can be used to reduce cleansing costs.

12.5.1 *Extraction with emulsions and solvents*

For creation of nanoparticles, a disseminated protein solution is subjected by sonication to remove the suspended particles and form emulsion. The primary w/o emulsion is stabilized by dispersing it in a second water phase with the help of a surfactant. Organic solvent is eliminated to form an emulsion in order to protect nanoparticles in buffer solution. Since emulsification process is unstable, it causes the bonding of particulates to lower systems free energy (Hong et al. 2020). Protein nanoparticles produced through this method may be chemically stabilized by including a crosslinking agent or thermally stabilized and purified by including a W/O emulsion to a warmed oil at 100°C or higher. EDC has been used as a chemical binding agent to stabilize the nano particles to obtain the required length of 300–700 nm, and the loading efficiency of the nano particles can be tuned through treatment with ultrasonic waves when generation of complex coagulants takes place.

12.5.2 *Nano spray drying*

This method is implemented in the treatment of liquid samples which contain the nano-particles which are complicated to treat. A hot gas mixture of nitrogen and carbon dioxide are used to drift the sprays of the liquid samples as they are sprayed into chambers from the nozzle (Fraga García et al. 2014). Nanoparticles are made into clusters at the electrodes which are placed at chamber's bottom where electrostatic force charges these spray droplets as they move down the chamber.

This is an additional step-by-step method that immediately and economically produces protein flakes on smaller scale. This synthesis of nanoparticles is beneficial due to the debris-sized scale that can be achieved using various parameters like sizing the scale of nozzle, and the manipulating the load at spot at which the debris are sprayed. In the case of protein nanoparticles, additional surfactants are required to stabilize polymer debris.

12.5.3 *Electrospray technique*

Electro spraying is an atomization technique. It has been used to regulate substances on the nano metric scale. This procedure involves the software of higher voltage to

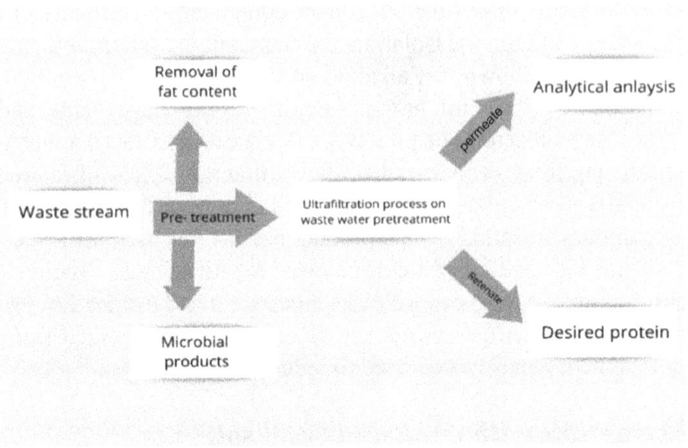

Fig. 5: Protein as recovery source from membrane technology.

the protein strategy (Hong et al. 2020) to be able to spray the circulation of the liquid through the nozzle for the form of the aerosol droplets. They include the collection of colloidal-sized protein nanoparticles.

12.5.4 *Self-assembly*

Protein micelles can be spontaneously generated while specific protein chains are dissolved in a reaction exceeding the critical micelle concentration (CMC) and the nanoscale aggregates are formed above the critical reaction temperature (CMT). The chains of micelles are stabilized by the process of bridging in solidification by a hydrophilic protein like albumin. Active amphiphilic ingredients are modified into hydrophilic proteins by the hydrophobic modification. They can self-assemble into micelle nanoparticles when added to an aqueous solution. Furthermore, the hydrophobic core can act as a conduit for active molecules. Protein micelle nanoparticles are synthesized mainly based on the specific first amino organization reaction of albumin and octaldehyde and can form nano micelle in core shape.

12.5.5 *Desolvation*

Desolvation is the most common technique used in the protein fabricated nanoparticles. In this method, solvents like ethanol and acetone act like reducing agents in order to reduce the drug-containing protein solutions. The reducing agents conduct the conformation of the protein and reduce its solubility, thereby leading to the formation of precipitates in the form of protein nanoparticles. The formation of fragments at certain stage in which the length of the particles will increase to the active stage. This could be achieved with the help of slow growth inside a variety of debris of the same length. Once the nanoparticles are formed, they are bridged using bridging agents, including GA. Protein attention, descale additive ratio, pH and temperature are the main factors influencing the effect on particle size. In particular, the high pH and attention to coffee protein can result in smaller nanoparticles. This

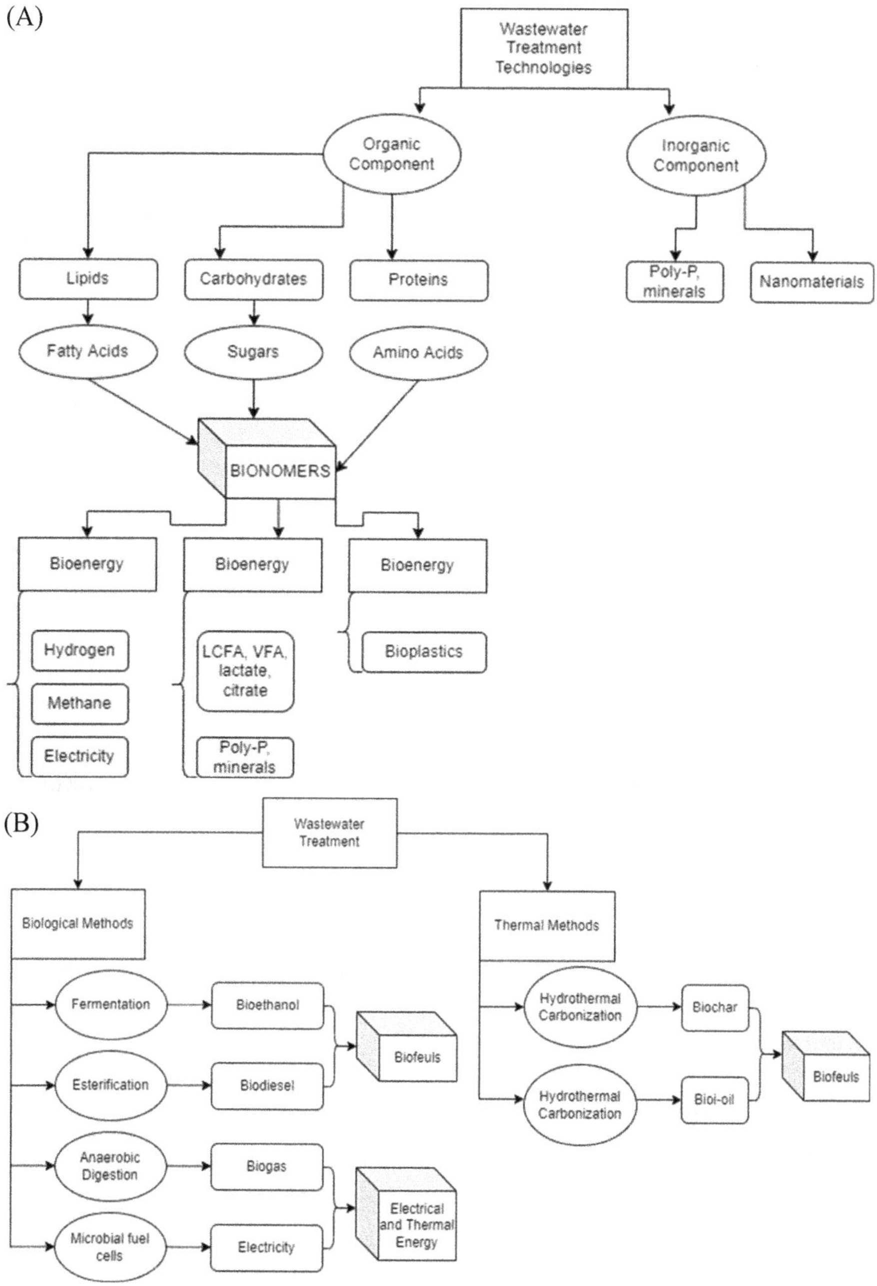

Fig. 6: (A) Flow chart of energy and resource recovery. (B) Flow chart of various technologies for energy recovery (Veluchamy et al. 2021).

technique involves the protein is concentrated with the help of a reducing agent to reduce the solubility of the protein. The technique is extensively used within side the era of nanoparticles which includes the use of albumin (Fraga García et al. 2014).

12.6 Recovery of supplementary materials

The distinction of technology in recovery of important products is achievable with nano-materials. It brings one closer to the ambitious target of a circular economic framework through commercial applications of such recoveries. Besides, it addresses the environmental challenges associated with wastewater treatment and pressure of consumption, especially in water-intensive sectors such as chemical industries, irrigation and associated cross-sectors while ensuring the redemption of value additive materials and their co-benefits (Veluchamy et al. 2021).

This endeavour requires modelling wastewater treatment plants with a multi-pronged process design that recovers valuable materials in addition to the treatment process. The above is illustrated in Fig. 6A, where treatment extracts components, ultimately resulting in the production of high-value products such as methane, hydrogen, bio-plastics, etc., and 6B, where the process design culminates into biofuels, electrical and thermal energy. Both physical as well as chemical treatment methods have a great potential in such process design of treatment plants (Yadav et al. 2020).

12.7 Conclusion

Nanotechnology provides a technologically and economically feasible solution to wastewater treatment and reuse. Although the public health toxicity and environmental aspects of the technology are still being debated, it has so far proved promising in protein recovery from wastewater. Nano particles are difficult to separate from treated solution, which may result in loss of nano particles. Nano particles in protein recovery are an addition for present treatment of water; the recovered protein material can be further utilised according to their properties, especially in drug delivery systems.

Recovered nanoparticles of protein can be characterized on the basis of their size, morphology, effective surface area for chemical reactions and entrapment capacity. A number of protein recovery methods with nanoparticles are still being researched for commercialization of this technology. The toxicity and bioaccumulation risks of nanomaterials for protein recovery from wastewater are currently being researched to understand the impact of public health and environment.

Abbreviations

GO	-	Graphite Oxide
RSM	-	Response Surface Methodology
ROS	-	Reactive Oxygen Species
UF	-	Ultra Filtration
POU	-	Point of Use
RNA	-	Ribonucleic Acid
CNT	-	Carbon based Nano Adsorbents
SCP	-	Single Cell Protein
PSB	-	Photosynthetic Bacteria
WCO	-	Molecular Weight Cut-Off

mg/L	-	Milligram per litre
MF	-	Microfiltration
NH4	-	N Ammonium
NF	-	Nano filtration
COD	-	Chemical oxygen demand
Nm	-	Nanometre
MEUF	-	Micellar enhanced ultrafiltration
PEUF	-	Polymer enhanced ultrafiltration

References

Abdelbasir, S.M. and Shalan, A.E. 2019. An overview of nanomaterials for industrial wastewater treatment. Korean Journal of Chemical Engineering. 36(8): 1209–1225. https://doi.org/10.1007/s11814-019-0306-y.

Ahn, Y.C., Park, S.K., Kim, G.T., Hwang, Y.J., Lee, C.G., Shin, H.S. et al. 2006. Development of high efficiency nanofilters made of nanofibers. Current Applied Physics, 6(6 SPEC. ISS.). 1030–1035. https://doi.org/10.1016/j.cap.2005.07.013.

Baruah, A., Chaudhary, V., Malik, R. and Tomer, V.K. 2018. 17-Nanotechnology based solutions for wastewater treatment. In Nanotechnology in Water and Wastewater Treatment: Theory and Applications. Elsevier Inc. https://doi.org/10.1016/B978-0-12-813902-8.00017-4.

Bouaifi, M., Hebrard, G., Bastoul, D. and Roustan, M. 2001. A comparative study of gas hold-up, bubble size, interfacial area and mass transfer coefficients in stirred gas-liquid reactors and bubble columns. Chemical Engineering and Processing. 40(2): 97–111. https://doi.org/10.1016/S0255-2701(00)00129-X.

Chen, C., Tsyusko, O.V., McNear, D.H., Judy, J., Lewis, R.W. and Unrine, J.M. 2017. Effects of biosolids from a wastewater treatment plant receiving manufactured nanomaterials on Medicago truncatula and associated soil microbial communities at low nanomaterial concentrations. Science of the Total Environment. 609: 799–806. https://doi.org/10.1016/j.scitotenv.2017.07.188.

Chollangi, A. and Hossain, M.M. 2007. Separation of proteins and lactose from dairy wastewater. Chemical Engineering and Processing: Process Intensification. 46(5): 398–404. https://doi.org/10.1016/j.cep.2006.05.022.

Elorm Obotey Ezugbe, S.R. 2020. Membrane Technologies in Wastewater Treatment_ A Review _ Enhanced Reader.pdf.

Engelhardt, S., Sadek, A. and Duirk, S. 2018. Rejection of trace organic water contaminants by an Aquaporin-based biomimetic hollow fiber membrane. Separation and Purification Technology, 197(December 2017): 170–177. https://doi.org/10.1016/j.seppur.2017.12.061.

Fane, A.G.T., Wang, R. and Jia, Y. 2011. Membrane and desalination technologies. In Membrane and Desalination Technologies (Vol. 13). https://doi.org/10.1007/978-1-59745-278-6.

Feng, X., Imran, Q., Zhang, Y., Sixdenier, L., Lu, X., Kaufman, G. et al. 2019. Precise nanofiltration in a fouling-resistant self-assembled membrane with water-continuous transport pathways. Science Advances. 5(8): 1–12. https://doi.org/10.1126/sciadv.aav9308.

Fraga García, P., Freiherr Von Roman, M., Reinlein, S., Wolf, M. and Berensmeier, S. 2014. Impact of nanoparticle aggregation on protein recovery through a pentadentate chelate ligand on magnetic carriers. ACS Applied Materials and Interfaces. 6(16): 13607–13616. https://doi.org/10.1021/am503082s.

Halim, N.S.A., Wirzal, M.D.H., Hizam, S.M., Bilad, M.R., Nordin, N.A.H.M., Sambudi, N.S. et al. 2021. Recent development on electrospun nanofiber membrane for produced water treatment: a review. Journal of Environmental Chemical Engineering. 9(1): 104613. https://doi.org/10.1016/j.jece.2020.104613.

Hong, S., Choi, D.W., Kim, H.N., Park, C.G., Lee, W. and Park, H.H. 2020. Protein-based nanoparticles as drug delivery systems. Pharmaceutics. 12(7): 1–28. https://doi.org/10.3390/pharmaceutics12070604.

Hülsen, T., Hsieh, K., Lu, Y., Tait, S. and Batstone, D.J. 2018. Simultaneous treatment and single cell protein production from agri-industrial wastewaters using purple phototrophic bacteria or microalgae—A comparison. Bioresource Technology. 254(January): 214–223. https://doi.org/10.1016/j.biortech.2018.01.032.

Jain, K., Patel, A.S., Pardhi, V.P. and Flora, S.J.S. 2021. Nanotechnology in wastewater management: A new paradigm towards wastewater treatment. Molecules. 26(6). https://doi.org/10.3390/molecules26061797.

Jhaveri, J.H. and Murthy, Z.V.P. 2016. A comprehensive review on anti-fouling nanocomposite membranes for pressure driven membrane separation processes. Desalination. 379: 137–154. https://doi.org/10.1016/j.desal.2015.11.009.

Jiang, C., Wu, Z., Li, R. and Liu, Q. 2011. Technology of protein separation from whey wastewater by two-stage foam separation. Biochemical Engineering Journal. 55(1): 43–48. https://doi.org/10.1016/j.bej.2011.03.005.

Joshi, A., Mehta, K., Shah, H., Joshi, U., Sharma, A. and Shah, M.P. 2021. Chapter 34 - Broad spectrum application of nanotechnology for wastewater treatment. 715–738. https://doi.org/10.1016/C2019-0-04851-0.

Khuntia, S., Majumder, S.K. ad Ghosh, P. 2015. Quantitative prediction of generation of hydroxyl radicals from ozone microbubbles. Chemical Engineering Research and Design, 98: 231–239. https://doi.org/10.1016/j.cherd.2015.04.003.

Kulkarni, A.A. and Joshi, J.B. 2005. Bubble formation and bubble rise velocity in gas-liquid systems: A review. Industrial and Engineering Chemistry Research. 44(16): 5873–5931. https://doi.org/10.1021/ie049131p.

Lannutti, J., Reneker, D., Ma, T., Tomasko, D. and Farson, D. 2007. Electrospinning for tissue engineering scaffolds. Materials Science and Engineering C. 27(3): 504–509. https://doi.org/10.1016/j.msec.2006.05.019.

Malik, M.A., Wani, M.Y. and Hashim, M.A. 2012. Microemulsion method: A novel route to synthesize organic and inorganic nanomaterials. 1st Nano Update. Arabian Journal of Chemistry. 5(4): 397–417. https://doi.org/10.1016/j.arabjc.2010.09.027.

Manikandan, S., Karmegam, N., Subbaiya, R., Karthiga Devi, G., Arulvel, R., Ravindran, B. et al. 2021. Emerging nano-structured innovative materials as adsorbents in wastewater treatment. Bioresource Technology. 320(PB): 124394. https://doi.org/10.1016/j.biortech.2020.124394.

Miettinen, T., Ralston, J. and Fornasiero, D. 2010. The limits of fine particle flotation. Minerals Engineering. 23(5): 420–437. https://doi.org/10.1016/j.mineng.2009.12.006.

Munshi, F. 2019. Forward Osmosis for Algae Dewatering and Electrical Field-driven Membrane Fouling Mitigation. Electronic Theses and Dissertations. https://stars.library.ucf.edu/etd/6393.

Nasrollahzadeh, M., Sajjadi, M., Iravani, S. and Varma, R.S. 2021. Green-synthesized nanocatalysts and nanomaterials for water treatment: Current challenges and future perspectives. Journal of Hazardous Materials. 401(July 2020): 123401. https://doi.org/10.1016/j.jhazmat.2020.123401.

Nnaji, C.O., Jeevanandam, J., Chan, Y.S., Danquah, M.K., Pan, S. and Barhoum, A. 2018. Engineered nanomaterials for wastewater treatment: Current and future trends. In Fundamentals of Nanoparticles: Classifications, Synthesis Methods, Properties and Characterization. Elsevier Inc. https://doi.org/10.1016/B978-0-323-51255-8.00006-9.

Pedersen, P.A., Bjørkskov, F.B., Alvisse, S. and Hélix-Nielsen, C. 2018. From channel proteins to industrial biomimetic membrane technology. Faraday Discussions. 209: 287–301. https://doi.org/10.1039/c8fd00061a.

Pendergast, M.M. and Hoek, E.M.V. 2011. A review of water treatment membrane nanotechnologies. Energy and Environmental Science. 4(6): 1946–1971. https://doi.org/10.1039/c0ee00541j.

Reneker, D.H. and Yarin, A.L. 2008. Electrospinning jets and polymer nanofibers. Polymer. 49(10): 2387–2425. https://doi.org/10.1016/j.polymer.2008.02.002.

Roy, A. 2014. Nanotechnology in Industrial Wastewater Treatment. In Water Intelligence Online (Vol. 13). https://doi.org/10.2166/9781780406886.

SantosMiranda, M.E., Marcolla, C., Rodriguez, C.A., Wilhelm, H.M., Sierakowski, M.R., BelleBresolin, T.M. and Alves de Freitas, R. 2006. I . The role of N-carboxymethylation of chitosan in the thermal stability and dynamic. Polym Int. 55(May 2007): 961–969. https://doi.org/10.1002/pi.

Singh, N., Petrinic, I., Hélix-Nielsen, C., Basu, S. and Balakrishnan, M. 2018. Concentrating molasses distillery wastewater using biomimetic forward osmosis (FO) membranes. Water Research. 130: 271–280. https://doi.org/10.1016/j.watres.2017.12.006.

Song, W., Lang, C., Shen, Y.X. and Kumar, M. 2018. Design Considerations for Artificial Water Channel-Based Membranes. Annual Review of Materials Research, 48(March): 57–82. https://doi.org/10.1146/annurev-matsci-070317-124544.

Vasita, R. and Katti, D.S. 2006. Nanofibers and their applications in tissue engineering. International Journal of Nanomedicine. 1(1): 15–30. https://doi.org/10.2147/nano.2006.1.1.15.

Veluchamy, C., Loganath, R., Sharma, D., Gowd, S.C., Rajendran, K. and Varma, V.S. 2021. Recovery of value-added materials from wastewater. In Current Developments in Biotechnology and Bioengineering. BV. https://doi.org/10.1016/b978-0-12-821009-3.00014-2.

Vullev, V.I. 2011. From biomimesis to bioinspiration: What?s the benefit for solar energy conversion applications? Journal of Physical Chemistry Letters. 2(5): 503–508. https://doi.org/10.1021/jz1016069.

Wagh, P. and Escobar, I.C. 2019. Biomimetic and bioinspired membranes for water purification: A critical review and future directions. Environmental Progress and Sustainable Energy. 38(3). https://doi.org/10.1002/ep.13215.

Wen, Y., Yuan, J., Ma, X., Wang, S. and Liu, Y. 2019. Polymeric nanocomposite membranes for water treatment: a review. Environmental Chemistry Letters. 17(4): 1539–1551. https://doi.org/10.1007/s10311-019-00895-9.

Xia, L., Andersen, M.F., Hélix-Nielsen, C. and McCutcheon, J.R. 2017. Novel commercial aquaporin flat-sheet membrane for forward osmosis. Industrial and Engineering Chemistry Research. 56(41): 11919–11925. https://doi.org/10.1021/acs.iecr.7b02368.

Yadav, B., Chavan, S., Atmakuri, A., Tyagi, R.D. and Drogui, P. 2020. A review on recovery of proteins from industrial wastewaters with special emphasis on PHA production process: Sustainable circular bioeconomy process development. Bioresource Technology, 317(August), 124006. https://doi.org/10.1016/j.biortech.2020.124006.

Yaqoob, A.A., Parveen, T., Umar, K. and Ibrahim, M.N.M. 2020. Role of nanomaterials in the treatment of wastewater: A review. Water (Switzerland), 12(2). https://doi.org/10.3390/w12020495.

Zhang, Y., Wu, B., Xu, H., Liu, H., Wang, M., He, Y. and Pan, B. 2016. Nanomaterials-enabled water and wastewater treatment. NanoImpact, 3–4, 22–39. https://doi.org/10.1016/j.impact.2016.09.004.

13

Biosorbents for Wastewater Treatment

Amandeep Singh[1,*] and *Jyothy G. Vijayan*[2]

13.1 Introduction

Wastewater treatment is a course through which contaminants are removed from wastewater or sewage and rejoined into water cycle. Once returned to water cycle, treated water makes a satisfactory influence on ecosystem of environment, and it is further reused for different purposes, such reusability of water is termed as water reclamation. Wastewater treatment process is carried out in a wastewater treatment plant (WWTP). Wastewater, on the basis of origin, is of two kinds: domestic wastewater (also called municipal wastewater) and industrial wastewater (also called industrial effluent), and both kinds of wastewater are handled by a WWTP. Treatment plant designed for the treatment of domestic wastewater is known as a sewage treatment plant (STP) or water resource recovery facility (WRRF). For industrial WWTP, a dedicated plant is required because it contains hazardous chemicals, heavy metal ions, dyes, oils, etc. Wastewater treatment course usually takes place in following sequences- phase separation (saponification, sedimentation, etc.), oxidation (biochemical, chemical, etc.), and polishing (adsorption, absorption adjusting, carbon filtering, etc.). Types of WWTP, on the basis of contaminated water, are industrial WWTP, STP, leachate treatment plants, and agricultural WWTP. The key intention of WWTP is to make wastewater be able enough to be disposed/ reused, in safe manner. However, prior to employing any treatment method, options for disposal or reuse must be decided in order to select correct treatment process. Disposal of treated water means to let it restore into water bodies. In order to dispose into the ocean, the concerned environmental treaty necessities should be met. Reuse of treated wastewater means to bring it into daily water requirements. Reuse of wastewater is a better option in order to dilute water stress in cities, and it promotes consciousness among people and also diminish the pollutants. Furthermore, reused

[1] Department of Polymer Science and Technology, University of Calcutta, Kolkata, India.
[2] Department of Chemistry, M. S. Ramaiah University of Applied Sciences, Bengaluru, India.
* Corresponding author: aspst_rs@caluniv.ac.in

water can also be utilized in various activities, for instance, landscape as well as crop irrigation, for recreational purposes, groundwater recharge, etc. In several developing nations, WWT projects are either deficient or if they are there, then not well functioned. Afghanistan and Pakistan are disposing untreated hazardous wastewater into local waterbodies, for instance, the Kabul River takes in nearly 1 m^3 of untreated wastewater per second (Khan and Khan 2019).

At the worldwide level, approximately 52% of wastewater is being treated, properly (Jones et al. 2021). However, rates of wastewater treatment are vastly unequal among countries. For instance, countries with significantly enough income treat about 74%, whereas countries with comparatively small income handle merely about 4.2% of their wastewater. Proper WWT practices need to be efficiently implemented, executed, and improved across the world in order to reduce water pollution to environment and attain water quality improvements. Therefore, a United Nations body named Sustainable Development Goal 6 has framed a target named 'Target 6.3' which expresses as follows: "By the year of 2030, improve water quality by reducing pollution, eliminating dumping, and minimizing release of hazardous chemicals and materials, halving proportion of untreated wastewater and substantially increasing recycling and safe reuse globally".

In India, regulations and guidelines regarding wastewater treatment are governed by central government agencies, namely Ministry of Environment Forest and Climate Change (MoEF&CC), Ministry of Housing and Urban Affairs (MoHUA), and newly formed Ministry of Jal Shakti (Schellenberg et al. 2020). Apart from these, water and sanitation related policies are also regulated under 'National Environment Policy 2006' and 'National Sanitation Policy 2008' schemes. At a state level, local administration along with municipality corporations take accountability and responsibility of sewage disposal as well as of construction and maintenance of sewerage infrastructure. Efforts of states get support from several schemes run by Central Government of India such as 'Jawaharlal Nehru National Urban Renewal Mission (JNNURM)', 'National River Conservation Plan (NRCP)', 'National Lake Conservation Plan (NLCP)', etc. The Government of India has established various initiatives through Ministry of Environment and Forest that inspire industries to establish common facilities to commence wastewater treatment at industry level. Contaminated wastewater, either domestic discharge or industrial effluent, consists numerous hazardous heavy metal and dyes. The toxicological aspects of heavy metals and dyes are given below.

13.1.1 Toxicological aspects of heavy metals

Human activities generate and dispose of post-consumer wastes consisting organic dyes and heavy metals to the water resource, as a result of which water bodies are getting polluted. Generally, a metal possessing density greater than 5 gcm^{3-} is considered as heavy metal. Heavy metals are further categorized into two groups, namely essential heavy metals and non-essential heavy metals. Heavy metal necessary for the growth of cells is termed as essential heavy metals such as Zn, Ni, Cu, etc. A very low concentration of essential metals is required for human body functioning. Concentration of metals beyond a certain limit cause harmful effects to living organisms. Metals that are not required for cell growth or for any biological functioning are known as non-essential metals such as Pb, Cd, Hg, etc. Non-essential metals are toxic for living organisms even at very low concentration. Heavy metals are reaching at hazardous levels than other toxic effluents (Khan and Jhung 2013). Various industries such as printing, photography, electrolysis, leather, aerospace, etc., dispose left over waste which contains heavy metals into water resources (Fein et al. 1997). Heavy metals are non-biodegradable and hazardous to human health even when present in small amount. Metals such as chromium, lead, zinc, arsenic, mercury, cobalt, and cadmium get accumulated into living organism and causes severe diseases. The surface water supplies as well as groundwater supplies consisting heavy metals are considered a greater inorganic contaminant associated with ecosystem because of their penetration into aquatic environment as well as their toxic nature towards all forms of living organisms. The presence of such metals even in undetectable and diluted quantities makes water contaminated. Such metals are either detected in their free state (if not treated for any biodegradation methods) or get adhered with various salt compounds. Through any of the process, mineralization of heavy metal ions is not possible. Aside the mere environmental concern, methodological as well as technological prospects of heavy metal eradication from industrial effluent should also be taken care of. Although heavy metals are present naturally in soil in minute amounts, their amount and presence in water bodies vary from region to region. If somehow such metals get concentrated, then they cause a serious danger. For instance, As and Cd are found to be creating cancers in human body. Hg causes mutations and genetic damage whereas Cu, Pb, and Hg are responsible for causing brain and bone defects. Adverse effects of four heavy metals are mentioned below.

Chromium. Humans get exposed to Cr and its compounds through eating, drinking, breathing, or through skin contact. Generally, concentration of Cr in drinking water is very low but contaminated water may comprise beyond expected level of dangerous Cr(VI) or hexavalent chromium. Chromium contaminated water is known to develop numerous health issues like skin rashes, kidney and liver damage, respiratory problems, lung cancer, weakened immune systems, alteration of genetic material, etc. Health threats related to exposure of Cr depend upon its oxidation state. Metal form of chromium Cr(III) possesses comparatively low toxicity than hexavalent form Cr(VI). There are life-threatening consequences of water containing hexavalent form and can cause dermatitis, ulcerations, and allergic skin problems. National Toxicology Program (NTP) is an inter-agency program run by Department of Health and Human Services (United States) has claimed on basis of an appropriate number of evidences that hexavalent Cr compounds like lead chromate, calcium chromate, zinc chromate, strontium chromate, and chromium trioxide have carcinogenic nature.

Mercury. It is considered one of the most toxic metals present in environment. Once it inscribes with food chain through consumption of polluted water, it gets progressively accumulated into humans and animals. Generally, mercury loaded polluted water is discharged into environment from various industries such as paints, oil refining, chlor-alkali, pulp and paper, rubber processing and fertilizer, pesticides, batteries, fluorescent light tubes and high intensity street lamps, thermometers, cosmetics and pharmaceuticals, etc. Methyl mercury present in industrial effluent causes abnormalities in children called teratogenic effects, an illness related to nervous system. In this illness, patient suffers from mental retardation, cerebral palsy, convulsions, and some other complications. On prolonged exposure, Hg laded industrial discharge causes genetic defects due to breakage in chromosome as well as interference in cell division process that leads to irregularity in chromosomal distribution. Furthermore, impairment of pulmonary function, dyspnoea, kidney function, and chest pain are also caused by mercury intakes. The destructive consequence of methyl mercury on aquatic ecosystem and humans were seen in Minamata pandemic during 1969–1973 in Japan.

Nickel. Electroplating is a process by which surface finishing followed by metal deposition is carried out to prolong the lifespan of containers. Various metals are used for this purpose such as Ni, Cu, Zn, Cr, etc. The selection of metal depends upon specific requirement of articles. On washing of electroplated tanks or articles, significant quantities of metal ions leech out into effluent. An appropriate amount of Ni(II), that is sufficient enough for water pollution and contamination, is found in effluents coming out from silver refineries, electroplating plant, battery industries, etc. Ni(II) in higher concentration may cause the cancer of lungs, nose, and bone. However, an acute poisoning of Ni(II) may also leads to hazardous consequences such as headache, dizziness, nausea and vomiting, chest pain, tightness of chest, dermatitis, dry cough and shortness of breath, rapid respiration, cyanosis, and extreme weakness.

Iron. Iron is found in two forms in industrial discharges: soluble ferrous iron (Fe^{2+}) form and insoluble ferric particulate iron (Fe^{3+}) form. The elemental presence of Fe in natural water is found due to dissolution of minerals and rocks, landfill leachate sewage, acid mine drainage, and due to industrial discharge. Generally, Fe is found in water in insoluble ferric particulate iron form. The existence of Fe at concentrations greater than 0.1 mgL^{-1} harms the gills of fishes. Under controlled mechanism, ferrous sulphate present in water becomes unstable and gets precipitated as insoluble ferric hydroxide that further gets settled down as a rust-coloured silt, but intake in higher concentration leads to haemochromatosis, a tissue damage phenomenon due to accumulation of iron particles.

13.1.2 Toxicological aspects of dyes

Dyes are carcinogenic, mutagenic, and toxic in nature even at micro molar concentration (Ngah et al. 2011). Dyes are characterized by complex molecular structure, high stability, and different substituents. Dye based industrial effluents comprise one of the utmost challenging water pollutants to be processed. Such effluents have high chemical as well as biological oxygen demands, chemical toxicity, and suspended solids. Dye containing effluents reduce light penetration that greatly affects photosynthetic process in aquatic organisms and it is toxic to aquatic organism due to existence of various chemical species such as metals, aromatics, chlorides, etc. (Fu and Viraraghavan 2001, Robinson et al. 2001). Generally, dyes possess complex aromatic molecular structures and synthetic origin that stabilizes dyes, and hence become difficult to biodegrade. The dyes for textile purpose are treated with chemicals and lights in order to make them more resistant towards fading. Textile dyes are made resistant towards high temperatures as well as enzyme degradation so that textile can be washed with detergent without any harm. The treatments of dyes to make them more resistant towards various activities leads to slow down the biodegradation of dyes. Broadly, dyes are categorized into three groups, namely cationic dyes (basic), anionic dyes (direct, acid, reactive), and nonionic dyes (disperse). Typically, azo groups or anthraquinone chromophores are present in anionic as well as non-ionic dyes. The toxic amines are formed in effluent by reductive cleavage of azo linkages, whereas anthraquinone-based dyes are most resilient towards biodegradation because of having fused aromatic structures, thus exhibiting colour for prolonged period in wastewater. The azo-based chromophores get combined with various reactive groups such as difluorochloro pyrimidine, vinyl sulfone, trichloro pyrimidine, and chlorotriazine in reactive dyes. These dyes vary from others in terms of their binding properties to textile fibres (cotton) by covalent bonds. Reactive dyes are used widely in textile industries due to their promising properties such as water fast, bright colour, and easy handling. Various effluents from dye industry are considered as very challenging type of wastewaters for the treatment because they have high chemical oxygen demand (COD), high biological oxygen demand (BOD), turbidity, solids suspension, toxic components, and colour. Out of all types of dyes, a basic dye possesses high intensity and brilliance; thus, such dye is very noticeable even if present in very low amount. Therefore, colour removal from contaminated water becomes chief objective not only due to toxicity but also due to visibility of dyes. Generally, metal complex dyes consist of Cr metal, which is considered as a potential carcinogenic metal. The disperse dye is non-ionizable into the aqueous medium. Few such dyes possess a tendency to bioaccumulate. Conventional wastewater treatment systems are not found to be sufficient for treating dye containing wastewater due to their chemical stability and low biodegradability. The utilization of dyes has attracted ample attention due to its toxic effects. Dyes are found to be creating serious health concern including carcinogenesis, mutagenesis, chromosomal fractures, teratogenicity, and respiratory toxicity. A study of mutagenic activity using salmonella/microsome assay showed that dye containing textile wastewater effluents have highest percentage (67%) of mutagenic activity (McGeorge et al. 1985). Textile effluent stands at second position out of eight industrial sectors in terms of toxicity (Costan et al. 1993). The calculation of LC_{50} (lethal concentration 50%, which means amount required to kill 50% of the population) values of various commercial dyes subjected to fishes at different time intervals was carried out earlier in 1980s. A survey carried out by ETAD (Ecological and Toxicological Association of the Dyestuffs Manufacturing Industry, Switzerland) on 4000 dyes suggested that about 90% of tests were having LD_{50} (lethal dose 50%, which means 50% of the population if killed in a given period of time) values more than 2×10^3 mg/kg. Basic and diazo direct dyes showed highest rates of toxicity. The sub-chronic exposure of benzidine-based dyes up to 13 weeks caused hepatic neoplastic nodules and hepatocellular carcinomas in rodents even in very short duration. Histopathological changes in rats exposed to textile wastewater showed dropping the numbers of Leydig cells as well as germs that results in impaired spermatogenesis.

The traditional methods for the removal of dyes and metals from the polluted water are highly expensive and parallelly create large amount of sludge and toxic wastes that further need a careful disposal (Kanamarlapudi et al. 2018). For the elimination of non-biodegradable dyes from aqueous streams, adsorption process seems to be a potential technique using activated carbons as an adsorbent due to its effectiveness and versatility. Generally, activated carbons are produced from high carbon content materials and it has an excellent adsorption efficiency due to its porous structure. Adsorption is considered as the most effective method among the conventional methods (Reddy and Gambrell 1987, Rafati et al. 2016). Different bioreactors are also used to remove contaminants from huge volume of effluents. Industrialization is a major factor behind the contamination of water, where lakes and river bodies are overloaded with large number of toxic substances. These contaminants cause accumulation of dyes and pigments. Therefore, biomass-based adsorption process named biosorption is considered as a significant favourable alternative for treatment of water contaminants. Adsorption is a physiochemical process in which certain material (adsorbent) offers their surface and allows to passively concentrate and bind various contaminants (adsorbates) onto its cellular structure. The biological substrate on the surface of which biosorption occurs is known as biosorbent whereas particles, atoms, molecules, or ions that adhere onto surface of biosorbent by intermolecular attractions are known as biosorbate materials. Charcoal, silica, gel, and alumina gel are taken as good adsorbents because they have highly porous structures and have large surface area.

Animal, plant, and microbial biomass based adsorbents are termed as biosorbent (Vasudevan et al. 2001). Biosorption can also be considered as an ability of any biomaterial to accumulate various contaminations such as heavy metals, dyes, oils, etc., from wastewater through physico-chemical hierarchy or metabolically mediated method. Various biomasses are being used in various forms since lone time for environment cleaning. However, biosorbents are yet to be explored in different dimensions due to its potentiality to be an economical, eco-friendly, sustainable, and green choice for eradicating toxic dyes/hazardous heavy metals from polluted wastewater (Fard et al. 2011). Biosorption reduces overall cost of treatment based on a fact that regeneration of biosorbent could be done at very low cost. Biosorption procedure requires a biosorbent (solid phase) along with a solvent (liquid phase). The water purification by biosorption process is highly favoured because biosorbents are economic and have good recyclability. The cost, availability, sustainability, and recyclability of the substance to be used for any purpose are major factors to be considered for circular economy (Singh et al. 2021a).

The biosorption process is affected by different factors including pH of wastewater, operation temperature, pollutant concentrations, dosage of biosorbent, agitation fastness, etc. (Oren and Kaya 2006, Gao et al. 2009, Seredych et al. 2009). Apart from the research papers, lots of patents have also been published in the biosorption field. Several patents demonstrate the ways to improve the sorption capacity of bio-based sorption material by surface modifications and immobilizations of biosorbents as shown in Table 1.

Table 1: Registered patents related to biosorption process.

Patent No.	Title	Year	Patent country	References
3725291	Sorbent and method of manufacturing same	1973	Czechoslovakia	Serbus et al. 1973
4021368	Process of treating mycelia of fungi for retention of metals	1977	Czechoslovakia	Nemec et al. 1977
4067821	Method of treating a biomass	1978	Czechoslovakia	Votapek et al. 1978
4293333	Microbiological recovery of metals	1981	USA	Drobot 1981
4289531	Process for recovering precious metals	1981	USA	Lechavelier and Drobot 1981
4320093	Separation of uranium by biosorption	1981	Canada	Volesky and Tsezos 1981
4690894	Treatment of microorganisms with alkaline solution to enhance metal uptake properties	1987	USA	Brierley et al. 1987
4701261	Process for the separation of metals from aqueous media	1987	England	Gibbs et al. 1987
4769223	Biosorbent for gold	1988	Canada	Volesky and Kuyucak 1988
4898827	Metal recovery	1990	USA	Brierley et al. 1990
5055402	Removal of metal ions with immobilized metal ion-binding microorganisms	1991	USA	Greene et al. 1991
5152969	Processes to recover and reconcentrate gold from its ores	1992	USA	Kleid et al. 1992
5084389	Bioadsorption composition and process for production thereof	1992	Canada	Lakshmanan and McCready 1992
5460791	Method for adsorbing and separating heavy metal elements by using a tannin adsorbent and method of regenerating the adsorbent	1995	Japan	Shirato and Kamei 1995
5538645	Process for the removal of species containing metallic ions from effluents	1996	Israel	Yannai and Meshulam 1996
5578547	Bead for removing dissolved metal contaminants	1996	USA	Summers and Gress 1996
5648313	Method for production of adsorption material	1997	Germany	Pohl 1997
5789204	Biosorbent for heavy metals prepared from biomass	1998	South Korea	Kogtev et al. 1998

Table 1 contd. ...

...Table 1 contd.

Patent No.	Title	Year	Patent country	References
5750065	Adsorption of PCBs using biosorbents	1998	USA	Kilbane 1998
5976847	Hydrophilic urethane binder immobilizing organisms having active sites for binding noxious materials	1999	USA	Hermann 1999
6395143	Biosorption system	2002	England	McHale and Bustard 2002
6402953	Adsorption means for radionuclides	2002	Russia	Gorovoj and Kosyakov 2002
6579977	Biosorbents and process for producing the same	2003	Germany	Pieschel et al. 2003
6786336	Composite biosorbent for treatment of waste aqueous system(s) containing heavy metals	2004	USA	Boddu and Smith 2004
20060070949	Process and plant for the removal of metals by biosorption from mining or industrial effluents	2006	Chile	Tadic et al. 2006
20070202588	Petroleum biosorbent based on strains of bacteria and yeast	2007	Russia	Khabibullina et al. 2007
20080169238	Biosorption system produced from biofilms supported in faujasite (FAU) Zeolite, process obtaining it and its usage for removal of hexavalent chromium (Cr(VI))	2008	Portugal	Simões Campos Tavares and Pontes Correia Neves 2008
7658849	Use of *Rhizopus stolonifer* (Ehrenberg) *Vuillemin* in methods for treating industrial wastewaters containing dyes	2010	Italy	Prigione et al. 2010a
7790031	Use of *Cunninghamella elegans lendner* in methods for treating industrial wastewaters containing dyes	2010	Italy	Prigione et al. 2010b
7935257	Use of *Rhizomucor pusillus* (lindt) *schipper* in methods for treating industrial wastewaters containing dyes	2011	Italy	Prigione et al. 2011
20110269169	*Pseudomonas alcaliphila* MBR and its application in bioreduction and biosorption	2011	China	Li et al. 2011
7951578	Bacterial strain for metal biosorption process	2011	Chile	Cotoras and Viedma 2011
20150353381A1	Porous nanocomposite polymer hydrogels for water treatment	2015	USA	Rodrigues 2015

13.2 Types of biosorbents and their efficiency

A large variety of bio-based adsorbents are used to remove contaminants from wastewater. The main characteristics of an ideal biosorbent includes high affinity towards contaminants, cost-effective, easy availability, and should have a significant desorption efficiency. Different biology based bioadsorbents are described below.

13.2.1 *Microorganisms-based*

Microorganism includes bacteria, algae, and fungi, which are capable of removing pollutants from wastewater (Vendruscolo et al. 2017). Such microbial forms are potential biosorption candidates for adsorption process. Generally, agricultural and industrial wastes are utilized as biosorbent in order to adsorb various adsorbates such as heavy metal ions and dye particles. The microbial forms are considered eco-friendly as well as cost-effective due to non-laborious culture, rapid growth, and ability of high yielding.

Fungi possess significant binding properties due to presence of inbuilt metal binding groups. Fungus biosorbent is considered as the most economic and ecofriendly biosorbent. It is easy to grow and modify to develop as biosorbent. Cell wall is rich in polysaccharides which promotes more metal binding sites during biosorption. It is cheaper and easily produced and is less sensitive towards variation in nutrients, pH, temperature, and aeration (Wang and Chen 2009). Yeast is the best sorbent material among all fungus. The cultivation of fungus is comparatively easy and it is yielded at large scale to be purposely used as biosorbent for the removal of metal ions. Fungus biomass is effortlessly available at reasonable cost and can be cultured on leftovers of the fermentation industries. Due to their filamentous nature, fungi are non-laborious to distinguish by simple techniques such as filtration. Additionally, a fungus is comparatively less responsive towards process temperature, pH, process aeration, nutrition variations, etc.

Bacteria are used to develop different resistance mechanism for the removal of contaminants from wastewater. Bacteria are made up of polysaccharides which detoxify metals/dyes from contaminated water. Advantages of bacteria as biosorbate is because of its large surface area to volume ratio and efficient active sorption sites. Bacterial biosorbents have capsule like layer on cell wall which helps to detoxify metal ions from wastewater. Such biosorbents have high surface area to volume ratio. Bacteria provide more active sorption site and are highly efficient in bio adsorption. Gram positive and negative bacteria are involved in biosorption process of organic dyes and divalent ions. Compared to living biomass, dead microbial biomass is preferred due to the absence of nutrients and BOD/ COD monitoring.

Algae and their derivatives are used as biosorbent due to their unique characteristics such as their abundant availability, high surface area, less sludge wastage, and metal regeneration capacity. Algal biomass based biosorbents possess adequate sorption capacity, abundant availability, significantly higher surface area to volume ratio, little sludge disposal, excellent efficiency in metal regeneration and discovery, and economic and ecofriendly characteristics. Main family includes red and green algae which possess different functional groups and shows higher biosorption. Algae as biosorbate is extremely competent for dyes and metal removal. Algae cell wall mainly contains cellulose, alginic acid, and sulfated polysaccharides which possess more carboxyl groups which helps in biosorption. Biosorption efficiencies of various fungi, bacteria, and algae biomass towards different metals and dyes are shown in Table 2.

13.2.2 Biomass-based biosorbents

13.2.2.1 Industrial byproducts

Low cost materials from industry are effectively used for treatment of wastewater (Aksu and Tezer 2005). Food industry is a major one which disposes large quantities of waste material. Other industries like steel, aluminium, paper, fertilizer, food, mining, etc., also discard waste which need quick action. Important aspects associated with biosorption process include selectivity, significantly higher efficiency, cost-effective, excellent removal efficiency. The biomaterials that are either plentiful (such as sea weeds) or are leftovers of the industries (such as fermentation wastes, activated sludge process wastes, etc.) can be utilized as biosorbent materials. Utilization of dead microbial cells for biosorption is advantageous for WWT as such cells are inactive towards the toxic wastes, continuous supply of nutrients is not required, and is regenerated as well as reused for several cycles. Dead cells may be stored or used for extended periods at room temperature without putrefaction occurring. Their operation is easy and their regeneration is simple. Moreover, dead cells exhibit the accumulation of pollutants onto their surfaces to a greater extent than living cells. The textile dyes differ significantly in compositions and degree of interactions that bank on nature of dye, preparation capacity, type of biomass, specific surface properties of biomass, and process parameters such as operational temperature, pH of wastewater, ionic strength of impurities, etc.

13.2.2.2 Agricultural waste materials

Removal of pollutants from wastewater is also focused on agricultural waste which contains cellulose and lignin as major parts. It possesses functional groups like amine, carbonyl, phenolic, alcoholic, etc. Agro-wastes are highly effective for heavy metals and organic dye remediation. Agricultural residues possess negligible or zero financial worth and their discarding is considered as a challenging task. The utilization of agricultural residue waste in dye and metal ion removal from wastewater is a good choice. Various agricultural waste materials are being used for eradication of distinct dyes and heavy metal ions from wastewater at different conditions. Agricultural residue comprises *Curcuma longa* (turmeric) leaves, *Saccharum Officinarum* (sugarcane) leaves, *Cocos nucifera* shell, rice husk, straw, saw dust, oil palm trunk fibre, durian (*Durio zibethinus*) peel, guava leaves, almond shell, *Citrus grandis* (pomelo) peel, broad bean peel, peanut hull, *Citrullus lanatusrind* (watermelon), etc. In another study, batch and continuous tests to separate Zn and Cu(II) from wastewater by using tea wastes as adsorbent material were conducted (Amarasinghe and Williams 2007). Study claimed a maximum removal of Zn and Cu metals by maintaining pH between

Table 2: Biosorption of metals and dyes through fungal, bacterial, and algal biomass.

Culture	Metals/dyes	Removal	References
Fungal biomass			
Pencillium Aspergillus	Pb	21%	Kuroda and Ueda 2003
Pencillium sp.	Ag	50%	Kotrba et al. 2011
Chryosporium	Cd	3%	Say et al. 2001
Pencillium Chrysogenum	Cu	12%	Deng and Ting 2005
Pencillium Digitatum	Ni	4%	Zulfadhly et al. 2001
Pencillium sp.	Zn	0.2%	
Aspergillus Niger	Cu	29%	Chen et al. 2005
Pencillium oxalium	Reactive blue 19	91%	Zhang et al. 2007
Aspergillus Niger	Polar red	94%	Abd El-Rahim et al. 2003
Rhizopus Oxyzae	Rhodamine B	90%	Das et al. 2006
P. Chrysoporium	Direct dye	100%	Robinson 2005
Irpex flavus	Coracryl pink	100%	Chander and Arora 2007
Aspergillus Niger	Synazol	88%	Patel and Suresh 2008
Rhizopus Stolonifer	Direct red 80	100%	Prigione et al. 2008
Rhizopus Nigaricus	Reactive green	86%	Kumari and Abraham 2007
Funalia Trogii	Astrazone blue	48%	Asma et al. 2006
Bacterial biomass			
Pseudomnas Putida	Pb	27%	Uslu and Tanyol 2006
Streptomyces Rimosus	Zn	30%	Mishra 2014
Bacillus Firmus	Cu	38%	Salehizadeh and Shojaosadati 2003
Pseudomonas sp.	Cr	95%	Ziagova et al. 2007
Micrococcus Luteus	U	39%	Nakajima et al. 2001
Pseudomonas luteala	Active red	37%	Hu 1994
Treptomyces rimosus	Methyl blue	86%	Nacera and Aicha 2006
Streptomyces BW 130	Azo-reactive dye	29%	Zhou and Banks 1993
Streptomyces BW 130	Azo copper red	73%	
Mixed anaerobic culture	Diazolinked chromophores	85%	Knapp and Newby 1999
Algal biomass			
Ascophyllum nodosum	Pb	1.3–2.3 mg/g	Romera et al. 2006
Chlorella miniate	Ni	0.24 mg/g	
Chondrus crispus	Cd	0.83 mg/g	
Fucus vesiculosus	Pb	1.1–2.9 mg/g	
Padina sp.	Cd	0.53 mg/g	
Sargassum sp.	Cu	1.08 mg/g	
Azola rongpong	Acid green	31%	
Spirogyra sp.	Synazol	85%	Padmesh et al. 2005
Cosmarium sp.	Malachite green	74%	Khalaf 2008
Chorella Vulgaris	Remazol black	52%	Daneshvar et al. 2007

5 and 6. Recently, more studies were conducted using novel agriculture waste based adsorbents due to low cost, high abundance, and easy availability. Besides leaves and barks, cereals and shells of dry fruits can also be used as natural adsorbents (Gupta 2009). Usually, adsorption capacity of such bioadsorbents is not significant; thus, some surface modifications of bioadsorbents through chemical and physical techniques are highly needed. The chemical modifications of solid surface include acid, base, metal, and metal oxide treatments. The physical modifications comprise microwave treatments and heat (Srivastava et al. 2015). Generally, shells of dry fruits are pre-treated by thoroughly washing with pressurized water to remove surface impurities and coloration and thereafter thermally-treated at various temperatures in order to activate as natural adsorbents. Different shells such as walnut, pistachio, and almond as bioadsorbent material to generate activated carbons nanoparticles having larger surface areas were also used (Kazemipour et al. 2008). In a study, shells of *Coix lacryma-jobi,* a perennial tropical plant of poaceae grass family, as bioadsorbent to remove Cu from solutions having metallics concentrations as 10, 30, 50, 70, 100 and 150 mg L^{-1} were used (de Luna et al. 2015). Adlai shells were pre-treated by thoroughly washing with distilled water and dried at 100°C for a whole day and then pulverized and sieved. Additionally, chemical and physical treatment using nitric acid solution and carbonization at 350°C were carried out for activations. The results showed that adlai shells turned to be an efficient bioadsorbent for Cu(II) removal from wastewater. Biochar is an excellent bioadsorbent for removing heavy metal ions from wastewater (Patra et al. 2017). Usually, biochar is obtained by applying a thermal treatment called pyrolysis of biomass rich in carbon such as agricultural and forest waste, municipal solid waste, industrial wastes, papers, tires, bones, etc. Adsorption capacities of different biochar towards Cu(II) ions was found in a range of 0.4 to 52.1 mg g^{-1}. Another study also claimed that biochar obtained from wood and corn straw exhibits efficient bioadsorbent capacity for removal of Cu and Zn from wastewater (Chen et al. 2011). Pulverization of biochar followed by pyrolysis enhances adsorption capacity of biochar to remove Cu(II) ion (Tong et al. 2011). Furthermore, biochar to be used as bioadsorbent can also be obtained from switchgrass (300°C) and pine wood (700°C) to remove U and Cu(II) from contaminated water, respectively (Kumar et al. 2011, Liu et al. 2010a). Different biochar materials such as corn straw, orange waste, rice husk, olive pomace, compost, cow manure, pig manure, dairy manure, hardwood, peanut straw, soybean straw, canola straw, switchgrass, softwood and pinewood have been used to remove Cu(II) (Mohan et al. 2014). Cow manure, pig manure, and peanut straw showed Q_{max} as 90 mg g^{-1}. Thermal treatment of bioadsorbents generally enhances adsorption capacity. However, a large amount of heat is required to obtain such natural bioadsorbents. Thus, novel methodologies are needed to be employed to avoid pre-treatments in order to get an optimum working efficiency of bioadsorbents.

13.2.2.3 Lignocellulose

The main characteristics of lignocellulosic material in biosorption includes chemical composition, surface area of material, availability of functional groups, surface morphology of lignocellulosic material, and porosity. Lignocellulose contains cellulose, hemicellulose, lipid, protein, etc. Lignin is considered as a key storage of pollutant. Rich lignin substance causes greater affinity towards pollutant of organic nature. Lignin possesses different functional groups like hydroxyls, carboxyl, amines, etc., which enhances sorption of pollutant from wastewater. Cost-effective natural biosorbents (such as sawdust, date pits, lignin) can be development and used to treat wastewater consisting heavy metal ions and other poisonous materials (Albadarin et al. 2011). Lignin is formed by phenylpropane units and is a complex structural heterogeneous three-dimensional aromatic polymer. Lignin residue can be produced in large quantity from effluent of paper mills at negligible cost. Lignin possesses as high a surface area as180 m^2/g; therefore, it is considered as a potential candidate to be used as a natural bioadsorbent material for removing various pollutants from wastewater (Celik and Demirbas 2005, Mohan et al. 2006). Cost of waste disposal can be reduced by using lignin as a natural bioadsorbent as it offers an inexpensive substitute to prevailing commercial activated carbon. The Dubinin-Radushkevich (DR) value was found to be 12.7 kJ/mol which shows that biosorption process of Cr(VI) onto lignin surface takes place by the development of an ion-exchange mechanism. Results revealed that the removal of Cr(VI) from wastewater took place due to adsorption of Cr(VI) as well as reduction of Cr(VI).

13.2.2.4 *Chitin/Chitosan*

Chitosan has numerous applications in various fields because of its properties such as abundance, easy availability, biocompatibility, biodegradability, and inexpensiveness. Recently, chitosan is often preferred as a bioadsorbent for removal of pollutants. The chitosan-molecularly imprinted polymers (MIPs) having significant absorption capacity and are being used to remove heavy metal ions such as Cu^{2+}, Hg^{+}, Mn^{2+}, etc. Inexpensive chitin and chitosan polymers are progressively used for wastewater treatment (Guibal et al. 2006, Huang et al. 2004). A study was conducted to know the effects of chitosan properties onto adsorption and it claimed that anion present in the dyes can be taken away (Crini 2005). Moreover, it has also been presented that reduction efficiency gets increased up to 90%, on treatment of wastewater with chitosan polymers. Efficiency of chitosan based bioadsorbents was found to be improved than some commercial purifying agents such as polychloride and alums. Removal efficiency of heavy metal ions from wastewater primarily depends upon molecular weight (Mw) of chitosan and on amino distribution. Flocculation as well as condensation effect of various organic substances is significantly enhanced by maintaining pH close to neutral value (pH 7) and taking chitosan of low molecular weight as well as of high deacetylation degree (Kurita and Iwakura 1979). Coagulation-flocculation and adsorption process are thoroughly related to deacetylation degree, thus chitosan is more effective than chitin in order to separate metals, anionic dyes, and polychlorinated biphenyls (PCBs), whereas chitin is extra effective in order to treat petrochemically contaminated water to eradicate polycyclic aromatic hydrocarbons. Along with deacetylation degree and molecular weight, the purity of pigment in wastewater also determines separation rate of pollutants because purity directly affects the availability of amido groups. Thus, selection of a chitinous material for pollutant depends upon characteristics of adsorbent polymer material (deacetylation degree, crystallinity, molecular weight) as well as on characteristics of wastewater (pH, ionic strength). Consequently, chitosan of high deacetylation degree and low crystallinity results in an excellent removing efficiency in WWT. Recently, MIPs are being widely practiced for wastewater treatment because of several advances such as easy synthesis, appropriate stability, and having plentiful combining sites. Initially, chitosan is blended with molecular template of heavy metal under ultrasonication, followed by addition of some elution solution in order to rinse the template molecules that imparts porosity to the MIPs. Such polymers possess chemical and steric memory towards template molecules and therefore acts as binding sites in order to rebind template molecules. On coming in contact with heavy metal ions in wastewater, such polymers get imposed with them through binding sites; thus, heavy metal gets collected and flocculated. The degree of imposition depends upon intermolecular forces. The efficiency of chitosan-MIPs towards heavy metal removal depends upon molecular weight and deacetylation degree of chitosan. The effects of molecular weight of chitosan on rebinding properties of chitosan-MIPs were analyzed using three different samples of chitosan of molecular weight 29, 47, and 92 kDa, and results showed that chitosan of molecular weight 47 kDa accomplished highest rebinding capacities (Fu et al. 2007). Ag(I) ion were attained from wastewater with chitosan-MIPs by developing a coating of thiocarbamide onto surface of Fe_3O_4 magnetite (Fan et al. 2011). Numerous micro balloon structures were developed through coating onto surface of chitosan-MIPs which increases absorption rate towards wastewater contaminants as high as 90% due to increase in surface area. Molecular structure of chitosan-MIPs possesses large number of free amino groups under acidic condition and later they get protonated, which brings positive charge onto surface of chitosan; as a result of that, such chitosan-MIPs become capable of adsorbing negatively charged dye from wastewater. However, amino groups of chitosan are able to be protonated at decreased pH but it weakens the adsorbent properties. Recently, carboxylic and hydroxyl groups are being adhered onto surface of chitosan in order to enhance adsorption capacity towards dyes and metals present in wastewater. Chitin is a natural mucopolysaccharide synthesized from the exoskeleton of crab and other anthropods. Main compounds of chitin are deacetylated derivative called glucosamine or chitosan. Chitin/chitosan is highly hydrophobic in nature and is insoluble in water. Chitosan is synthesized from N-acetyl-D-glucosamine and D-glucosamine units. Surface area of chitin/chitosan is less compared to lignocellulose. Chitosan is chemically modified by different methods to improve adsorption efficiency for water treatment. Surface modification helps chitosan in the adsorption of specific organic pollutant (Li et al. 2008, Benguella and Benaissa 2002). Adsorption of some dyes by chitosan as biosorbent is listed in Table 3.

Table 3: Adsorption of some dyes by chitosan.

Dye	Chitosan	Adsorption efficiency mg/g	References
Reactive blue 222	Flakes	199	Wu et al. 2001
Reactive red 222	Bead	1106	
Basic blue 3	Powder	167	Grini and Badot 2008
Acid green 25	Crab shell	645	Wong et al. 2004
Methyl orange	Powder	130	Morais et al. 2007

13.2.2.5 *Cellulose*

Cellulose is considered as one among the world's most renewable, ample, and organic biopolymer resource which is available in different forms of biomasses like trees, plants, and bacteria. Generally, physical adsorption arises due to feeble van der Waals forces and it happens at lower temperature due to the creation of multilayers of adsorbate onto the adsorbent that get declined with the temperature. Due to unique characteristics, nanocellulose such as cellulose nanocrystal (CNC or NCC), cellulose nanofibers (CNF) also called nanofibrillated cellulose (NFC), and bacterial nanocellulose, have attracted scholars in the past. It shows properties like nano scale dimensions and higher mechanical characteristics. Due to the attachment of hydroxyl group onto surface of nanocellulose, it can be modified with other functional groups to attain expected surface modification. Many reactions like esterification, oxidation or polymer grafting can be used effectively to functionalize nanocellulose and to improve dispersion of nanocellulose into different polymer matrices. One of the most common techniques for fibres is electrostatic adsorption. In this technique, fibres are subjected to certain gas at optimum temperature during carbonization that makes fibres lose their crystal packaging which imparts the porosity (Rombaldo et al. 2014). The presence of carboxylic groups onto cellulose fibre surface makes them negatively charged; therefore, cations present in wastewater get adsorbed onto surface of bioadsorbent. Such adsorption processes are not selective as they undergo through a mechanism that comprises outer sphere surface complexation (Loganathan et al. 2013). This mechanism includes the migration of ions instead of transfer of electrons during redox reaction (Song et al. 2016). The dissolution of the cellulose into ionic liquid loses its crystalline nature; therefore, the excess amount of water that surrounds cellulose gets engrossed into the ionic solution. Thereafter, the crystalline segments of cellulose start to accumulate and act as a crosslinking point for formation of ion gel as depicted in Fig. 1. In another investigation, an ionic liquid was used to develop cellulose and chitin based natural composite to act as biosorbents for heavy metal adsorption from wastewater (Sun et al. 2009).

13.2.2.6 *Alginate compounds*

Alginate containing adsorption methods seem to be excellent techniques, especially in case of As metal separation. Usually, in media formation, adsorbents such as hydroxide, iron oxide, nano zero valent iron (nZVI), magnetite, goethite, minerals, ZrO_2, etc., are infused in alginate beads. Characteristic feathers of alginate such as rough and large surface area, adequate biocompatibility, amorphous, and high water permeability advance the adsorption efficiency of As and ions. It consists of easier regeneration process as compared to traditional adsorbents. Recently, research is being carried about laboratory-scale applications of alginate beads along with their encapsulation and impregnation with adsorbents in order to remove arsenic metal and ions from wastewater. The removal efficiencies of alginate-based biosorbents towards arsenic was reported maximum up to 95% at wide range of concentrations from 10 ppb to 1,000 ppb and pH from 3.0 to 7.5. Performance of different bioadsorbents coated or impregnated into alginate beads is shown in Table 4.

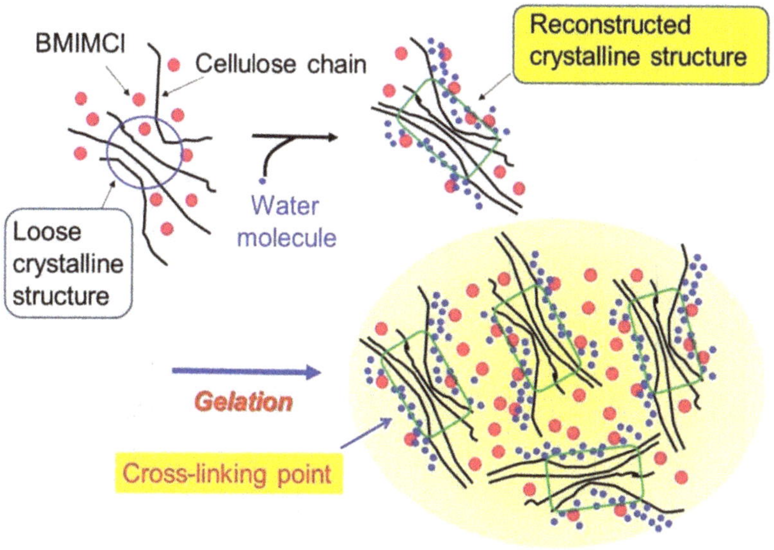

Fig. 1: A possible mechanism for the formation of cellulose ion gel. Reproduced with due permission (Takada and Kadokawa 2015). Copyright MDPI 2015.

13.2.2.7 Peat

Peat is a dark brown coloured partly fossilized plant matter, and is developed in poorly oxygenated wetlands. Such lands have greater rate of plant matter accumulation as compared to decomposition rate. Peat is a biomaterial, containing lignin as well as cellulose as main elements. The separation of pollutants with the help of peat seems to be an excellent method. Being a polar in nature, peat adsorbs various metals in huge amount onto surface and thus becomes competitive with other bioadsorbents. As a natural bioadsorbent, peat is able extract non-ferrous metals from contaminated wastewater, as well as polyvalent metal cations. High chelation capacity of peat allows to develop methods for the separation of Cd, Fe, Cu, Sb, Cr, Pb, Hg, Zn, Ni, Ag, phosphates, and organic compounds, for instance, detergents, oils, and dyes. The chelation property along with establishment of complexes by chemical bonds during adsorption process are believed as main interactions between peat and polar groups. The chelation capacity of peat towards metallic cations depends upon characteristics of metallic ions and existence of polar functional groups such as –OH and –CHO. An order of chelation capacity of cations towards peat was obtained as $Ni^{2+} < Cu^{2+} < Ba^{2+} < Pb^{2+} < Fe^{3+}$. The peat has shown 99% and 93–96% removal efficiency towards Cu and Zn, respectively. Peat is able to adsorb Cd, Pb, Cu, and Zn sulfides with significant efficiency from polluted wastewater, and also decreases Hg concentrations. Peat is very economic and easily available biosorbent. It is made up of lignin and cellulose which are highly porous in nature. It has limitations like poor chemical stability, hydrophilicity, and tendency to shrink. High cost regeneration related to activated carbon and other adsorbents leads to develop low cost adsorbents. It means a material which is available in nature or as a waste or a by-product of another industry could be utilized. Also, peat requires less pre-treatment methodology.

Table 4: Alginate based bio adsorbents and their adsorption efficiency.

Adsorbent	Adsorption efficiency (%)/Q_{max}	Remarks	References
Alginate-based magnetic nanocomposites	▪ 1.10 mg/g (45.9%) for As(V) ▪ 1.04 mg/g (32.3%) for As(III)	▪ Beads have lower adsorption efficiency than magnetic NCs ▪ Desorption is not carried-out	Luna-Pineda et al. 2009
Ca-alginate/activated carbon composite beads	▪ 66.7 mg/g for As(V) at 30 C in 60 min	▪ Less adsorption at lower pH ▪ Desorption is not carried-out	Hassan et al. 2014
Hydrous iron oxide (HIO) loaded alginate beads	▪ 47.8 mg/g for As(III) ▪ 55.1 mg/g for As(V) at 6–9 pH	▪ Efficiency increases with high HIO loading ▪ Not efficient at lower and higher pH ▪ Efficient to remove As(V) and As(III)	Sigdel et al. 2016
Fe(III) treated alginate beads	▪ 94% for As(V) ▪ 400 μg/L with 20 beads at pH 4 for 120 hr	▪ Sluggish sorption kinetics ▪ Desorption is not carried-out	Min and Hering 1998
Iron-coated Ca-Fe alginate beads	▪ 7.2 μg for As/g of wet alginate bead ▪ 1.8 μg for As/mg of Fe	▪ Pre-oxidation step is essential to increase the separation ▪ Effective on 1d coating period	Zouboulis and Katsoyiannis 2002
Iron-doped calcium beads	▪ 96.6% for As(V) at pH 3 ▪ 40% for As(III) at pH 8	▪ > 70% As(V) is separated from beads ▪ Efficient at conc 20–50 mg/kg ▪ Max adsorption at pH 3 for 20 minutes	Banerjee et al. 2007
Magnetite encapsulated by Ca-alginate	▪ 6.75 mg/g for As(V)ions	▪ Pre-oxidation of As(III) is required ▪ Max adsorption is at acidic pH	Lim and Chen 2007
Adsorption onto alginate and carboxymethyl cellulose beads	▪ 4.39 mg/g for As(V)	▪ Adsorption declined at > 26 C ▪ Adsorption capacity remains unchanged post bead's desorption	Tiwari et al. 2008
Microencapsulated particles of ferric hydroxide encapsulated in Ca-alginate	▪ 3.8 mg/g ▪ 99% removal efficiency of As(V) and As (III) at initial conc 300 μg/L	▪ Safe disposal of bead is required ▪ Ferric hydroxide particle sized 1–10 microns	Sarkar et al. 2010
Ca–Fe(III) alginate beads	▪ 0.364 mg/g for As(V) ▪ 0.117 mg/g for As(III) ▪ 80% at pH 2.0 for As(V) ▪ 54% at pH 4.0–8.0	▪ Sorption of As(V) lowers at high pH ▪ As(V) desorbed up to 100% ▪ As(III) desorbed up to 84.60% using H_2SO_4	Sanchez-Rivera et al. 2013

Iron(III) oxide-hydroxide/ chloride, mineral akageneite	▪ 99.1% adsorption efficiency	▪ High adsorption capacity in bare akageneite nanorods ▪ High akageneite conc increases removal efficiency	Cho et al. 2014
NZVI entrapped calcium (Ca)–alginate beads used for As adsorption	▪ 99.9% adsorption efficiency	▪ Pre-oxidation of As(III) is required ▪ NZVI trapped into bead ▪ Not loose reactivity ▪ Bead sized 3.4 mm	Bezbaruah et al. 2014
Fe(III) crosslinked alginate nanoparticles	▪ 0.0553 mg/g ▪ 69.12%	▪ NPs' separation from treated water is difficult ▪ Adsorption lowers with high As conc ▪ Bead sized 31–43 nm	Singh et al. 2014
Iron bearing hydroxide mineral goethite (α- FeO(OH)) impregnated Ca alginate (Cal-Alg-Goe) beads	▪ 30.44 mg/g ▪ 95%	▪ Adsorption lowers at lower and higher pH ▪ Optimum goethite conc is 18% ▪ Mineral particles sized 60–900 nm	Basu et al. 2015
Water treatment residuals (WTR) alginate beads	▪ 3.4 mg/g for As(III) ▪ 2.9 mg/g for As(V) at 3.0–7.5 pH	▪ Pre-treatment required ▪ Capsulation of WTRs in alginate beads lowers adsorption rate	Ocinski et al. 2016
Hydrous zirconium oxide (HZO) immobilized alginate beads	▪ 32.3 mg/g for As(III) ▪ 28.5 mg/g for As(V)	▪ Bare HZO shows max capacity ▪ Cation/ anion removal	Kwon et al. 2016
HIO loaded alginate beads	▪ 13.8 mg/g for As(V) post 168 hr	▪ Adsorption high at low pH ▪ Adsorption declines at pH more than 7 ▪ As and Cu removal	Jung et al. 2017

13.2.2.8 Clay

It is the most abundant naturally available biosorbent. It possesses high adsorption properties like porosity, large surface area and low cost. It is more economical than activated carbon. It possesses high cationic exchange property, for instance, montmorillonite, mica, kaolinite, etc. The adsorption of Pb ions through clay is a subject of numerous studies in recent time. The capacities of isothermal adsorption for several clay from various geographical locations of the entire world were calculated. Clay from South Africa showed comparatively higher capacity as 62 mg/g towards lead adsorption (Potgieter et al. 2006), whereas the USA originated phosphatic clay showed as 32 mg/g towards Pb ion (Singh 2001). The adsorption behaviour of Nigerian kaolinite clay towards Pb ions was analyzed at different temperatures and concentrations and it has been found that the elevated temperature enhances the adsorption capacity from 3.9 to 9.0 mg/g (Unuabonah et al. 2007a), though these capacities stand lower in front of clays from South Africa and America. The adsorption capacity enhancement with temperature proposes that the Pb adsorbed onto kaolinite clay surface is an endothermic process. Adsorption relies on the pH—as it gets higher, adsorption enhances up to certain level. Thus, the effect of pH onto adsorption must be analyzed prior to the measurement of adsorption isotherm (Unuabonah 2007b). The hectorite and vermiculite clay from USA and Spain, respectively, are changeable to thiol-functionalized hectorite as well as thiol-functionalized vermiculite. The adsorption of Pb ions onto thiol-functionalized vermiculite clay was found to be higher (Q_{max} 33 mg/g) than on thiol-functionalized hectorite clay (Q_{max}10 mg/g) (Diaz et al. 2007).

13.2.2.9 Zeolite

The occurrence of natural zeolites is in abundance and are considered as low cost resources. They are crystalline in nature and are hydrated aluminosilicates, having a skeleton construction, consisting pores filled with alkali metallic cations, alkaline earth metallic cations, and water. It possesses highly porous three dimensional structure with greater surface area. As zeolites possess higher ability of cation-exchange as well as molecular sieve properties, these materials are extensively utilized as adsorbents in WWT. Chemical modification of zeolites causes superior adsorption capacity. Different natural-based zeolites materials exhibit variable capacity towards cations-exchange phenomenon, for instance, ammonium and heavy metal ions. A few natural zeolite materials present adsorption capacity towards anions and organics. The modification of nature-based zeolites is carried-out by different techniques including ion exchange method, acid treatment, and surfactant functionalization method. Modifications make natural zeolites more efficient to achieve higher adsorption capacity towards anions, and organics.

13.3 Adsorption capacity of biosorbents towards organic herbicides, pesticides and phenolic compounds

Organic herbicides such as molinate and paraquat are adsorbed by different biosorbent with maximum adsorption capacity. Adsorbent with efficient regeneration is highly recommended. Adsorption of pesticides like Dichlorodiphenyltrichloroethane (DDT), andrin, malathion, Dichlorodiphenyldichloroethane (DDD), and methyl parathion by different biosorbents is recently reported. The adsorption capacities of biosorbents towards pesticides is found to be lower as compared to the high cost commercial adsorbents. The acid treatment and pyrolysis methods are used to enhance the capacities of biosorbents. Many biosorbents are reported to be efficient for the separation of phenolic compounds. It includes nut shells, rice husk, cellulose, chitin, chitosan etc. Optimal conditions for adsorption of phenolic compounds were found near to pH 7. Adsorption capacities of biosorbents towards organic herbicides, pesticides, and phenolic compounds are shown in Table 5.

Table 5: Efficiencies of different biosorbents towards herbicides, pesticides, and phenolic compounds

Adsorbate	**Adsorbent**	Q_{max}	**References**
Adsorption of herbicides			
Molinate	Activated carbon	113	Silva et al. 2004
Paraquat	Rice husk	293	Hsu et al. 2009
Prometon	Coconut carbon	26	Moreno et al. 2010
Propazine	Mineral activated carbon	26	
Atrazine	Wood charcoal	0.8	Lopez-Tellez et al. 2011
Adsorption of pesticides			
4,4 DDT	Wood sawdust	4-70	Boussahel et al. 2009
DDD	Bagasse fly ash	$8 \times 10^{3-}$	Zolgharnein et al. 2011
Malethion	Bagasse fly ash	$2 \times 10^{3-}$	
DDE	Bagasse fly ash	$7 \times 10^{3-}$	
Methyl parathion	Rice husk	101	Ahmad et al. 2010
Adsorption of phenolic compounds			
4-Chlorophenol	Chitosan	322	Kumar et al. 2010
2,4-dichlorophenol	Pomogranite peel	66	Soto et al. 2011
Nonyl phenol	Chitosan	56	
4-Nitrophenol	Activated carbon	480	Liu et al. 2010b
Phenol	Rice husk	14	Mbui et al. 2002
Phenol	Banana peel	689	Achak et al. 2009
o-Chloro phenol	Activated carbon	97	Nadavala et al. 2009
p-Nitrophenol	Chitosan	25	Li et al. 2009

13.4 Characteristics of biosorption

Adsorption-based WWT is a promising method due to little cost of operation, high order of selectivity, lower degree of toxic sludge generation, and optimum efficiency. Adsorption method is extensively practiced to adsorb metals ions, industrial toxic effluents, various atmospheric pollutants, pesticides, fertilizers, wastes from nuclear reaction, etc. In spite of numerous practical research results obtained from biosorption, the proper commercialization is yet to achieve. The scientific community should work in the directions to commercialize various biosorption techniques. However, several national/international policies along with industry's requirements while solving technical/scientific problems to scale-up the technologies may restrict the proper commercialization of biosorption techniques. In a few cases, for instance, the commercial application of activated carbon makes adsorption process expensive. Therefore, any cost-effective adsorbents (such as agricultural wastes-based) are used for the same purpose. Similarly, discarded solid waste materials such as sugarcane bagasse, rice husks, coffee residue, orange peels, sawdust, yeast biomass, and rice husks are used for removal of heavy metals, fertilizers, pesticides, toxic industrial effluents, nuclear wastes, and air pollutants. Cost-effectiveness, high capacity of adsorption, occurrence of different functional groups, and a higher chance of metal recovery makes biosorbent the most suitable candidate for remediation of environmental pollutants. Cost-effectiveness and easy preparation of biosorbents has resulted in establishment of a mature and widespread adoption of biosorption technology. Biosorption employs a widespread spectrum of mechanisms to regulate concentrations of heavy metals, toxic industrial dyes, fertilizers, pesticides, atmospheric pollutants, and waste management in environment. High surface area, large pore volume, and occurrence of ligands and functional groups in biosorbent make them efficient for WWT.

13.5 Mechanism of biosorption

Mechanism of biosorption is diverse in nature. Using mechanism, one can control the removal of pollutant and efficiency of biosorbent. Physorption and chemisorption give more details about the mechanism. Physisorption involves interaction between surface of adsorbent and adsorbate by Van der Waals force. The process is exothermic and spontaneous, and thus Gibbs free energy is negative, whereas chemisorption includes ionic exchange, chemical precipitation, and electrostatic interaction. Mechanism of biosorption is very simple and it requires binding of sorbates onto sorbent. It involves physical, chemical, binding, chelation, reduction, precipitation, and complexation. The main points for mechanism include availability, type of binary sites, factors influencing process (pH, temperature, concentration of sorbent, and sorbate), and structure as well as nature of biosorbent. Several types of mechanisms associated with adsorption process are discussed below.

13.5.1 Physical adsorption

Process by which the ions get shifted to solid phase from liquid phase includes subsequent three steps: first, boundary layer mass transfer across liquid film that surrounds the particle; second, mass transport or internal diffusion amongst the particle boundary; and third, the adsorption within particle and on external surface (Foots et al. 1976). Two pathways through which adsorption occurs onto surface are: first, physical adsorption that is rapid as well as reversible and it occurs due to non-specific attraction forces such as Van der Waals forces, and second, electrostatic adsorption that is rapid as well as largely reversible and occurs due to coulombic attraction forces between adsorbing phase and charged solute species. Biosorption of U, Cd, Zn, Cu, and Co can be carried out by dead biomasses of algae, fungi, and yeasts through electrostatic interactions between microbial cell walls and metal ions present in wastewater. The electrostatic interactions are considered to be accountable for Cu adsorption by alga *Chiarella vulgaris* and bacterium *Zoogloea ramigera*, also for Cr adsorption by fungi *Aspergillus niger* and *Ganoderma lucidum*.

13.5.2 Transport across the microbial cell wall

Various ions get transported in cells through diffusion process across the microbial cell wall. The transport of cations largely depends upon permeability of cell membrane. Migration of various heavy metals through microbial cell membranes is called intracellular accumulation and it is facilitated by same transportation mechanism that used to carry metabolically essential ions such as K, Mg, and Na. Principally, biosorption process through any living organisms comprises of two phases: first, metabolism-independent binding process, in which metal binds up with cell walls, and second, metabolism-dependent intracellular uptake process, in which metal ion migrates through microbial cell membrane (Salman et al. 2014). Another study (Price et al. 2001) has presented that the mechanism of removal of Cu through *Aspergillus niger* is primarily because of an active metabolic reaction that leads to internal absorption of Cu. The accumulation of metals inside microbial cell results from bioaccumulation, a slow metabolic-dependent metal removal process. The adsorption and accumulation process of macrophytes cleans contaminated water body by removing heavy metals.

13.5.3 Ion exchange

Polysaccharides are one of the chief components in microbial cell wall where divalent metal ion gets exchanged by counter ions. Ion exchange is taken as an imperative method in biosorption as it describes several explanations regarding heavy metal removal experiments. Ion exchange includes the mechanism in which metal binds with biomass through precise mechanisms consisting of physical binding (London forces, electrostatic, Van der Waals forces) and chemical binding (ionic and covalent) (Uluozlu et al. 2008, Li et al. 2010). The biosorption of Cd(II) and Pb(II) metals onto *L. scrobiculatus* and *A. rubescens* biomasses happens because of chemical ion exchange process (Anayurt et al. 2009, Sari and Tuzen 2009). Binding of metal cation to available site of biomasses takes place. Higher concentration of biosorbent materials results in a higher separation of Cd(II) and Pb(II) metal cations due to plentiful availability of sites for sorption onto the biomass surfaces as well as exchanging ions (Vimala and Das 2009). Significance of electrostatic attraction relies on types as well as amount of sites available on biomass surface and whether such sites are ionized/occupied by protons or other ions. Therefore, relevance of attractions indirectly depends upon *p*H and *p*Ka values of attached group. The $-NH_2$ groups are positively charged in protonated form and become neutral on deprotonation, whereas $-COOH$, $-SO_4^{2-}$, and $-PO_4^{3-}$ groups are neutral on protonated, and becomes negatively charged on deprotonated. Alkaline treatment is needed to separate proteins from the biomass and to contribute to deacetylate chitin based material present in cell wall of microorganisms, whereas alcohol treatment may not associate with accessibility and availability of sorption sites but encourage denaturizing of protein. Declining pH of metal solution during adsorption designates discharging of protons during metal uptake that indicates a mechanism of ion exchange on carboxylic functions (Syecova et al. 2006). A strong ionic balance between adsorbed acidic agent and metal ions onto biomass gets established during metal sorption on biomasses of *Eichhornia crassipes*, *Vallisneria spiralis*, and *Pistia stratiotes*. Furthermore, adsorption of targeted metal ions from multi-metal solution stays unaffected in presence of other metal ions which helps to understand the metal adsorption mechanism, that is, ion exchange mechanism, from a multi metal system such as contaminated industrial wastewater (Verma et al. 2008).

13.5.4 Complexation

The removal of metals from wastewater is also carried out by a complex formation on cell surface by establishing reactions between the active functional groups present on cell wall and metals. Some of the functional groups such as carboxyl, amino, thiol, hydroxy, phosphate, and hydroxy-carboxyl get attached with heavy metal ions through coordination bonds. Complexation mechanism plays a significant role in metal–ligand and sorbate–sorbent interactions. A coordination compound is a multiatomic molecule consisting of one or several central metal cations and surrounded by a number of negative or neutral charges ligands. Coordinate compounds are either neutral or positively charged. Number of coordinating atoms in ligands attached directly to central atom is known as coordination number and can be greater than valence of central metal atom. Complexation is an only mechanism associated with accumulation of Ca, Mg, Cd, Zn, Cu, and Hg by *Pseudomonas syringae*. Microorganisms produce several organic acids such as citric, oxalic, gluonic, fumaric, lactic, and malic acids, and these acids chelate toxic metals and form metallo-organic molecules. Such organic acids help to solubilize metal compounds and leach out of metals from the surfaces. Metal complexes become more complicated by –COOH groups present in the microbial polysaccharides. Metal sequestration is a complexation or coordination of a central heavy metal with a multidentate ligand existing onto surface of algae that can be considered as a polyfunctional macromolecule. The binding of various metallic ions with biomaterials through several functional groups relies on ionic properties including ionization potential, electronegativity, redox potential of metals, and ionic radius. As per the final *p*H assessment, main mechanism behind Cu removal was found to be due to the development of insoluble complexes between Cu^{2+} and anionic species, i.e., CO_3^{2-}, SO_4^{2-}, and OH^- that occur onto surface of biosorbent (Sarioglu et al. 2009).

13.5.5 Precipitation

Metals might get accumulated inside microbial cells or onto cell walls through sorption–precipitation process at elevated pH. Precipitation of metals can either be dependent on or independent of cellular metabolism. Metal removal process from cell wall is related with an active defense system of microorganisms. Precipitation occurs due to the interactions between target metals and compounds developed by microorganisms as a result of the defense system. The cellular metabolism-independent precipitation is a result of chemical reaction between metal and cell surface. In case of precipitation, uptake of metal can take place either in solution or on cell surface. Furthermore, precipitation can be a factor of cell metabolism if microorganism produces any compounds in the presence of toxic metals that favour the precipitation process. Precipitation may be independent of cell metabolism if it occurs after the chemical interaction between cell surface and metal. The plausible mechanism of bio adsorption by different biomass based biosorbents of heavy metals, and involvement of various functional bonds are shown in Table 6.

13.6 Factors affecting biosorption process

There are several factors which have a strong influence on biosorption process such as *p*H, temperature, initial biosorbent concentration etc. Some of such factors are discussed in detail below.

13.6.1 Influence of pH

The pH of solution is an important factor in biosorption. At low pH (acidic), hydronium ions are thoroughly accompanying to biosorbent's active sites. There is a lower number of hydronium ions at higher *p*H which causes an increase in uptake of contaminant in biosorption. General range of *p*H between 2.5 to 6.0 for metal uptake is optimum. Beyond the certain limit, the metal uptake ability of biosorbent becomes compromised.

13.6.2 Effect of temperature

For an efficient removal of metal ions from the environment, an optimum temperature is needed to be examined. Temperature is associated with the thermodynamics of biosorption process. Temperature deals with positive or negative effects on biosorption process at different intervals. It can cause a change in sorption efficiency of biosorbent. Rise in temperature increases biosorption rate of biosorbents but can cause structural damage to biosorbent. Generally, the biosorption process is employed at temperature range of 20 to 35°C. Temperature beyond 45°C damages the proteins which in turn affects the metal uptake process.

13.6.3 Initial concentration of pollutant dyes

Efficiency of biosorption increases with increase in pollutant concentration and then reaches a saturation value. Higher efficiency at low metal concentration or low dye concentration is due to the fact that at such condition, complete interaction of ions with available binding site takes places which causes high efficiency rate.

Table 6: Predominant mechanism of biosorption by various types of biomasses.

Biomass	**Heavy metal**	**Functional groups**	**Mechanism**	**References**
Yarrowia lipolytica (strains NCIM 3589 and 3590)	Cr(VI)	–COOH, –OH, amide	Ion exchange	Bankar et al. 2009
Rhizopus arrhizus (a harmful fungus)	Cr(VI)	–COOH, $-NH_2$	Electrostatic attraction	Shroff and Vaidya 2012
Agaricus bisporus (a species of mushroom)	Cr(VI)	C–H, –CN, $-SO_3$	Physical adsorption	Ertugay and Bayhan 2008
Spirulina species (a blue-green algae)	Cr(III), Cd(II), Cu(II)	–COOH, –OH, $-PO_4^{3-}$	Physical adsorption, ion exchange	Chojnacka et al. 2005
Yarrowia lipolytica (a non-pathogenic fungus)	Ni(II)	–COOH, –OH, –CO–, $-NH_2$	Ion exchange	Shinde et al. 2012
Acidophilic (bacterium)	Cd(II)	–COOH, –OH, $-NH_2$	Electrostatic reaction and complex formation	Chakravarty and Banerjee 2012
Mesorhizobium amorphae (a mesophilic bacterium)	Cu(II)	O–H, N–H, C–H, C=O, –NH, –CN, C–N, C–O, C=C, C≡C, amide	Precipitation	Mohamad et al. 2009
Pseudomonas plecoglossicida (a motile bacterium)	Cd(II)	–NH, –OH, –CH, –CONH	Ionic exchange, electrostatic interaction	Guo et al. 2012
Cyanobacteria microcystis	Sb(III)	–COOH, –OH, $-NH_2$	Complexation	Wu et al. 2012
Inactivated fungal biomass	Cr(VI)	–COOH, $-PO_4^{3-}$, –SH, $-NH_2$	Physical adsorption, ion exchange, complexation and electrostatic attraction	Ramrakhiani et al. 2011
Alkali-modified sewage sludge	Cd(II)	–COOH, –OH	Ion exchange, electrostatic attraction and complexation	Hu et al. 2012
Aerobic granules and bacterial alginate	Pb(II)	–COOH	Ion exchange	Wang and Li 2012
Fruiting body of jelly fungus	Cd(II), Cu(II), Pb(II)	–COOH, –OH, $-NH_2$, $-PO_3^{2-}$, C–N–C	Ion exchange and surface complexation	Huang et al. 2012
Xanthate-modified magnetic chitosan	Pb(II), Cu(II), Zn(II)	–SH, $-NH_2$	Complexation	Zhu et al. 2012

Table 6 contd. ...

...*Table 6 contd.*

Biomass	Heavy metal	Functional groups	Mechanism	References
Magnetic PS-EDTA resin	Cr(VI)	$-NH_2$, $-NH-$	Chelating electrostatic interaction	Mao et al. 2012
Cyanobacteria Microcystis	Sb(III)	$-COOH$, $-OH$, $-NH_2$	Complexation	Sun et al. 2011
Barley straw	Cu(II), Pb(II)	$-COOH$, $-OH$	Ion exchange, complexation	Pehlivan et al. 2012
Rice straw	Cd(II)	$-COOH$, $-OH$, C=C	Ion exchange, chelation	Ding et al. 2012
Green tomato husk	Fe(II), Mn	$-COOH$, $Ph-O^-$	Ion exchange, complexation, precipitation	Garcia-Mendieta et al. 2012
Raw and treated agave bagasse	Cd(II), Pb(II) Zn(II)	$-COOH$	Ion exchange, complexation	Velazquez et al. 2013
Soybean	Cr(III), Cu(II)	$-COOH$, $-OH$	Ion exchange, chelation, precipitation	Witek-Krowiak et al. 2013
Date pits/tea-waste	Cr(VI)	$-COOH$, $-NH_2$, $-O-CH_3$	Complexation	Albadarin et al. 2013
Watermelon rind	Cu, Zn, Pb	$-COOH$, $-NH_2$	Ion exchange and micro-precipitation	Liu et al. 2012

13.6.4 Initial concentration of pollutant metals

The initial concentration of pollutant metals offers a significant motivation to subdue mass transfer resistance of metal between aqueous phases and solid. Amount of metal to be adsorbed by biomass adsorbent gets enhanced on increasing initial concentration of metals. The optimum percentage of metal removal is achieved at low initial concentration of metal. Therefore, at a given concentration of biomass, metal separation surges with increase in initial concentration of metal in pollutant solution.

13.6.5 Nature of biosorbent

Metal and dye removal from wastewater is described in various methods such as freely suspended microbial cells, immobilization of microbial cells, and biofilms. Metal uptake phenomenon is changeable between physical, chemical, or by combining both treatments. The physical treatment consists of autoclaving, drying, boiling, sonication, whereas chemical treatment involves the use of acid or alkali to enhance biosorption capacity. Deacetylated fungal cells affect structure of chitin resulting in formation of chitosan-glycan complexes which results in high metal affinities (Wang and Chen 2009). Age of biosorbent and growth of medium components during biosorption result in alteration of cell wall composition, cell size, and EPS formation.

Generally, biosorbent generates binding sites for biosorption. Dosage of biosorbent has a strong role in biosorption. Increase in the dosage of biosorbent increases the surface area, which in turn increases the number of binding sites in the process.

13.6.6 Concentration of biomass

Concentration of biomass is directly related with metal uptake (Abbas et al. 2014). It has been found that electrostatic bonding among the cells plays a vital part in the process of metal removal. At a given equilibrium, biomass adsorbs large number of metal ions at low cell densities as compared to high densities (Gourdon et al. 1990). As metal uptake relies on biding sites, higher biomass concentration increases adsorption efficiency due to increase in binding sites (Malkoc and Nuhoglu 2005).

13.6.7 Metal affinity to biosorbent

The physical as well as chemical pre-treatment affects permeability. Furthermore, the biomass surface charges produce metal binding groups onto surface, which would further be accessible for the binding. Such accessibility is operated by pre-treatment of biomass with alkalis, acids detergents, and heat, in order to increase amount of metal uptake.

13.6.8 Effect of contact time

Time required to gain highest biosorption relies on the type of biosorbent, type of pollutant, and their combined effect. The rate of biosorption is high at initial stage due to vacant sites available for biosorption. But with time, the biosorption rate gets decreased due to enhancement in the percentage saturation by pollutant that remains in the solution.

13.6.9 Effect of agitation speed

The agitation speed increases the biosorption efficiency of biosorbent by decreasing the resistance in mass transfer. But it can lead to destruction of physical nature of biosorbent medium. Speed is best for homogeneity for suspension with huge adsorption. High speed can cause vortex phenomenon which leads to loss of homogeneous nature of suspension.

13.6.10 Surface area to volume ratio

The surface area to volume ratio plays an important part in efficient removal of heavy metal from the medium. The surface area is a major concern, especially in case of the biofilms. The binding efficiency of metallic ions to the microbial cell wall has already been previously reported (Gadd et al. 1985). Even though the intracellular metal adsorption process is an energy-consuming process, still microorganisms prefer it over wall adsorption.

13.7 Types of pollutants and application of biosorbents

Biomaterials and microorganism are capable of adsorbing contaminants from wastewater. Wastewater is processed by physical and chemical treatment for effective removal of pollutants. Conventional methods are very costly and cause more disposal of sludge. The old techniques are ineffective to remove contaminants and are expensive and less adaptable. Table 7 shows about various types of pollutants and their characteristics.

Biosorption is highly useful over traditional techniques. The efficiency of biosorbent is explored at small scales. But a difference is seen in real laboratory environment at a comparatively large scale, i.e., contaminant treatment industries. Few commercial processes are also developed. It can be related to the diversity of pollutants and adsorption background. Biosorbents are optimized/standardized for a special effluent. The cost and feasibility in terms of large-scale applications should also be analyzed. High efficiencies of metal removal at lower costs and minimum labour when compared to other traditional treatments can assure industry/state to adopt biosorption. However, there is a lack of field experiments. Field studies require

Table 7: Types of pollutants and their characteristics.

Heavy metal ions	▪ It is mainly from industries like electroplating, leather, dye manufacturing, mining, jewelry making, etc. ▪ Main challenge is its high cost operation, high maintenance, and high sludge disposal. ▪ Continuous disposal of water causes increase in metal ion concentration. ▪ Use of low cost adsorbent is most viable for removal.
Organic dyes	▪ Water pollution is caused by textile, printing, cosmetic, and leather industries. ▪ Raw agricultural waste material can be processed to make adsorbents for removal of organic dyes. ▪ Low cost adsorbent with maximum adsorption capacity can be used to remove organic dyes from wastewater.
Toxic industrial wastes	▪ Toxic dyes from textiles, printing, cosmetic industries cause toxicity in water. ▪ High adsorption capacity adsorbent preferred for removal of such toxic substances.
Fertilizer/pesticides	▪ Excess applications of pesticides and fertilizers in agricultural and pesticide industry causes water pollution. ▪ It causes the concentration of chemical nutrient rich in water causes eutrophication activated. ▪ Activated carbon can be used as an effective adsorbent. ▪ Carbonaceous adsorbent is used as an alternative for removal of fertilizer from wastewater.
Atmospheric pollutant	▪ Chemical, petrochemical, and industrial waste causes water pollution. ▪ Volatile organic pollutant is the main pollutant which causes atmospheric pollution. ▪ VOC causes depletion of ozone layer. ▪ Carbon adsorption is used for VOC emission control.

capital, manpower, coordination and infrastructure. State should assist researchers to not only fund but also to concern the issues related to the industries. The general indifference of the industry toward waste treatment may be an issue. The state is the centre for providing, informing, and facilitating availability of biomass from different sources to different polluting units. This kind of effort will develop a sustainable wastewater environment. New technology based startups based on biosorption need to be developed. Strict norms and scrutiny against waste disposal can convince industry to view waste treatment as a necessary investment rather than an avoidable overhead cost. Novel efforts should be initiated for the biosorption to be applied in small laboratories as well as in large centralized treatment plants. Many techniques such as surface methodology and general algorithm can be used for optimizing the suitable process. Modeling should also be carried-out in solutions with different metals and organic matter simulating real wastewater parameters and conditions. The pilot and field studies need to be employed to compare biosorption with conventional techniques. Also, the use of computer-based simulations or modeling can reduce the number of processes.

13.8 Techniques to enhance biosorption

Biosorbents obtained from untreated biomass feedstock show comparatively lesser sorption capacity. Different methods and treatments are used for enhancement of biosorption capacity. Some of such methods are described below.

13.8.1 Nano-bioadsorbents

Combining bioadsorbents with nanoparticles form nano bioadsorbents which is a nanohybrid and highly efficient for biosorption. This process helps to increase surface area. Here, main limitation is agglomeration of ultra-fine powder to large particles which leads to decrease in surface area and catalytic performance of material. Use of nanoparticles enhances activity and reduces toxicity. An optimized blending of nanotechnology with material science imparts a useful insight for functional nanocomposite materials (Singh et al. 2017, Pattanayak et al. 2021). The nanoparticles/nanofillers embedded nanocomposites are widely used in almost all domains of material science (Singh et al. 2020, Kumar et al. 2015). Therefore, some surface modifications of bioadsorbents are also carried out to make such materials antimicrobial and conductive in nature to serve various applications. Silver nanoparticles (AgNPs) are one of the most compatible and widely used antimicrobial agents to be incorporated into polymer matrix through different methods to impart antimicrobial property (Kumar et al. 2013, Singh et al. 2018a).

13.8.2 Biosorption by microbial biosorbents

Bioremediation is an advanced method for removal and recovery of pollutants from wastewater and it involves using living organism to reduce or recover pollutants into less toxic form. It involves activities by algae, bacterial, fungi, etc. Bioremediation can function naturally or can be through addition of electron acceptor, nutrient or other factors. Bacteria have been used as biosorbents owing to the ability to grow under controlled condition and resistance to extensive environmental conditions.

13.8.3 Pyrolysis

Pyrolysis is a thermal decomposition of materials at elevated temperatures in an inert atmosphere. It involves a change of chemical composition. Biomaterial under limited oxygen environment gets converted into char. It is done either by slow pyrolysis or fast one.

13.8.4 Hydrothermal carbonization

Hydrothermal carbonization is a thermochemical process for pre-treatment of high moisture content biomass under hot compressed water, making it relevant for diversified purposes. It is accomplished in a closed reactor at a temperature range of 180–280°C under pressure (2–6 MPa) for 5 to 240 min. This method is used for biomasses like bamboo, coconut fibre, peanut shell, etc. In this process, the biomaterial is disintegrated by varying the temperature.

13.9 Limitations of biosorption

Difficulties for the application of biosorption suggest people to implement hybrid technology for pollutant removal. The advancement of biosorption process needs analysis in the direction of modelling, regeneration, and immobilization of biosorbent and also pilot study of real industrial wastewater (Michalak et al. 2013). Application of biosorption is affected by several factors such as biosorption capacity, cost of biosorbent, availability of biosorbent, ease of regeneration, ease with different reactor configuration, etc. The main challenges consist high cost and operational difficulties to synthesize the adsorbent materials. Even though a vast number of biosorbents are available, still it is essential to develop more efficient, economic, and selective biosorbents. Different mathematical models are used for a single metal biosorption process; however, elaboration of new, improved, and simplified mathematical models for description of multi-sorbate systems is required. In order to achieve best performance of biosorption process, it is necessary to identify the biosorption mechanism in relation to class of biosorbents. However, a huge number of publications and patents are available, but process is still at the laboratory scale. The technology should be commercialized for its real time utility. In order to use biosorption technology at industrial scale, economic analyses are required to be carried out and overall cost of sorbent and biosorption process should be investigated before employing to wastewater treatment.

13.10 Future aspects

Biosorbents are very efficient to remove different types of dyes and heavy metal ions. Biosorption will provide most economic biosorbents. Many techniques, especially drying, autoclaving, and chemical treatment, are required to enhance the biosorption capacity of adsorbent. There is a high demand to synthesize biosorbents in a simpler, and less costly way with high sorption efficiency. The use of non-viable fungi, bacteria, and algae biomass is more feasible and effective along with excellent efficiency of biosorption. However, biosorption efficiency depends upon characteristics of biosorbent, biosorbate, and process variables. Many effluents coexist in an industrial effluent and biosorption under actual wastewater treatment needs to be evaluated. There is still a challenge to use non-viable immobilized biomass. Various methods are being used to reduce chemicals in the wastewater treatment process, activated carbon being the most commonly used one. However, activated carbon is an expensive and unsustainable option. Fortunately, cheaper alternatives are being developed by 3D (three-dimensional) printing methods (Singh et al. 2018b, 2021b, 2021c, 2022).

13.11 Conclusions

This chapter details different biosorbents which are easily available for removal of pollutants, especially heavy metals and dyes, from wastewater. Pollutants such as organic dyes and heavy metal ions being released during industrial operations are among the major causes of soil and water contamination. Persistent pollution is a threat for all the living organisms as it creates mutagenic and toxic consequence. Several methods exist to treat contaminant wastewater, but biosorption technique is highly appreciable. Biosorption is an eco-friendly, sustainable, economic, and a green chemistry process where pollutants are removed by taking up dead or inactive biomaterial as biosorbents. Biosorbents act as an efficient material for eliminating the toxic wastes such are fertilizers, effluents waste from industries, pesticides, pollutants associated with atmosphere, etc. Selection of biosorbents and its implementation for wastewater treatment and soil remediation requires more efforts. In biosorption, toxic sludge is obtained in minor quantity as compared to conventional technique. The wastewater treatment by chemical and physical methods is more expensive and such methods produce a significant quantity of chemical waste. Various biobased adsorbents are used in biosorption technique such as chitin, clay, cellulose, chitosan, microbial biomass, and agricultural wastes. Such bio adsorbents can be effectively used to remove dyes, heavy metal ions, toxic industrial effluents, and pesticides/chemical fertilizers from wastewater. Selection of a suitable biosorbent and its proper implementation need more effort. An effective futuristic approach needs to be developed for selection of industrial effluent, operating conditions, and efficient mechanism of removal of contaminants from wastewater through treatment. Adsorption and related bioremediation require high selectivity, efficiency, low cost, and minimal toxic disposal. Key factor in biosorption is regeneration of biosorbent which needs to be economic. Presence of different functional groups in biosorbent makes them selective and specific for biodegrading dye from wastewater. More effects are needed to industrialize and commercialize biosorption by proper selection of suitable biosorbent based on market analysis and pilot scale studies with actual contaminated water and full scale monitoring and demonstration system. Efficiency of adsorption potential of biosorbent can be enhanced by surface treatment, functionalization, and oxidation process. This chapter summarizes the process and mechanism of biosorption along with different methods to enhance the efficiency of biosorbents and physico chemical condition to remediate heavy metal ions and organic dyes from wastewater. In addition to that, optimum application utilities of various biosorbents, nano biosorbents, and different biomass-based biosorbents for bioremediation are also explored. To conclude, the selection of biosorbents is the best choice for wastewater treatment because such adsorbents are abundant, easily available, ecofriendly, biodegradable, and are green chemistry materials. It has huge application at domestic and industrial scale to reduce environmental pollution by treating domestic and industrial wastewater by adsorption method.

References

Abbas, S.H., Ismail, I.M., Mostafa, T.M. and Sulaymon, A.H. 2014. Biosorption of heavy metals: a review. J Chem Sci Technol. 3: 74–102.

Abd El-Rahim, W.M. and Moawad, H. 2003. Enhancing bioremoval of textile dyes by eight fungal strains from media supplemented with gelatine wastes and sucrose. Journal of Basic Microbiology: An International Journal on Biochemistry, Physiology, Genetics, Morphology, and Ecology of Microorganisms. 43: 367–375.

Achak, M., Hafidi, A., Ouazzani, N., Sayadi, S. and Mandi, L. 2009. Low cost biosorbent "banana peel" for the removal of phenolic compounds from olive mill wastewater: Kinetic and equilibrium studies. Journal of Hazardous Materials. 166: 117–125.

Ahmad, T., Rafatullah, M., Ghazali, A., Sulaiman, O., Hashim, R. and Ahmad, A. 2010. Removal of pesticides from water and wastewater by different adsorbents: a review. Journal of Environmental Science and Health, Part C. 28: 231–271.

Aksu, Z. and Tezer, S. 2005. Biosorption of reactive dyes on the green alga Chlorella vulgaris. Process Biochemistry. 40: 1347–1361.

Albadarin, A.B., Ala'a, H., Al-Laqtah, N.A., Walker, G.M., Allen, S.J. and Ahmad, M.N., 2011. Biosorption of toxic chromium from aqueous phase by lignin: mechanism, effect of other metal ions and salts. Chemical Engineering Journal. 169: 20–30.

Albadarin, A.B., Mangwandi, C., Walker, G.M., Allen, S.J., Ahmad, M.N. and Khraisheh, M. 2013. Influence of solution chemistry on Cr (VI) reduction and complexation onto date-pits/tea-waste biomaterials. Journal of Environmental Management. 114: 190–201.

Amarasinghe, B.M.W.P.K. and Williams, R.A. 2007. Tea waste as a low cost adsorbent for the removal of Cu and Pb from wastewater. Chemical engineering journal. 132: 299–309.

Anayurt, R.A., Sari, A. and Tuzen, M. 2009. Equilibrium, thermodynamic and kinetic studies on biosorption of Pb (II) and Cd (II) from aqueous solution by macrofungus (*Lactarius scrobiculatus*) biomass. Chemical Engineering Journal. 151: 255–261.

Asma, D., Kahraman, S., Cing, S. and Yesilada, O. 2006. Adsorptive removal of textile dyes from aqueous solutions by dead fungal biomass. Journal of Basic Microbiology. 46: 3–9.

Banerjee, A., Nayak, D. and Lahiri, S. 2007. Speciation-dependent studies on removal of arsenic by iron-doped calcium alginate beads. Applied Radiation and Isotopes, 65: 769–775.

Bankar, A.V., Kumar, A.R. and Zinjarde, S.S. 2009. Removal of chromium (VI) ions from aqueous solution by adsorption onto two marine isolates of *Yarrowia lipolytica*. Journal of Hazardous Materials. 170: 487–494.

Basu, H., Singhal, R.K., Pimple, M.V. and Reddy, A.V.R. 2015. Arsenic removal from groundwater by goethite impregnated calcium alginate beads. Water, Air, & Soil Pollution. 226: 1–11.

Benguella, B. and Benaissa, H. 2002. Effects of competing cations on cadmium biosorption by chitin. Colloids and Surfaces A: Physicochemical and Engineering Aspects. 201: 143–150.

Bezbaruah, A.N., Kalita, H., Almeelbi, T., Capecchi, C.L., Jacob, D.L., Ugrinov, A.G. and Payne, S.A. 2014. Ca–alginate-entrapped nanoscale iron: arsenic treatability and mechanism studies. Journal of Nanoparticle Research. 16: 1–10.

Boddu, V.M., and Smith, E.D. 2004. Composite biosorbent for treatment of waste aqueous system(s) containing heavy metals. US Patent # 6,786,336.

Boussahel, R., Irinislimane, H., Harik, D. and Moussaoui, K.M. 2009. Adsorption, kinetics, and equilibrium studies on removal of 4, 4-DDT from aqueous solutions using low-cost adsorbents. Chemical Engineering Communications. 196: 1547–1558.

Brierley, J.A., Brierley, C.L., Decker, R.F. and Goyak, G.M. 1987. Treatment of microorganisms with alkaline solution to enhance metal uptake properties. US Patent # 4,690,894.

Brierley, J.A., Brierley, C.L., Decker, R.F. and Goyak, G.M. 1990. Metal recovery. US Patent # 4,898,827.

Celik, A. and Demirbaş, A. 2005. Removal of heavy metal ions from aqueous solutions via adsorption onto modified lignin from pulping wastes. Energy Sources. 27: 1167–1177.

Chakravarty, R. and Banerjee, P.C. 2012. Mechanism of cadmium binding on the cell wall of an acidophilic bacterium. Bioresource Technology. 108: 176–183.

Chander, M. and Arora, D.S. 2007. Evaluation of some white-rot fungi for their potential to decolourise industrial dyes. Dyes and Pigments. 72: 192–198.

Chen, H.Z., Xu, J. and Li, Z.H. 2005. Temperature control at different bed depths in a novel solid-state fermentation system with two dynamic changes of air. Biochemical Engineering Journal. 23: 117–122.

Chen, X., Chen, G., Chen, L., Chen, Y., Lehmann, J., McBride, M.B. and Hay, A.G. 2011. Adsorption of copper and zinc by biochars produced from pyrolysis of hardwood and corn straw in aqueous solution. Bioresource Technology. 102: 8877–8884.

Cho, K., Shin, B.Y., Park, H.K., Cha, B.G. and Kim, J. 2014. Size-controlled synthesis of uniform akaganeite nanorods and their encapsulation in alginate microbeads for arsenic removal. RSC Advances. 4: 21777–21781.

Chojnacka, K., Chojnacki, A. and Gorecka, H. 2005. Biosorption of Cr^{3+}, Cd^{2+} and Cu^{2+} ions by blue–green algae *Spirulina sp.*: kinetics, equilibrium and the mechanism of the process. Chemosphere. 59: 75–84.

Costan, G., Bermingham, N., Blaise, C. and Ferard, J.F. 1993. Potential ecotoxic effects probe (PEEP): a novel index to assess and compare the toxic potential of industrial effluents. Environmental Toxicology and Water Quality. 8: 115–140.

Cotoras, D. and Viedma, P. 2011. Bacterial strain for a metal biosorption process. US Patent # 7,951,578.

Crini, G. 2005. Recent developments in polysaccharide-based materials used as adsorbents in wastewater treatment. Progress in Polymer Science. 30: 38–70.

Daneshvar, N., Ayazloo, M., Khataee, A.R. and Pourhassan, M. 2007. Biological decolorization of dye solution containing Malachite Green by microalgae *Cosmarium* sp. Bioresource Technology. 98: 1176–1182.

Das, S.K., Bhowal, J., Das, A.R. and Guha, A.K. 2006. Adsorption behavior of rhodamine B on rhizopus oryzae biomass. Langmuir. 22: 7265–7272.

de Luna, M.D.G., Flores, E.D., Cenia, M.C.B. and Lu, M.C. 2015. Removal of copper ions from aqueous solution by adlai shell (Coix lacryma-jobi L.) adsorbents. Bioresource Technology. 192: 841–844.

Deng, S. and Ting, Y.P. 2005. Fungal biomass with grafted poly (acrylic acid) for enhancement of Cu (II) and Cd (II) biosorption. Langmuir. 21: 5940–5948.

Diaz, M., Cambier, P., Brendlé, J. and Prost, R. 2007. Functionalized clay heterostructures for reducing cadmium and lead uptake by plants in contaminated soils. Applied Clay Science. 37: 12–22.

Ding, Y., Jing, D., Gong, H., Zhou, L. and Yang, X. 2012. Biosorption of aquatic cadmium (II) by unmodified rice straw. Bioresource Technology. 114: 20–25.

Drobot, W. 1981. Microbiological recovery of metals. US Patent # 4,293,333.

Ertugay, N. and Bayhan, Y.K. 2008. Biosorption of Cr (VI) from aqueous solutions by biomass of *Agaricus bisporus*. Journal of Hazardous Materials. 154: 432–439.

Fan, L., Luo, C., Lv, Z., Lu, F. and Qiu, H. 2011. Removal of Ag^+ from water environment using a novel magnetic thiourea-chitosan imprinted Ag^+. Journal of Hazardous Materials. 194: 193–201.

Fard, R.F., Azimi, A.A. and Bidhendi, G.N. 2011. Batch kinetics and isotherms for biosorption of cadmium onto biosolids. Desalination and Water Treatment. 28: 69–74.

Fein, J.B., Daughney, C.J., Yee, N. and Davis, T.A. 1997. A chemical equilibrium model for metal adsorption onto bacterial surfaces. Geochimica et Cosmochimica Acta. 61: 3319–3328.

Foots, V.J., Mckey, G. and Healy, J.J. 1976. The removal of acid dye from effluent using natural adsorbent. Wat. Res. 10: 1061–1070.

Fu, G., Zhao, J., Yu, H., Liu, L. and He, B. 2007. Bovine serum albumin-imprinted polymer gels prepared by graft copolymerization of acrylamide on chitosan. Reactive and Functional Polymers. 67: 442–450.

Fu, Y. and Viraraghavan, T. 2001. Fungal decolorization of dye wastewaters: a review. Bioresource Technology. 79: 251–262.

Gadd, G.M. and White, C. 1985. Copper uptake by *Penicillium* ochro-chloron: influence of pH on toxicity and demonstration of energy-dependent copper influx using protoplasts. Microbiology. 131: 1875–1879.

Gao, Z., Bandosz, T.J., Zhao, Z., Han, M. and Qiu, J. 2009. Investigation of factors affecting adsorption of transition metals on oxidized carbon nanotubes. Journal of Hazardous Materials. 167: 357–365.

García-Mendieta, A., Olguín, M.T. and Solache-Ríos, M. 2012. Biosorption properties of green tomato husk (*Physalis philadelphica* Lam) for iron, manganese and iron–manganese from aqueous systems. Desalination. 284: 167–174.

Gibbs, D.F., Greenhalgh, M.E., Watson, J.H.P., Yeo, R.G. and Ellwood, D.C. 1987. Process for the separation of metals from aqueous media. US Patent # 4,701,261.

Gorovoj, L.F. and Kosyakov, V.N. 2002. Adsorption means for radionuclides. US Patent # 6,402,953.

Gourdon, R., Bhende, S., Rus, E. and Sofer, S.S. 1990. Comparison of cadmium biosorption by gram-positive and gram-negative bacteria from activated sludge. Biotechnology Letters. 12: 839–842.

Greene, B., McPherson, R.A., Darnall, D.W. and Gardea-Torresdey, J.L. 1991. Removal of metal ions with immobilized metal ion-binding microorganisms. US Patent # 5,055,402.

Grini, G. and Badot, P.M. 2008. Application of chitosan, a natural aminopolysaccharide, for dye removal form aqueous solutions by adsorption processes using batch studies: A review of recent literature. Prog Polym Sci. 33: 39–447.

Guibal, E., Van Vooren, M., Dempsey, B.A. and Roussy, J. 2006. A review of the use of chitosan for the removal of particulate and dissolved contaminants. Separation Science and Technology. 41: 2487–2514.

Guo, J., Zheng, X.D., Chen, Q.B., Zhang, L. and Xu, X.P. 2012. Biosorption of Cd (II) from aqueous solution by *Pseudomonas plecoglossicida*: kinetics and mechanism. Current Microbiology. 65: 350–355.

Gupta, V.K. 2009. Application of low-cost adsorbents for dye removal–a review. Journal of Environmental Management. 90: 2313–2342.

Hassan, A.F., Abdel-Mohsen, A.M. and Elhadidy, H. 2014. Adsorption of arsenic by activated carbon, calcium alginate and their composite beads. International Journal of Biological Macromolecules. 68: 125–130.

Hermann, P. 1999. Hydrophilic urethane binder immobilizing organisms having active sites for binding noxious materials. US Patent # 5,976,847.

Hsu, S.T., Chen, L.C., Lee, C.C., Pan, T.C., You, B.X. and Yan, Q.F. 2009. Preparation of methacrylic acid-modified rice husk improved by an experimental design and application for paraquat adsorption. Journal of Hazardous Materials. 171: 465–470.

Hu, J.L., He, X.W., Wang, C.R., Li, J.W. and Zhang, C.H. 2012. Cadmium adsorption characteristic of alkali modified sewage sludge. Bioresource Technology. 121: 25–30.

Hu, T.L. 1994. Decolourization of reactive azo dyes by transformation with *Pseudomonas luteola*. Bioresource Technology. 49: 47–51.

Huang, H., Yuan, Q. and Yang, X. 2004. Preparation and characterization of metal–chitosan nanocomposites. Colloids and surfaces B: Biointerfaces. 39: 31–37.

Huang, H., Cao, L., Wan, Y., Zhang, R. and Wang, W. 2012. Biosorption behavior and mechanism of heavy metals by the fruiting body of jelly fungus (*Auricularia polytricha*) from aqueous solutions. Applied Microbiology and Biotechnology. 96: 829–840.

Jones, E.R., van Vliet, M.T., Qadir, M. and Bierkens, M.F. 2021. Country-level and gridded estimates of wastewater production, collection, treatment and reuse. Earth System Science Data. 13: 237–254.

Jung, W., Park, Y.K., An, J.S., Park, J.Y. and Oh, H.J. 2017. Sorption of arsenic and heavy metals using various solid phase materials. International Journal of Environmental Science and Development. 8: 71.

Kanamarlapudi, S.L.R.K., Chintalpudi, V.K. and Muddada, S. 2018. Application of biosorption for removal of heavy metals from wastewater. Biosorption. 18: 69.

Kazemipour, M., Ansari, M., Tajrobehkar, S., Majdzadeh, M. and Kermani, H.R. 2008. Removal of lead, cadmium, zinc, and copper from industrial wastewater by carbon developed from walnut, hazelnut, almond, pistachio shell, and apricot stone. J. Hazard Mater. 150: 322–327.

Khabibullina, F., Archegova, I., Shubakov, A., Sharapova, I., Romanov, G., Chernov, I. et al. 2007. Petroleum biosorbent based on strains of bacteria and yeast. US Patent # 0,202,588.

Khalaf, M.A. 2008. Biosorption of reactive dye from textile wastewater by non-viable biomass of *Aspergillus niger* and *Spirogyra* sp. Bioresource Technology. 99: 6631–6634.

Khan, N.A. and Jhung, S.H. 2013. Effect of central metal ions of analogous metal-organic frameworks on the adsorptive removal of benzothiophene from a model fuel. Journal of Hazardous Materials. 260: 1050–1056.

Khan, T. and Khan, H. 2019. Environmental sustainability of grey water footprints in Peshawar Basin: Current and future reduced flow scenarios for Kabul River. International Journal of Agricultural and Biological Engineering. 12: 162–168.

Kilbane, J.J. 1998. Adsorption of PCB's using biosorbents. US Patent # 5,750,065.

Kleid, D.G., Kohr, W.J. and Thibodeau, F.R. 1992. Processes to recover and reconcentrate gold from its ores. US Patent # 5,152,969.

Knapp, J.S. and Newby, P.S. 1999. The decolourisation of a chemical industry effluent by white rot fungi. Water Research. 33: 575–577.

Kogtev, L.S., Park, J.K., Pyo, J.K., and Mo, Y.K. 1998. Biosorbent for heavy metals prepared from biomass. US Patent # 5,789,204.

Kotrba, P., Mackova, M., Fišer, J. and Macek, T. 2011. Biosorption and metal removal through living cells. In Microbial Biosorption of Metals (pp. 197–233). Springer, Dordrecht.

Kumar, N.S., Suguna, M., Subbaiah, M.V., Reddy, A.S., Kumar, N.P. and Krishnaiah, A. 2010. Adsorption of phenolic compounds from aqueous solutions onto chitosan-coated perlite beads as biosorbent. Industrial & Engineering Chemistry Research. 49: 9238–9247.

Kumar, P., Singh, A.D., Kumar, V. and Kundu, P.P. 2015. Incorporation of nano-Al_2O_3 within the blend of sulfonated-PVdF-*co*-HFP and Nafion for high temperature application in DMFCs. RSC Advances. 5: 63465–63472.

Kumar, S., Loganathan, V.A., Gupta, R.B. and Barnett, M.O. 2011. An assessment of U (VI) removal from groundwater using biochar produced from hydrothermal carbonization. Journal of Environmental Management. 92: 2504–2512.

Kumar, V. and Singh, A. 2013. Polypropylene clay nanocomposites. Reviews in Chemical Engineering. 29: 439–448.

Kumari, K. and Abraham, T.E. 2007. Biosorption of anionic textile dyes by nonviable biomass of fungi and yeast. Bioresource Technology. 98: 1704–1710.

Kurita, K.T.S. and Iwakura. Y. 1979. Studies on chitin. VI. Binding of metal cations. J. Appl. Polym. Sci. 23: 511–515

Kuroda, K. and Ueda, M. 2003. Bioadsorption of cadmium ion by cell surface-engineered yeasts displaying metallothionein and hexa-His. Applied Microbiology and Biotechnology. 63: 182–186.

Kwon, O.H., Kim, J.O., Cho, D.W., Kumar, R., Baek, S.H., Kurade, M.B. and Jeon, B.H. 2016. Adsorption of As (III), As (V) and Cu (II) on zirconium oxide immobilized alginate beads in aqueous phase. Chemosphere. 160: 126–133.

Lakshmanan, V. I. and McCready, R.G.L. 1992. Bioadsorption composition and process for production thereof. US Patent # 5,084,389.

Lechavelier, H.A. and Drobot, W. 1981. Process for recovering precious metals. US Patent # 4,289,531.

Li, C.B., Hein, S. and Wang, K. 2008. Biosorption of chitin and chitosan. Materials Science and Technology. 24: 1088–1099.

Li, D., He, X., Tao, Y. and Wang, X. 2011. Pseudomonas alcaliphila MBR and its application in bioreduction and biosorption. US Patent # 0,269,169.
Li, H., Lin, Y., Guan, W., Chang, J., Xu, L., Guo, J. and Wei, G. 2010. Biosorption of Zn(II) by live and dead cells of *Streptomyces ciscaucasicus* strain CCNWHX 72-14. Journal of Hazardous Materials. 179: 151–159.
Li, J.M., Meng, X.G., Hu, C.W. and Du, J. 2009. Adsorption of phenol, p-chlorophenol and p-nitrophenol onto functional chitosan. Bioresource Technology. 100: 1168–1173.
Lim, S.F. and Chen, J.P. 2007. Synthesis of an innovative calcium-alginate magnetic sorbent for removal of multiple contaminants. Applied Surface Science. 253: 5772–5775.
Liu, C., Ngo, H.H. and Guo, W. 2012. Watermelon rind: agro-waste or superior biosorbent? Applied Biochemistry and Biotechnology. 167: 1699–1715.
Liu, Q.S., Zheng, T., Wang, P., Jiang, J.P. and Li, N. 2010a. Adsorption isotherm, kinetic and mechanism studies of some substituted phenols on activated carbon fibers. Chemical Engineering Journal. 157: 348–356.
Liu, Z., Zhang, F.S. and Wu, J. 2010b. Characterization and application of chars produced from pinewood pyrolysis and hydrothermal treatment. Fuel. 89: 510–514.
Loganathan, P., Vigneswaran, S., Kandasamy, J. and Naidu, R. 2013. Defluoridation of drinking water using adsorption processes. Journal of Hazardous Materials. 248: 1–19.
López-Téllez, G., Barrera-Díaz, C.E., Balderas-Hernández, P., Roa-Morales, G. and Bilyeu, B. 2011. Removal of hexavalent chromium in aquatic solutions by iron nanoparticles embedded in orange peel pith. Chemical Engineering Journal. 173: 480–485.
Luna-Pineda, T., Ortiz-Rivera, M., Perales-Pérez, O. and Román-Velázquez, F. 2009. Removal of arsenic from aqueous solutions with alginate based-magnetic nanocomposites. NSTI-Nanotech. 2: 395–398.
Malkoc, E. and Nuhoglu, Y. 2005. Investigations of nickel (II) removal from aqueous solutions using tea factory waste. Journal of Hazardous Materials. 127: 120–128.
Mao, N., Yang, L., Zhao, G., Li, X. and Li, Y. 2012. Adsorption performance and mechanism of Cr (VI) using magnetic PS-EDTA resin from micro-polluted waters. Chemical Engineering Journal. 200: 480–490.
Mbui, D.N., Shiundu, P.M., Ndonye, R.M. and Kamau, G.N. 2002. Adsorption and detection of some phenolic compounds by rice husk ash of Kenyan origin. Journal of Environmental Monitoring. 4: 978–984.
McGeorge, L.J., Louis, J.B., Atherholt, T.B. and McGarrity, G.J. 1985. Mutagenicity analyses of industrial effluents: results and considerations for integration into water pollution control programs. In Short-term bioassays in the analysis of complex environmental mixtures IV (pp. 247–268). Springer, Boston, MA.
McHale, A. and Bustard, M. 2002. Biosorption system. US Patent # 6,395,143.
Michalak, I., Chojnacka, K. and Witek-Krowiak, A. 2013. State of the art for the biosorption process- a review. Applied Biochemistry and Biotechnology. 170: 1389–1416.
Min, J.H. and Hering, J.G. 1998. Arsenate sorption by Fe (III)-doped alginate gels. Water Research. 32: 1544–1552.
Mishra, V. 2014. Biosorption of zinc ion: a deep comprehension. Applied Water Science. 4: 311–332.
Mohamad, O.A., Hao, X., Xie, P., Hatab, S., Lin, Y. and Wei, G. 2009. Biosorption of copper (II) from aqueous solution using non-living *Mesorhizobium amorphae* strain CCNWGS0123. Microbes and Environments. 27: 1202170359.
Mohan, D., Pittman Jr., C.U. and Steele, P.H. 2006. Single, binary and multi-component adsorption of copper and cadmium from aqueous solutions on Kraft lignin—a biosorbent. Journal of colloid and interface science. 297: 489–504.
Mohan, D., Sarswat, A., Ok, Y.S. and Pittman Jr., C.U. 2014. Organic and inorganic contaminants removal from water with biochar, a renewable, low cost and sustainable adsorbent–a critical review. Bioresource Technology. 160: 191–202.
Morais, W.A., Fernandes, A.L.P., Dantas, T.N.C., Pereira, M.R. and Fonseca, J.L.C. 2007. Sorption studies of a model anionic dye on crosslinked chitosan. Colloids and Surfaces A: Physicochemical and Engineering Aspects. 310: 20–31.

Moreno, R.F., López, C.J., Galvín, M.R., Cordón, M.M. and Mellado, R.J. 2010. On the removal of s-triazine herbicides from waters using commercial low-cost granular carbons. Journal of the Serbian Chemical Society. 75: 405–412.

Nacèra, Y. and Aicha, B. 2006. Equilibrium and kinetic modelling of methylene blue biosorption by pretreated dead streptomyces rimosus: Effect of temperature. Chemical Engineering Journal. 119: 121–125.

Nadavala, S.K., Swayampakula, K., Boddu, V.M. and Abburi, K. 2009. Biosorption of phenol and o-chlorophenol from aqueous solutions on to chitosan–calcium alginate blended beads. Journal of Hazardous Materials. 162: 482–489.

Nakajima, A., Yasuda, M., Yokoyama, H., Ohya-Nishiguchi, H. and Kamada, H. 2001. Copper biosorption by chemically treated Micrococcus luteus cells. World Journal of Microbiology and Biotechnology. 17: 343–347.

Nemec, P., Prochazka, H., Stamberg, K., Katzer, J., Stamberg, J., Jilek, R. et al. 1977. Process of treating mycelia of fungi for retention of metals. U.S. Patent # 4,021,368.

Ngah, W.W.S., Teong, L.C. and Hanafiah, M.A.K.M. 2011. Adsorption of dyes and heavy metal ions by chitosan composites: A review. Carbohydrate Polymers. 83: 1446–1456.

Ociński, D., Jacukowicz-Sobala, I. and Kociołek-Balawejder, E. 2016. Alginate beads containing water treatment residuals for arsenic removal from water—formation and adsorption studies. Environmental Science and Pollution Research. 23: 24527–24539.

Ören, A.H. and Kaya, A. 2006. Factors affecting adsorption characteristics of Zn2+ on two natural zeolites. Journal of Hazardous Materials. 131: 59–65.

Padmesh, T.V.N., Vijayaraghavan, K., Sekaran, G. and Velan, M. 2005. Batch and column studies on biosorption of acid dyes on fresh water macro alga *Azolla filiculoides*. Journal of Hazardous Materials. 125: 121–129.

Patel, R. and Suresh, S. 2008. Kinetic and equilibrium studies on the biosorption of reactive black 5 dye by Aspergillus foetidus. Bioresource Technology. 99: 51–58.

Patra, J.M., Panda, S.S. and Dhal, N.K. 2017. Biochar as a low-cost adsorbent for heavy metal removal: A review. Int. J. Res. Biosci. 6: 1–7.

Pattanayak, P., Papiya, F., Kumar, V., Singh, A. and Kundu, P.P. 2021. Performance evaluation of poly (aniline-co-pyrrole) wrapped titanium dioxide nanocomposite as an air-cathode catalyst material for microbial fuel cell. Materials Science and Engineering: C. 118: 111492.

Pehlivan, E., Altun, T. and Parlayici, S. 2012. Modified barley straw as a potential biosorbent for removal of copper ions from aqueous solution. Food Chemistry. 135: 2229–2234.

Pieschel, F., Lange, E., Camacho, J., Fieseler, C. and Ikier, H. 2003. Biosorbents and process for producing the same. US Patent # 6,579,977.

Pohl, P. 1997. Method for production of adsorption material. US Patent # 5,648,313.

Potgieter, J.H., Potgieter-Vermaak, S.S. and Kalibantonga, P.D. 2006. Heavy metals removal from solution by palygorskite clay. Minerals Engineering, 19: 463–470.

Price, M.S., Classen, J.J. and Payne, G.A. 2001. *Aspergillus niger* absorbs copper and zinc from swine wastewater. Bioresource technology. 77: 41–49.

Prigione, V., Varese, G.C., Casieri, L. and Marchisio, V.F. 2008. Biosorption of simulated dyed effluents by inactivated fungal biomasses. Bioresource Technology. 99: 3559–3567.

Prigione, V. P., Varese, G. C., Casieri, L., Voyron, S., Bertolotto, A., and Filipello Marchisio, V. 2010a. Use of *Rhizopus stolonifer* (Ehrenberg) Vuillemin in methods for treating industrial wastewaters containing dyes. US Patent # 7,658,849.

Prigione, V. P., Varese, G. C., Casieri, L., Voyron, S., Bertolotto, A., and Filipello Marchisio, V. 2010b. Use of *Cunninghamella elegans* lendner in methods for treating industrial wastewaters containing dyes. US Patent # 7,790,031.

Prigione, V. P., Varese, G. C., Casieri, L., Voyron, S., Bertolotto, A., and Filipello Marchisio, V. 2011. Use of *Rhizomucor pusillus* (lindt) schipper in methods for treating industrial wastewaters containing dyes. US Patent # 7,935,257.

Rafati, L., Ehrampoush, M.H., Rafati, A.A., Mokhtari, M. and Mahvi, A.H. 2016. Modeling of adsorption kinetic and equilibrium isotherms of naproxen onto functionalized nano-clay composite adsorbent. Journal of Molecular Liquids. 224: 832–841.

Ramrakhiani, L., Majumder, R. and Khowala, S. 2011. Removal of hexavalent chromium by heat inactivated fungal biomass of *Termitomyces clypeatus*: Surface characterization and mechanism of biosorption. Chemical Engineering Journal. 171: 1060–1068.

Reddy, K.S. and Gambrell, R.P. 1987. Factors affecting the adsorption of 2, 4-D and methyl parathion in soils and sediments. Agriculture, Ecosystems & Environment. 18: 231–241.

Robinson, T., McMullan, G., Marchant, R., and Nigam, P. 2001. Remediation of dyes in textile effluent: A critical review on current treatment technologies with a proposed alternative. Biores Technol. 77: 247–255.

Robinson, T. 2005. Studies on the production of enzymes by white-rot fungi for the decolourisation of textile dyes. Enzyme Microb. Tech. 36: 17–24.

Rodrigues, D.F. 2015. Porous nanocomposite polymer hydrogels for water treatment. US Patent # 0,353,381,A1.

Rombaldo, C.F., Lisboa, A.C., Mendez, M.O. and Coutinho, A.R. 2014. Brazilian natural fiber (jute) as raw material for activated carbon production. Anais da Academia Brasileira de Ciências. 86: 2137–2144.

Romera, E., Gonzalez, F., Ballester, A., Blazquez, M.L. and Munoz, J.A. 2006. Biosorption with algae: a statistical review. Critical Reviews in Biotechnology. 26: 223–235.

Salehizadeh, H. and Shojaosadati, S.A. 2003. Removal of metal ions from aqueous solution by polysaccharide produced from *Bacillus firmus*. Water Research. 37: 4231–4235.

Salman, H.A., Ibrahim, M.I., Tarek, M.M. and Abbas, H.S. 2014. Biosorption of heavy metals: a review. J. Chem. Sci. Tech. 3: 74–102.

Sánchez-Rivera, D., Perales-Pérez, O. and Román, F.R. 2013. Removal of inorganic arsenic oxyanions using Ca–Fe (III) alginate beads. Desalination and Water Treatment. 51: 2162–2169.

Sarı, A. and Tuzen, M. 2009. Kinetic and equilibrium studies of biosorption of Pb (II) and Cd (II) from aqueous solution by macrofungus (*Amanita rubescens*) biomass. Journal of hazardous materials. 164: 1004–1011.

Sarioglu, C.M., Guler, U.A. and Beyazit, N. 2009. Removal of copper from aqueous solutions using biosolids. Desalination. 239: 167–174.

Sarkar, P., Pal, P., Bhattacharyay, D. and Banerjee, S. 2010. Removal of arsenic from drinking water by ferric hydroxide microcapsule-loaded alginate beads in packed adsorption column. Journal of Environmental Science and Health Part A. 45: 1750–1757.

Say, R., Denizli, A. and Arıca, M.Y. 2001. Biosorption of cadmium (II), lead (II) and copper (II) with the filamentous fungus *Phanerochaete chrysosporium*. Bioresource Technology. 76: 67–70.

Schellenberg, T., Subramanian, V., Ganeshan, G., Tompkins, D. and Pradeep, R. 2020. Wastewater discharge standards in the evolving context of urban sustainability–The case of India. Frontiers in Environmental Science. 8: 30.

Serbus, C., Hora, K., Rezae, J., Pribil, S., Marvan, P., Krejdirik, L. et al. 1973. Sorbent and method of manufacturing same. U.S. Patent # 3,725,291.

Seredych, M., Lison, J., Jans, U. and Bandosz, T.J. 2009. Textural and chemical factors affecting adsorption capacity of activated carbon in highly efficient desulfurization of diesel fuel. Carbon, 47: 2491–2500.

Shinde, N.R., Bankar, A.V., Kumar, A.R. and Zinjarde, S.S. 2012. Removal of Ni (II) ions from aqueous solutions by biosorption onto two strains of *Yarrowia lipolytica*. Journal of Environmental Management. 102: 115–124.

Shirato, W., and Kamei, Y. 1995. Method for adsorbing and separating heavy metal elements by using a tannin adsorbent and method of regenerating the adsorbent. US Patent # 5,460,791.

Shroff, K.A. and Vaidya, V.K. 2012. Effect of pre-treatments on the biosorption of Chromium (VI) ions by the dead biomass of *Rhizopus arrhizus*. Journal of Chemical Technology & Biotechnology. 87: 294–304.

Sigdel, A., Park, J., Kwak, H. and Park, P.K. 2016. Arsenic removal from aqueous solutions by adsorption onto hydrous iron oxide-impregnated alginate beads. Journal of Industrial and Engineering Chemistry. 35: 277–286.

Silva, M., Fernandes, A., Mendes, A., Manaia, C.M. and Nunes, O.C. 2004. Preliminary feasibility study for the use of an adsorption/bio-regeneration system for molinate removal from effluents. Water Research. 38: 2677–2684.

Simões Campos Tavares, T. M. J., and Pontes Correia Neves, I. M. 2008. Biosorption system produced from biofilms supported in faujasite (FAU) zeolite, process obtaining it and its usage for removal of hexavalent chromium (Cr(VI)). US Patent # 0,169,238.

Singh, A., Kumari, K. and Kundu, P.P. 2017. Extrusion and evaluation of chitosan assisted AgNPs immobilized film derived from waste polyethylene terephthalate for food packaging applications. Journal of Packaging Technology and Research. 1: 165–180.

Singh, A., Khamrai, M., Samanta, S., Kumari, K. and Kundu, P.P. 2018a. Microbial, physicochemical, and sensory analyses-based shelf life appraisal of white fresh cheese packaged into PET waste-based active packaging film. Journal of Packaging Technology and Research. 2: 125–147.

Singh, A., Kumari, K. and Kundu, P.P. 2018b. Adaptation of 3D Printing Techniques in Bone Tissue Engineering: An Assessment of Its Need, Reliability, Validity, Sustainability, and Future Scope. In 3D Printing Technology. CRI Press India.

Singh, A., Banerjee, S.L., Dhiman, V., Bhadada, S.K., Sarkar, P., Khamrai, M. et al. 2020. Fabrication of calcium hydroxyapatite incorporated polyurethane-graphene oxide nanocomposite porous scaffolds from poly (ethylene terephthalate) waste: A green route toward bone tissue engineering. Polymer. 195: 122436.

Singh, A., Banerjee, S.L., Kumari, K. and Kundu, P.P. 2021. Recent innovations in chemical recycling of polyethylene terephthalate waste: a circular economy approach towards sustainability. In Handbook of Solid Waste Management Sustainability through Circular Economy (pp 1–28), Springer, Singapore.

Singh, A., Kumari, K. and Kundu, P. P. 2021. Nanocellulose biocomposites for bone tissue engineering. In Handbook of Nanocelluloses: Classification, Properties, Fabrication, and Emerging Applications (pp. 1–51). Cham: Springer International Publishing.

Singh, A., Vijayan, J.G. and Moodley, K.G. 2021. Surface functionalizations of nanocellulose for wastewater treatment. In Handbook of Nanocelluloses: Classification, Properties, Fabrication, and Emerging Applications (pp. 1–48). Cham: Springer International Publishing.

Singh, A., Kumari, K., & Kundu, P. P. 2022. Polyurethane Nanocomposites for Bone Tissue Engineering. In Engineered Nanomaterials for Innovative Therapies and Biomedicine (pp. 373–403). Cham: Springer International Publishing.

Singh, P., Singh, S.K., Bajpai, J., Bajpai, A.K. and Shrivastava, R.B. 2014. Iron crosslinked alginate as novel nanosorbents for removal of arsenic ions and bacteriological contamination from water. Journal of Materials Research and Technology. 3: 195–202.

Singh, S.P., Ma, L.Q. and Harris, W.G. 2001. Heavy metal interactions with phosphatic clay: sorption and desorption behavior. Journal of Environmental Quality. 30: 1961–1968.

Song, Y., Lei, S., Zhou, J. and Tian, Y. 2016. Removal of heavy metals and cyanide from gold mine wastewater by adsorption and electric adsorption. Journal of Chemical Technology & Biotechnology. 91: 2539–2544.

Soto, M.L., Moure, A., Domínguez, H. and Parajó, J.C. 2011. Recovery, concentration and purification of phenolic compounds by adsorption: A review. Journal of Food Engineering. 105: 1–27.

Srivastava, S., Agrawal, S.B. and Mondal, M.K. 2015. A review on progress of heavy metal removal using adsorbents of microbial and plant origin. Environmental Science and Pollution Research. 22: 15386–15415.

Summers, Jr., B.L. and Gress, L.B. 1996. Bead for removing dissolved metal contaminants. US Patent # 5,578,547.

Sun, F., Wu, F., Liao, H. and Xing, B. 2011. Biosorption of antimony (V) by freshwater cyanobacteria Microcystis biomass: chemical modification and biosorption mechanisms. Chemical Engineering Journal. 171: 1082–1090.

Sun, X., Peng, B., Ji, Y., Chen, J. and Li, D. 2009. Chitosan (chitin)/cellulose composite biosorbents prepared using ionic liquid for heavy metal ions adsorption. AIChE Journal. 55: 2062–2069.

Svecova, L., Spanelova, M., Kubal, M. and Guibal, E. 2006. Cadmium, lead and mercury biosorption on waste fungal biomass issued from fermentation industry. I. Equilibrium studies. Separation and purification Technology. 52: 142–153.

Tadic, D.C., Lozano, F.V., Zarzar Maza, M.E., and Viedma Elicer, P.L. 2006. Process for the removal of metals by biosorption from mining or industrial effluents. US Patent # 7,326,344.

Takada, A. and Kadokawa, J.I. 2015. Fabrication and characterization of polysaccharide ion gels with ionic liquids and their further conversion into value-added sustainable materials. Biomolecules. 5: 244–262.

Tiwari, A., Dewangan, T. and Bajpai, A.K. 2008. Removal of toxic As (V) ions by adsorption onto alginate and carboxymethyl cellulose beads. Journal of the Chinese Chemical Society. 55: 952–961.

Tong, X.J., Li, J.Y., Yuan, J.H. and Xu, R.K. 2011. Adsorption of Cu (II) by biochars generated from three crop straws. Chemical Engineering Journal. 172: 828–834.

Uluozlu, O.D., Sari, A., Tuzen, M. and Soylak, M. 2008. Biosorption of Pb (II) and Cr (III) from aqueous solution by lichen (Parmelina tiliaceae) biomass. Bioresource Technology. 99: 2972–2980.

Unuabonah, E.I., Adebowale, K.O. and Olu-Owolabi, B.I. 2007a. Kinetic and thermodynamic studies of the adsorption of lead (II) ions onto phosphate-modified kaolinite clay. Journal of Hazardous Materials. 144: 386–395.

Unuabonah, E.I., Olu-Owolabi, B.I., Adebowale, K.O. and Ofomaja, A.E. 2007b. Adsorption of lead and cadmium ions from aqueous solutions by tripolyphosphate-impregnated Kaolinite clay. Colloids and surfaces A: Physicochemical and Engineering Aspects. 292: 202–211.

Uslu, G. and Tanyol, M. 2006. Equilibrium and thermodynamic parameters of single and binary mixture biosorption of lead (II) and copper (II) ions onto *Pseudomonas putida*: effect of temperature. Journal of Hazardous Materials. 135: 87–93.

Vasudevan, P., Padmavathy, V., Tewari, N. and Dhingra, S.C. 2001. Biosorption of heavy metal ions. Journal of Scientific and Industrial Research. 60: 112–120.

Velazquez, L.H., Pavlick, A. and Rangel, J.R. 2013. Chemical characterization of raw and treated agave bagasse and its potential as adsorbent of metal cations from water. Ind. Crop. Prod. 43: 200–206.

Vendruscolo, F., da Rocha Ferreira, G.L. and Antoniosi Filho, N.R. 2017. Biosorption of hexavalent chromium by microorganisms. International Biodeterioration & Biodegradation. 119: 87–95.

Verma, V.K., Tewari, S. and Rai, J.P.N. 2008. Ion exchange during heavy metal bio-sorption from aqueous solution by dried biomass of macrophytes. Bioresource Technology. 99: 1932–1938.

Vimala, R. and Das, N. 2009. Biosorption of cadmium (II) and lead (II) from aqueous solutions using mushrooms: a comparative study. Journal of hazardous materials. 168: 376–382.

Volesky, B. and Tsezos, M. 1981. Separation of uranium by biosorption. US Patent # 4,320,093.

Volesky, B. and Kuyucak, N. 1988. Biosorbent for gold. US Patent # 4,769,223.

Votapek, V., Marval, E., Jilek, R. and Stamberg, K. 1978. Method of treating a biomass. U.S. Patent # 4,067,821.

Wang, J. and Chen, C. 2009. Biosorbents for heavy metals removal and their future. Biotechnology Advances. 27: 195–226.

Wang, L. and Li, Y. 2012. Biosorption behavior and mechanism of lead (II) from aqueous solution by aerobic granules (AG) and bacterial alginate (BA). Journal of Ocean University of China. 11: 495–500.

Witek-Krowiak, A. and Reddy, D.H.K. 2013. Removal of microelemental Cr (III) and Cu (II) by using soybean meal waste–unusual isotherms and insights of binding mechanism. Bioresource Technology. 127: 350–357.

Wong, Y.C., Szeto, Y.S., Cheung, W.H. and McKay, G. 2004. Pseudo-first-order kinetic studies of the sorption of acid dyes onto chitosan. Journal of Applied Polymer Science. 92: 1633–1645.

Wu, F., Sun, F., Wu, S., Yan, Y. and Xing, B. 2012. Removal of antimony (III) from aqueous solution by freshwater cyanobacteria Microcystis biomass. Chemical Engineering Journal. 183: 172–179.

Wu, F.C., Tseng, R.L. and Juang, R.S. 2001. Enhanced abilities of highly swollen chitosan beads for color removal and tyrosinase immobilization. Journal of Hazardous Materials. 81: 167–177.

Yannai, S., and Meshulam, G. 1996. Process for the removal of species containing metallic ions from effluents. US Patent # 5,538,645.

Zhang, X., Liu, Y., Yan, K. and Wu, H. 2007. Decolorization of anthraquinone-type dye by bilirubin oxidase-producing nonligninolytic fungus *Myrothecium* sp. IMER1. Journal of bioscience and bioengineering. 104: 104–110.

Zhou, J.L. and Banks, C.J. 1993. Mechanism of humic acid colour removal from natural waters by fungal biomass biosorption. Chemosphere. 27: 607–620.

Zhu, Y., Hu, J. and Wang, J. 2012. Competitive adsorption of Pb (II), Cu (II) and Zn (II) onto xanthate-modified magnetic chitosan. Journal of Hazardous Materials. 221: 155–161.

Ziagova, M., Dimitriadis, G., Aslanidou, D., Papaioannou, X., Tzannetaki, E.L. and Liakopoulou-Kyriakides, M. 2007. Comparative study of Cd (II) and Cr (VI) biosorption on *Staphylococcus xylosus* and *Pseudomonas* sp. in single and binary mixtures. Bioresource Technology. 98: 2859–2865.

Zolgharnein, J., Shahmoradi, A. and Ghasemi, J. 2011. Pesticides removal using conventional and low-cost adsorbents: a review. Clean–Soil, Air, Water. 39: 1105–1119.

Zouboulis, A.I. and Katsoyiannis, I.A. 2002. Arsenic removal using iron oxide loaded alginate beads. Industrial & Engineering Chemistry Research. 41: 6149–6155.

Zulfadhly, Z., Mashitah, M.D., and Bhatia, S. 2001. Heavy metals removal in fixed-bed column by the macro fungus *Pycnoporus sanguineus*. Environmental Pollution. 112: 463–470.

14

Bio-Solids' Recovery from Wastewater

S. Ranjitha,[1,*] *C. Sudhakar*[2] and *V. Aroulmoji*[3]

14.1 Introduction

Biosolids are defined as the residual solids generated from the processing of domestic wastewater that meet the regulatory requirements for recycling process. Generally, in most cases, biosolids can be used produced from domestic wastewater which consists of wastes solids in the form of liquid generated from residences, businesses, and institutions. Any form of the materials that enter from the municipal wastewater collection system subsequently modified into biosolids. It is well known that biosolids are rich in nutrients and are good source of natural fertilizer to stimulate plant growth or soil amendment to enhance the land. Normally, biosolids contain significant quantities of water, organic matter, nutrients, and trace elements. According to U.S Environmental Protection Agency, biosolids contain about 93–99% of water, solids, and dissolved matters present in the waste water or added during wastewater or biosolids treatment processes. Meanwhile, animal manures, untreated septage, municipal solid waste (MSW), untreated wastewater sludges, hazardous wastes, industrial waste, grit, and screenings removed during the initial wastewater treatment process are the substances that are not included under biosolids.

The presence of major biosolids in the wastewater treatment facility is encountered by the raw sewage that enters the unit that may be contaminated by the chemicals. Biosolids' production made damage and harm the environment because of right treatment while they're launched in air with out purification for you to treat them, we have to have enrolled a monitoring machine to deal with the majority cloth elimination and similarly, liberating the wastewater biosolids waste with

[1] Professor , Velalar College of engineering and Technology, Erode-638012.
[2] Assistant Professor, Department of Biotechnology, Mahendra Arts & Science College, Kalippatti. Namakkal-637501, Tamil Nadu.
[3] Center for Research and Development, Mahendra Engineering College (Autonomous), Mallasamudram-637503, Tamil Nadu, India.
* Corresponding author: ranjilotus31@gmail.com

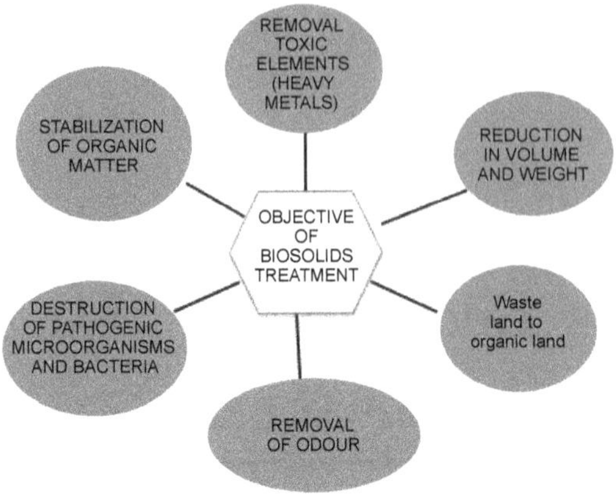

Fig. 1. Illustration of objectives of biosolids treatment.

none treatment might lead the goal of prevention of the purpose of environmental protection. Among the various treatment systems that can be employed to treat the wastewater solids and focus of the treatment will be the same which is used to convert the wastewater solids into a form that can be disposed of without giving any harm to the environment or creating inconvenience conditions. In this case, it does not matter what system or combination of systems is chosen for the treatment. The removal of the constituents during wastewater treatment cycling processes, largest production of biosolids is by volume, leading to a complex problem for process of disposal and treatment system. Typically, biosolids contain organic matter (protein, carbohydrates, fats, oils, greases, chemicals, etc.), pathogens and microorganisms (bacteria, viruses, parasites, etc.), heavy and toxic metals, and toxins that include pesticides, household, and industrial chemicals. All of these will be a risk and hazardous to humans and the environment. In order to overcome these problems, it is important to treat the biosolids in a proper manner. The following figure 1shows the objectives of biosolids treatment.

14.2 Biosolids production

The World Bank predicts that biosolid waste generation reached 1.14–1.73 times in rates of per capita by both developed and developing countries between the period of 1995 and 2025 due to high urbanization and industrialization, e.g., Shanghai in China generates 17,000 tonnes day^{-1} waste. The foremost case observe overview stated from important towns of growing international locations like Nepal, Pakistan, Guinea, Amman, and many others (World Bank 2008). Similarly, Indian population creates around 1,09,598 tonnes per day of MSW, which has been predicted to be 3,76,639 tonnes per day by the year 2025 reported by Hoornweg and Bhada-Tata (2012). With the current population of 1.28 billion, which includes 33% urban population, per year generation of MSW in India is expected to augment over 150 million tonnes by 2025 (Hanrahan et al. 2006) and 300 MT by 2047 (Pappu et al. 2007). The huge amount of biosolid waste requires wide areas of land for disposal of the contaminants. In the year 1997,

20.2 Km^2 area of land was used to manage 48 MT of waste and 169.6 Km^2 area of land would be required for disposal of 300 MT by 2047 (CPCB 2000, Pappu et al. 2007). With this trend of burgeoning population and increasing rate of waste generation, maintaining a pristine urban environment, for the future generations, appears to be a daunting task.

In the current scenario, of polluting natural resources and energy crisis, the importance and demand of developing a sustainable approach towards environmentally solid waste management system monitoring cannot be neglected, as has been reported by Pappu et al. (2007). The inappropriate disposal of solid wastes like biosolids and other biomedical wastes impose a serious threat to the environment, leading to problems like groundwater contamination, degradation of soil quality and air pollution which meets Covid-19 crisis like, and so on. In this mean time, different kind of approaches and analysis of safe biosolids waste disposal such as incineration, soil application, land filling and sea dumping have been explored all over the world with new environmental policy rules. Disposal methods like land filling and ocean dumping have their own problems and practical issues due to scarcity of land, pollution trouble and additionally don't make to reuse of the beneficial parts of biosolids. As a result, the united states and numerous european international locations have banned the ocean dumping of biosolids for the reason that 1991 and 1998. The growth of global population, and the generation of stable wastes like biosolids are bound to increase remarkably till now. Hence, according to the principles of waste management hierarchy, agricultural recycling of biosolids will be a more environmentally preferred option over the traditional disposal methods. Utilizing the potential of biosolids to recycle valuable plant nutrients and an effective soil amendment will not only help in sustainable management of this waste but also in minimizing the negatives associated with its traditional solid waste disposal.

The wastewater treatment can begin before the wastewater reaches the treatment plant. Generally, before the wastewater is released to the WWTP, the wastewater must be pretreated first in order to remove any hazardous contaminants including metals such as copper, lead, cadmium, and chromium and other pollutants. Over the past 20 years, pretreatment and pollution prevention programmes have reduced the level of metals and other pollutants going into WWTP. This can help to improve the quality of biosolids produced. Once the wastewater reaches the WWTP, it will undergo preliminary, primary, secondary, and tertiary treatment. Table 1 lists the types of wastewater treatment and the types of biosolids produced after each treatment. Basically, the quantity and characteristics of the biosolids produced at the WWTPs depend on three important factors that include the composition of wastewater, the type of wastewater treatment used, and the type of treatment applied to the biosolids. Generally, the total volumes of biosolids generated are influenced by the degrees of wastewater treatment. The higher the level of treatment is, the higher will be the concentrations of contaminants produced in the biosolids. This is because most of the components removed from the wastewater finally end up in the biosolids. Besides that, the addition of chemicals to precipitate solids, for example ferric chloride, lime, or polymers, can increase the concentrations of these chemicals in the biosolids produced at the end of the process. Indirect effects also can occur such as when alum (aluminium hydroxide) adsorbs trace metals such as cadmium to precipitate out of

Table 1: Classification of wastewater treatment and production of biosolids.

Types of Wastewater Treatment Level	***Types of Biosolids Produced***
Screening And Grit Removal (Preliminary Treatment)	
Wastewater screening removes coarse solids that can interfere with mechanical equipment. Grit removal separates heavy, inorganic, and sand-like solids that would settle in different channels and interfere with treatment processes.	The technique of screenings and grit are shaped as a biosolid waste and almost constantly land crammed and river location
Primary wastewater treatment	
Generally, the method involves gravity sedimentation of screened and degritted wastewater to remove SS before secondary treatment.	Biosolids produced at this stage usually contain 7–8% solids. Normally, water content can be reduced by thickening or dewatering cyclic process
Secondary wastewater treatment	
It depends on biological treatment process of suspended growth or fixed growth system. During the biological treatment, microorganisms, and nanoparticles are used to reduce BOD and remove SS.	Biosolids produced usually have a low solid content (0.5–3%). The products are more complicated to thicken and dewater compared to primary biosolids.
Tertiary wastewater treatment	
Common types of tertiary treatment include biological and chemical precipitations. The processes are used to remove nitrogen and phosphorus	Lime, polymers, iron, or aluminium salts used in tertiary treatment produce biosolids with varying water-absorbing characteristics.

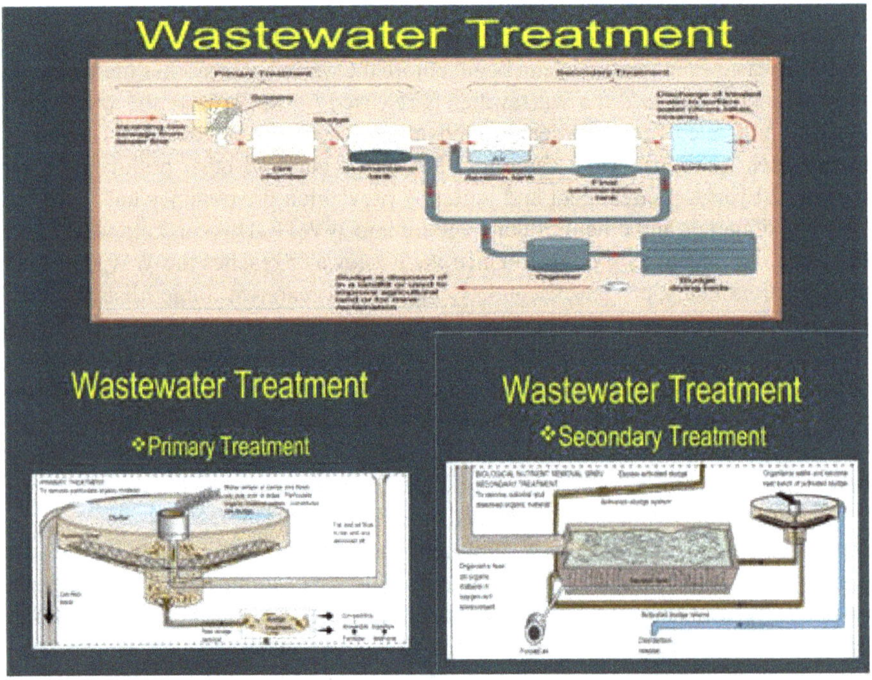

Fig. 2: Illustration of wastewater treatment from extraction of biosolids.

the wastewater and into the biosolids. Thus, the type of wastewater treatment or pretreatment used will affect the characteristics of biosolids and also can affect the types of biosolids treatment chosen.

14.3 Types of biosolids

14.3.1 Primary Biosolids

In global in most WWTPs machine observed by means of primary sedimentation technique extensively to remove comfortably settle capable biosolids from raw wastewater. It was discovered that the dry weight of primary biosolids is about 50% of the total sludge solids that was formed from a treatment plant equipped with primary sedimentation and a conventional activated sludge process for secondary treatment. Commonly, primary bio-solids are easier to manage to decompose compared to biological and chemical bio-solids and the reasons are as follows:

- ✓ Primary biosolids are readily thickened by gravity.
- ✓ Primary biosolids with low conditioning requirements can be mechanically dewatered rapidly.
- ✓ The dewatering device will produce a drier cake and it gives solid waste product materials than the biological and chemical biosolids.

Typically, primary biosolids' production is within the range of 100–300 mg/L of wastewater. There are two basic approaches to estimate primary biosolids' production that include: (1) computing the quantity of total suspended solids (TSS) entering the primary sedimentation tank and (2) assuming an efficiency of removal. Usually, estimates of 0.07–0.11 kg/capita/day of TSS are commonly used when site-specific data are not available. Meanwhile, the removal efficiency of TSS in the primary sedimentation tank is usually in the range of 50–65%.4

For estimating purposes, the removal efficiency of 60% is commonly used. This is subjected to the following conditions:

- ➢ The biosolids are mainly produced from a domestic wastewater treatment without major industrial loads.
- ➢ The biosolids did not contain chemical from the coagulation and flocculation process.
- ➢ No other biosolids have been added to the influent wastewater such as trickling filter biosolids.
- ➢ The biosolids did not contain major side streams from biosolids processing; for example, digester supernatant, elutriate, and filtrates or centrates and other biosolids such as waste-activated sludge (WAS).

14.3.2 Chemical biosolids

In wastewater treatment process, particularly in industrial wastewater treatment, chemicals such as ferric chloride, alum, lime, or polymer are used widely to precipitate the solids and to improve SS removal. Toxic chemicals, infectious organisms, and endotoxins or cellular material may all be present in biosolids. There are anecdotal reports attributing adverse health effects to biosolids exposures, ranging from relatively mild irritant and allergic reactions to severe and chronic health outcomes. Odours are a common complaint about biosolids, and greater consideration should be given to whether odours from biosolids could have adverse health effects. The addition of chemicals can result in the formation of chemical biosolids at the treatment plants. Most plants apply chemicals to secondary effluent and use tertiary clarifiers to remove the chemical precipitates. In certain cases, some treatment plants add the chemicals to a biological process. Thus, chemical precipitates are produced and mixed with the primary biosolids or biological biosolids. Chemicals can greatly influence the increasing impact of biosolids in this generation, and it depends on the chemicals used and the chemicals' addition rates. The followings are several types of precipitates that are produced and must be considered in measuring the total biosolids generation. EPA's 1993 chemical and pathogen standards for biosolids were based on the scientific and technical information available at that time and the expectation that the prescribed biosolids-management practices specified in the Part 503 rule would be effective in preventing harmful exposure to biosolids' constituents. To assure the public that the standards are protective of human health, it is important that EPA demonstrate that its chemical limits and pathogen-reduction requirements are supported by current scientific data and risk-assessment methods. Management practices (e.g., 10-meter setback from water bodies) are designed to control the potential risks; therefore, it is important to verify the effectiveness of the practices. In addition, EPA must demonstrate that the Part 503 rule is being enforced.

14.3.3 Biological biosolids

Generally, biological biosolids are produced from the treatment processes such as activated sludge, trickling filters, and rotating biological contactors. The metabolic and growth rates of microorganisms will affect the quantities and characteristics of biological biosolids. The quantity and quality of biosolids produced by the biological process are intermediate among that produced in non-number one structures and that produced in complete-primary structures in cases while high-quality displays or number one sedimentation tanks with high overflow quotes are used. Biological biosolids containing grit, plastics, papers, and fibres are produced at treatment plants lacking primary treatment. Normally, pure biological biosolids are produced at the treatment plants with primary sedimentation unit. The method of operation of the clarifiers will affect the concentrations and the volumes of biological biosolids produced. Typically, biological biosolids are more difficult to thicken and dewater compared to primary bio solids and chemical biosolids. Biosolids is a good source of organic matter as well as plant macro and micro-nutrients and in future can be substituted for expensive inorganic fertilization. Addition of treated biosolids to soil has been found to be beneficial to soil health, enriching soil with essential nutrient elements as also increasing the pH of the soil. Further, soil amendments with biosolids has been reportedly effective in increasing a number of agro morphological attributes as well as yield in different crop species. However, use of bio solids for commercial

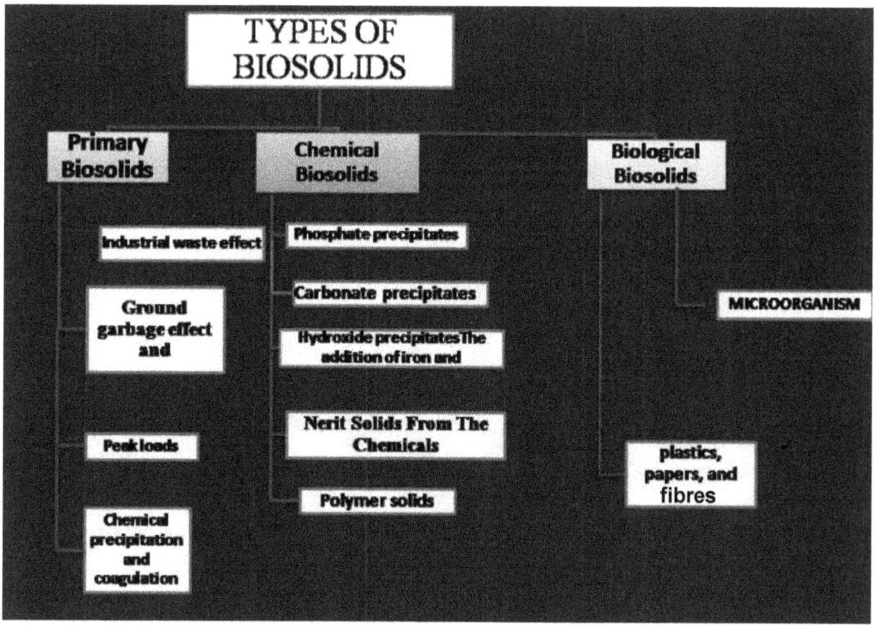

Fig. 3: Illustration of Types of Bio-solids Effluents.

agriculture has to be done cautiously. It is common knowledge that biosolids often contain toxic metal residues (heavy metals), as well as toxic organic residues, and indiscriminate use of it can be detrimental to the productivity of the soil as well as cause harm to the food chain. Moreover, the character of the bio solids changes over time and hence stringent and periodic monitoring of bio solids for agricultural use should be done. In general, a potential waste management depends on several tiers like disposal, recovery, recycle, reuse and prevention. This hierarchy is also suitable for managing biosolids (Fig. 3). It is evident that implication of biosolids induce agricultural productivity to a certain level, but the application of this waste in major food generation and supply chain still needs to be studied more.

14.4 Best Management Practices (BMPs) for land application of biosolids

The biosolids management system is designed to address such issues as storage or alternate waste management and methodology from rural to city for biosolids in particular time periods when application cannot take place due to unfavourable climate utrien. An increased number of high capacity wastewater treatment facilities generate large quantities of biosolids As a brief overview about the biosolids generation process, wastewater or sewage treatment plants receive wastewater and sewage from domestic, industrial, and agricultural sources. Table 2 shows the production and utilization of biosolids in some selected countries/region.
[Data adapted from—Singh and Sarkar 2015, Srivastava et al. 2016, Singh et al. 2017].

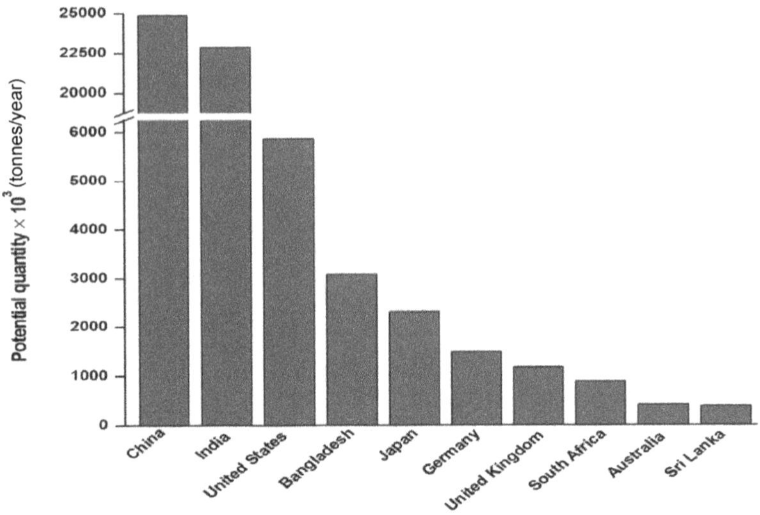

Table 2: Illustration of potential quantity of biosolids produced in selected countries; values were calculated based on biosolids generation assuming 50 g $person^{-1}$ day^{-1}.

Country/Region	Biosolids production (Million tonnes dry solid per year)	Biosolids utilization rate (%)	Main application
United Kingdom	1.05	85	**Land application, Energy recovery**
Australia	0.36	80	**Land application**
South Africa	1.0	80	**Land application**
India	2.3	80	**Land application**
Japan	2.2	74	**Energy recovery, Construction products**
Germany	2.3	60	**Land application, Energy recovery**
United States of America	17.8	55	**Land application**
European Union	9.0	40	**Land application**
Republic of Korea	1.9	6	**Land application, Construction products**
Singapore	0.12	0	–
Hong Kong, China	**0.3**	**0**	–

During the past three decades, in many growing nations faces with transformation of economies with unexpected drops made demand for meals because of uncontrolled increase of populace has caused enlargement of cropland, accompanied by means of shrinking primary forests and grassland regions. consistent with latest reports, in India, between 1880 and 2020, crop land place extended and forest insurance decreased fairly modifications with crop land vicinity has expanded from

92 to 140.1 million ha and wooded area land decreased (from 89 to 63 million ha) inside the past 50 years, worldwide rapid urbanization has been glaring that Migration in its numerous paperwork is the most important demographic factor effect on land utilization at collection of more than one centuries with historical proof fast economic increase is blended with a shift of agriculture land from to enterprise, infrastructure, street network, instructional institutions, smart metropolis development and home use international locations in East Asia, North america, and Europe have confronted all lost cultivated land and traditional organic farming techniques throughout their intervals of economic development. The dramatic growth and urbanization of Indian economy and market since economy reforms in 1990 have brought about a major loss of croplands which were converted to urban areas and transportation routes for industrial growth during 1990–2021.

The case have a look at approximately the land of Slovakia land tendencies had grown in many aspects development and in 2013 as parts of the entire region of Slovakia agricultural land blanketed 48.9% and woodland land 41.1%. The highest share of used agricultural land was represented by arable land (58.9%) followed by permanent grasslands (36.1%). The average amount of agricultural land per capita was 0.44 ha. In attention from the lengthy-time period factor of view, sluggish affore station and everlasting grassland conversion to forest land changed into mentioned wherein forest land multiplied from 67.7% in 1782 to eighty three 7% in 2006. Land software applications will dealing with the public competition as compared with technical constraints is the predominant reason for implementation constantly. Inspite of that, the proper land application of biosolids employ the Best management practices (BMPs) for protection of the environment and human health which are

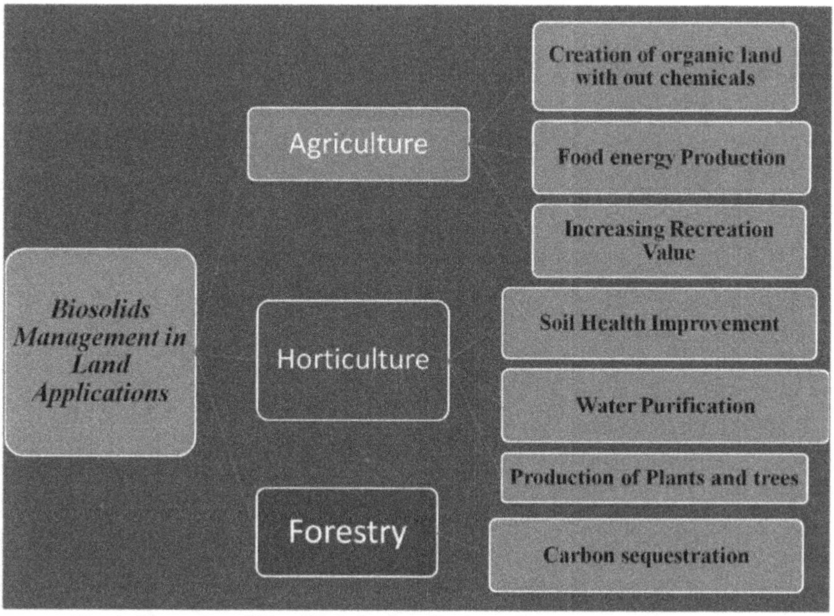

Fig. 4: Illustration of biosolids in land application utilization and implementation.

operating methods that ensure. The BMPs include agronomic loading rates, slope limitations, soil pH limitations, water conservation, organic farming, forest formation, public access restrictions, soil conservation practices, restrictions for saturated and frozen soils, protection of endangered species, and other site restrictions and how the nutrients will be provided by the biosolids. Taking account of both the initial content of trace elements in the biosolids and in the field is also one of the BMPs' requirements on the use of biosolids which given in illustration 4.

14.5 Concept of DPSIR framework

There are many different approaches created and used to analyze structure and dynamics of environmental issues and impacts on climate change worldwide in weather conditions and air pollution. Recently, Pressure- State- Response (PSR) framework was developed by OECD2 as a common framework system for environmental evaluation approach. Environmental problems and issues were taken in account with variables to show the cause-effect relationships between human abnormal activities against nature that exert pressure (P) on the environment, modify the conditions and care of the environment (R). The advanced PSR model was further improved with quality by the European Environmental Agency to become the Driving force- Pressure- State- Impact- Response (DPSIR) framework in order to provide a more complementary approach in analyzing environmental problems. The DPSIR framework is in providing the consistent dating among the origin and sequences of environmental troubles however in prospect recognize their dynamics is useful to focus on the connections between DPSIR elements. The DPSIR model

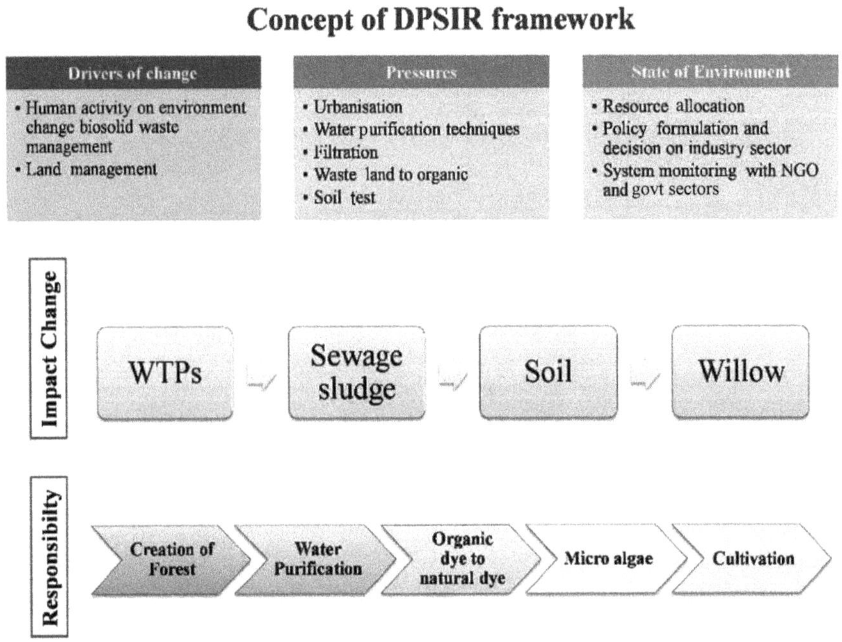

Fig. 5: Illustration of DSPIR Framework of Biosolid Management System.

was utilized used to identify the parameters of a series of core indicators and to establish the nature of interactions among different driving forces, pressures, states, impact issues, and positive responses from various organization, and thus to assess the agriculture and its impact on land usage, environment, and ecosystem. The sustainability indicators are tool makers that measure the sustainability and provide an opportunity to monitor whether a system is moving towards sustainability or not. However, these measurement indicators fail to provide the information about the progress of the wastewater management system. Drivers-Pressure-State-Impact-Response (DPSIR) framework given in illustration 5 was adopted in this study report as a way of using the sustainable indicators are used to identify the major issues in waste water to biosolid waste management system and to evaluate the continuous progress of the system towards sustainability in nature. Technology can affect labour market and operational processes on land. Demographic factors, such as increase and decrease of population, and migration patterns have a large impact on land use. Life-cycle features arise and affect rural as well as urban environments might be lead by way of the distinct stakeholders to monitor the gadget sustainable in phrases of meals safety.

14.5.1 Best management practices for biosolids

14.5.1.1 Site and soils

As per guidance of DES regulations (Env-Wq 800 Site selection), site plan is part of the site permitting process biosolids management system. The site plans and DES regulations address the focus on analysis of soil testing, setbacks from wells, surface waters, groundwater utilization, and control biosolids' application on steep slopes.

- ➢ Application sites in critical aquifer recharge areas require nutrient management controls system.
- ➢ Disclose of biosolids application on land that frequently floods or on websites with excessive water tables will prohibited and not permitted to use all through durations of high-water table.
- ➢ Attention the soil's base saturation to make imbalances of base metal salts contaminants (Ca, Mg, and k).
- ➢ Erosion manage dimension turned into discovered the cover vegetation for the autumn and winter, ought to be used with tilled soils in Soil variability test have to be analysed whilst choice of a website.

Biosolids should not be applied directly to the land within 300 feet of any private sector well and 500 feet from any community well or municipal water supply well system, and they should also not be applied during periods of high water table when biosolids are implemented from groundwater ought to be maintained two feet of separation at that time.

14.5.1.2 Crop considerations

The ideal crop will exhibit minimal uptake of metals and display significant benefit from the nutrients in biosolids. This would include non-agricultural land (e.g., landfills, turf, or roadside restoration) as well as field crops grown for animal feed (e.g., field corn grown for silage or grain, and forage crops).

Fall maintenance applications are preferred for forage crops (including hay, pasture, field corn, and small grains) and should be made before the ground freezes. The preferred application time is immediately after forage harvest. At least 45 days should be allowed between application and harvest or grazing to preserve animal acceptability. The ideal practice is to make top listing programs earlier to expected rainfall. Biosolids should be incorporated prior to tillage whenever possible. Application of biosolids should be avoided to land where food crops are grown. Leaf and root crops are the greatest accumulators of heavy metals. Careful management and monitoring of the soil and crop is critical (Hermanson et al. 1987).

14.5.1.3 Application planning

One of the unique worries associated with any biosolids that is to be land implemented pertains to the processes to reduce pathogens, one in all which is brief-term lime stabilization. This method involves raising the pH to 12 for a short period of time. Hydrated lime is often used to achieve this rapid pH elevation. Use of such stabilized biosolids on agricultural land, therefore, involves a high pH material, containing large amounts of readily available calcium. With repeated use, soil pH may increase above the desirable range, in turn affecting the availability of nutrients and the efficacy of such crop inputs as herbicides. It is possible that calcium, rather than nitrogen loading, will pose the most limiting factor in using biosolids and septage on agricultural land. Research data (McLean et al. 1983) show results that "strongly suggest that for maximum crop yields, emphasis should be placed on providing sufficient, but not excessive levels of each basic cation". Suggested soil cation saturation ratios of the cation exchange capacity (CEC) from many literature sources support a base saturation of 3–6% potassium, 12–13% magnesium, and 65–80% calcium. Soil pH should be maintained above 6.5 in order to limit the solubility of metals, but should not exceed 7.2.

- ✓ When possible, biosolids' spreading should be done when weather conditions allow rapid odour dissipation.
- ✓ Annual biosolids loading rates and supplemental fertilizer requirements are based on biosolids and soil analysis. Total available nitrogen in a given year should not exceed crop requirements. On sites with a high risk of soil erosion, biosolids applications should not provide phosphorus in excess of crop removal if soil test phosphorus levels are already high.
- ✓ Estimate the crop nutrient needs given the farm's yield goals.
- ✓ Calculate the agronomic application rate using the UNH Cooperative Extension Nutrient Management Worksheet for Biosolids (2015).
- ✓ The choice of land application equipment is determined by the physical characteristics of the biosolids. Equipment should be calibrated to ensure uniform accurate rates of application.
- ✓ Biosolids should be spread evenly over the ground and incorporated within 24 hours by discing or plowing whenever possible.

Fig. 6: Illustration of DSPIR Framework of Conversion of Biosolid Waste to Energy Products by adding Values with Protection.

- ✓ Heavy equipment should not be used when the soil is wet in order to minimize soil compaction.
- ✓ Biosolids should not be applied on soils that are frozen, excessively wet or covered with snow.
- ✓ Alkaline-stabilized biosolids should not be applied at rates greater than the soil liming requirements, and the nutrients present in the product (N, P) must be accounted for, even if the application is based solely on the liming value.

14.5.1.4 Monitoring and recordkeeping

Although the country permits the requirement for land application of biosolids, good stewardship requires a programme of site monitoring. DES currently requires annual soil testing to assess soil fertility and testing every five years to evaluate heavy metals loading to the soil. This can help assess the nutrient and metal loading effects during UNH Cooperative Extension, the application period, as well as after the site has reached its loading capacity. Major considerations include the potential loss of nitrogen and metals as non-point source pollutants and the development of nutrient imbalances which may have negative effects on long term crop production. Good recordkeeping is crucial to a successful land application programme which is provided in illustration 6. In the long-term production of any crop, field records are an indispensable tool in evaluating the effects of nutrient (fertilizers, manures and biosolids) applications on soil fertility and crop response.

In order to evaluate the effectiveness of biosolids applications, complete records (soil test results, recommendations, applications rates, nutrient analysis, etc.) are a necessary portion of every land application programme. Elements of an effective monitoring plan include:

- Complete soil analysis prior to each biosolids application through the UNH Soil Testing programme, or through another accredited soil testing programme. In addition to standard soil test results (pH, nutrient levels, organic matter), UNH Cooperative Extension recommends including an environmental package to monitor heavy metals.
- To refine nitrogen needs in annual row plants, makes use of of midseason tests (e.g., pre side dress nitrate take a look at, Adapt-N model, chlorophyll meters and regarding the better guard for quality the consuming water standards. The producers must examine concentration of nitrogen mineralization from soil organic matter that contributes to the crop's nitrogen fulfillment.
- Repeated application of biosolids is recommended to enhance dairy ration balancing and mineral analysis of feed crops from fields which have complete cropping records kept on individual fields to help analyse the effects of the land application programme on crop production.
- During the application of biosolids, the farmers are requested to maintain records of the source, quantity and quality of materials such as odours.

14.6 Biosolids' utilization and financial management structure

Solid waste management policy and implementation may vary from country to country based on urbanization and population. In most developing countries such as India, the local governments are responsible for bio solid waste management and may evolve over a period of time, depending on the variations in solid waste, political and administrative system, socioeconomic situation, and geo-climatic condition. Hence, it is useful to capture the evolving process with respect to laws, institutions, financial mechanisms, technology and infrastructure and stakeholder participation. Bio solid waste management system requires intensive use of environmentally with superior technology adopt in each level of development which can be commonly exported from advanced countries to growing nations. Hence, knowledge based technical transfer should cover not only advanced technology of waste treatment method but also management system, legal system, policy and investment in order to build a sustainable society with natural acceptance. This chapter reviewed and discussed about a wide range of knowledge held by developed countries such as Japan and needed by developing countries such as India requires in depth use of environmentally with advanced era undertake in each stage of development which may be normally exported from advanced international locations to growing international locations. The DPSIR framework may help more in current scenario.

A number of studies have looked at the potential for contamination of ground and surface waters from sewage biosolids and have found that it is not significant. It is important to evaluate the risks of biosolids land application in light of other common, widespread agricultural practices such as the land application of animal wastes. When put into this context, the risk of infectious disease to rural residents,

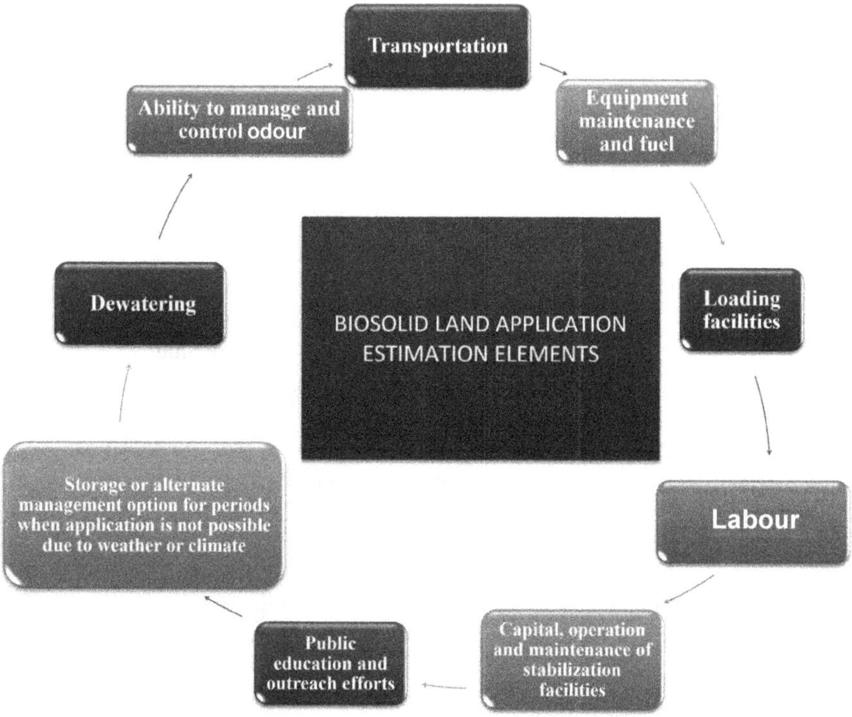

Fig. 7: Illustration of biosolid land estimation elements and cost effective techniques.

specifically from biosolids land application, appears to be relatively small. Science continues to move forward, but definitive answers will remain elusive. Currently, evidence supports ongoing land application of biosolids to recycle nutrients and organic matter as a benefit to individual farmers and society as a whole. Health professionals have to participate in the continual improvement of our water supply and wastewater treatment systems.

14.6.1 Identification of sources for finance

ULBs could also consider the opportunity of funds from bilateral and multilateral donors like ADB, KfW, the World Bank, etc., for soft loans and grants for infrastructure projects, after due approval from the State. States could also access these funds and provide them to the ULBs.

Local material recovery and recyclable sorting facilities, decentralized compost plants, bio methanation (biogas) plants, vermicomposting, windrow composting, and bin-composting are all easy to establish at the community or institutional level. Sizes can vary from small backyard composting to plants processing 3–20 tonnes per day (TPD) of organic waste. Material take-back and recycling facilities can also be established at the local community level to increase efficiency of collection.

- The extent of service provision by the ULBs is determined largely by the availability and allocation of finances to different services and functions. ULBs are empowered to derive their income from several sources such as taxes, fines, penalties, and remunerative enterprises. Apart from the above mentioned sources, ULBs also depend on grants and loans to meet their financial needs.

The traditional sources of financing MSWM activities include local taxes, e.g., property tax, water tax, conservancy tax, development fee, etc.; grants from state and central government, e.g., Swachh Bharat Mission, state finance commission grants; loans from capital market, government, and financial institutions, e.g., Housing and Urban Development Corporation (HUDCO) and National Bank for Agriculture and Rural Development (NABARD); loans from international agencies, e.g., Asian Development Bank (ADB), Japan Bank for International Cooperation (JBIC), German Development Bank (KfW) and the World Bank; municipal bonds or debentures' revenue from sale of products from waste processing plants owned by the ULB, tipping fees from the private operator.

14.6.2 Integrating the informal waste sector: policy directives

The integration of the informal waste sector into formal waste management systems is made possible through a set of formal or informal arrangements between waste pickers or organisations of waste pickers or organisations working with waste pickers and the local authorities, in their operational area. The integration process would typically result in the accrual of social benefits to wastepickers. Some of the salient features of the policies or regulations pertaining to municipal solid waste management (MSWM) as they relate to the informal waste management sector are the following:

- ✓ National Environment Policy, 2006. It acknowledges the informal waste sector and states, "Give legal recognition to, and strengthen the informal sector systems of collection and recycling of various material. In particular enhance their access to institutional finance and relevant technologies."
- ✓ National Action Plan on Climate Change, 2009. It stresses the need for giving legal recognition to the informal sector, which it recognizes as the "backbone of India's highly effective recycling system."
- ✓ National Labour Commission, 2002. It "recognises the useful role played by the scrap collectors both in helping recycling activities as well as in maintaining civic hygiene."
- ✓ Encouraging informal sector, NGO, and CBO through linkage to National Urban Livelihoods Mission.

The illegal activity measurement of providing identity cards, receipts for transactions, minimum wages when they are employed by contractors, health facilities, creation of welfare funds, prohibition of child labour from the activity. The Commission fully endorses the suggestions made by the United Nations Development Programme (UNDP) and the International Labour Organisation (ILO). SWM Rules, 2016 recognises the role of informal sector in waste management, and emphasises on establishing a system for integration of these waste collectors in order to facilitate their participation in biosolids.

14.6.3 *Developing capacities for implementing awareness-raising campaigns*

The following are some ways to develop capacities of urban local body (ULB) staff, non-government organisations (NGOs), self-help groups (SHGs), and other community based organisations (CBOs) for implementing awareness-raising campaigns:

- Indicate the issues to be covered as per social and cultural needs or requirements.
- Draft messages for selected media.
- Field-test draft material and finalize material with selective target groups (including women, youth, and children).
- Train the field workers or ULB staff.
- Train NGOs and SHGs in facilitating municipal solid waste management (MSWM).
- Monitor progress in the field.

14.6.4 *Public participation and awareness*

The SWM Rules, 2016 direct ULBs to create public awareness through information, education, and communication (IEC) campaigns and educate the waste generators to minimise waste and prohibit littering. Municipal authorities should organise awareness generation programmes promoting segregation of waste and recycling or reuse of segregated waste. The communities of people must be educated and well trained on biowaste segregation. ULBs should sensitize citizens to associate with environment and health hazards of improper biowaste management. Further, the citizens should be made aware of the need to pay user fees or charges for ensuring sustainability of the MSWM services. This process is most effective when led by the Chief Executive of the ULB, and prominent people are involved in the campaign to motivate the society at large. To ensure segregation at household level, along with the proper system of door-to-door collection and transportation, there must be sustained efforts by the authorities and strong leadership to motivate the citizens over a period of 15 days to one month along with the proper system (e.g., case study on Warangal). The practice of ULB through the NGOs, rotary clubs, CBOs, and other such organizations should conduct school-level awareness and education programmes focusing on source segregation, waste minimization through reduce, reuse, and recycle, and the importance of proper management of waste. Students should be made aware of increasing waste quantities and environmental impacts of unscientific disposal.

14.6.5 *Role of NGO*

During the recent years, NGOs' (non-governmental organizations) participation has turned into initiatives to work from local body to improve the social environmental issues and solution with systematic work completion. They have been playing an active role in organizing surveys and data studies in specified disciplines of social and technological sciences. In the field of biosolid management to land application, such data studies are helpful for identifying areas of commercial potential to attract young industrial entrepreneurs. They can play a vital role in separation of biowaste from waste water into various forms, its collection and handling techniques by the management system and successfully create awareness among the citizens about their rights and responsibilities towards bio solid waste and water purification. These organizations promote environmental education and awareness in schools and involve communities in the management of solid waste.

The NGO programmes' analysis:

- Create mass awareness, ensuring public participation in segregation of recyclable material and storage of waste at source.
- Provide employment through organizing door-to-door collection of waste.
- Ensure public participation in community based primary collection system.
- Encourage minimization of waste through in-house backyard composting, vermicomposting and biogas generation.
- Urban poverty is inextricably linked with waste. In India alone, over a million people find livelihood opportunities in the area of waste; they are engaged in waste collection (popularly known as rag picking) and recycling through well-organized systems. Substantial populations of urban poor in other developing countries also earn their livelihood through waste. It is important to understand issues of waste in this context. The informal sector dealing with waste is engaged in various types of work like waste picking, sorting, and recycling at the organized level, door-to-door collection, composting and recycling recovery.

14.6.6 Focus view on motion key factors view through recognition generation binded with technical aspects

In step with SWM regulations, in 2016 mandates ULBs to create public consciousness via statistics, education, and conversation (IEC) campaign to train the waste to energy product through MSWM activities maximum of software create the movement factors attention for focus generation with the following: ensure active participation of the community in reducing overall quantities of waste.

The different waste reduction strategies are given below:

1. Promotion of waste-reduction in community level with efficient training to manage the recyclable materials.
2. Awareness program on the environmental impact of using non-recyclable, non-reusable, and hazardous materials through NGOs and governmental agencies.
3. Training of the social welfare organizations and the citizen on the segregation of biowaste at the point of the generation.

4. The committee formed will conduct regular meetings among the ULB staff and representatives of market associations, NGOs, SHGs, social clubs and other stakeholders to ensure successful implementation of such programmes.
5. Ensure active participation of the community for successful implementation of primary and secondary collection systems.
6. Involve community in designing the primary collection system, e.g., in determining waste collection system and timings.
7. Generate awareness on bye-laws on waste collection and management system as well as user charges levied on different waste fractions.

14.7 Conclusions and perspectives

Land application of biosolids is a constructive method to recycle organic matter and nutrients, to improve physical, chemical, and biological properties of soils, and to re-establish vegetation and restoration of degraded ecosystem. However, precaution needs to be exercised when biosolids are applied again or at heavy application rates such as heavy metals, organic pollutants, and pathogens in biosolids, though at minimum concentration, may impose harmful effect to the environment and animal and human health with time. In the future, long-term monitoring is needed to evaluate the potential impacts of biosolids land application on soil quality and the environment including surface, soil and ground water, air, and plant/animal health as well as food quality. The protective regulation governing policies in international nations are anticipated based on scientific tracking and assessment information in biosolids control machine. There are many socio-cultural-regional-community issues which influence the bio solid waste management. The most prominent issue is inter-generational equity for the welfare of human and the social dimension includes the continued satisfaction of basic human needs, as well as higher-level of social cultural, and economic necessities. It is the responsibility of humans to protect nature from disaster and to design products, processes and technologies as well as systems balancing these multidimensions.

References

Brown, Chaney, Hallfrisch and Xue. 2003. Effect of biosolids processing on lead bioavailability in an urban soil. Journal of Env. Quality. 32: 100–108.

Chaney, Rufus L. September 1990. Twenty Years of Land Application Research. BioCycle. 31(9): 54–59.

Cheremisinoff, N.P. 2002. Handbook of Water and Wastewater Treatment Technologies,the United State of America: Butterworth-Heinemann.

Glawe, U., Visvanathan, C. and Alamgir, M. 2005. International conference on integrated solid waste management in Southeast Asiancities. 5–7 July, Siem Reap, Cambodia.

Hoornweg, Daniel and Bhada-Tata, Perinaz. 2012. What a Waste: A Global Review of Solid Waste Management. Urban development series; knowledge papers no. 15. World Bank, Washington, DC. © World Bank. https://openknowledge.worldbank.org/handle/10986/17388 License: CC BY 3.0 IGO.

Idris, A., Inanc, B. and Hassan, M.N. 2004. Overview of wastedisposal and landfills/dumps in Asian countries. Journal of MaterialCycles in Waste Management. 6: 104–110.

Jewell, W.J. and Seabrook. B.L. 1979 History of land application as a treatment alternative. Tech. Rep. EPA 430/9-79-012, Office of Water Program Operations, Washington, DC, USA.

Muga, H.E. and Mihelcic, J.R. 2008. Sustainability of wastewater treatment technologies.Journal of Environmental Management. 88: 437–447.

Puddephatt, Karen Joan. 2013. Determining the Sustainability of Land-Applying Biosolids to Agricultural Lands Using Environmentally-Relevant Terrestrial Biota. Ryerson University: Theses and dissertations, Paper 1579.

Spellman, F.R. 1997. Wastewater Biosolids to Compost, the United State of America: Technomic Publishing Company.

U.S. EPA. 2011. Water: sewage sludge (biosolids), http://water.epa.gov/polwaste/wastewater/treatment/biosolids/index.cfm.

Wang, H., Kimberley, M.O. and Schlegelmilch, M. 2003. Biosolids-derived nitrogen mineralization and transformation in forest soils. Journal of Environmental Quality. 32(5): 1851–1856.

Wang, L.K., Williford, C., Chen, W.Y. and Shammas, N.K. 2007. Land application of biosolids. *In*: Biosolids Engineering and Management.

World Bank. 1999. What a Waste: Solid waste management in Asia. Urban Development Sector Unit, East Asia and Pacific region.

15

Membranes in the Recovery of Oils and Lipids including Volatile Fatty Acids

Underlying Principles, Recent Advances and its Synergism with Nanotechnology

Tathagata Adhikary and *Piyali Basak**

15.1 Introduction

Oily wastewater from industrial effluents (e.g., petrochemical, mining, dye industry) and accidental oil spills pose a serious threat to the environment and also account for the scarcity of clean water (Yu et al. 2017). Oil spills block the sunlight to penetrate, reduce the percentage of dissolved oxygen in the water and hence disturb aquatic life (AG et al. 2013). Dyeing, being one of the largest wastewater producing industry, is responsible for 17–20% of the total industrial wastewater (Kant 2012). This draws the attention of researchers to develop efficient and cost-effective technologies in separating and recovering oils to meet the problems of freshwater scarcity and environmental hazards (Iglesias et al. 2007, Mairal et al. 2006). Traditional techniques employed in wastewater recycling for oil-water separation such as using gravity, centrifugation, coagulation and flocculation by chemical treatments, magnetic separations or bioremediation are being replaced by innovative alternatives involving multidisciplinary approaches (Al-Shamrani et al. 2002). These conventional methods of separation usually fail when the dimension of oil droplets in an emulsion is less than 20 μm, hence suitable for layered oil-water mixtures. Moreover, energy consumption is high and the process may add secondary pollution (due to additional chemical usage) (Ahmad et al. 2011, Kwon et al. 2012). The use of membranes stands out to be a promising solution in separating oil-water mixture or surfactant-stabilized emulsions due to its simplicity in functioning, low energy consumption

School of Bioscience and Engineering, Jadavpur University, West Bengal, India.
* Corresponding author: piyalibasak@gmail.com

and the ease of scaling up (Drioli and Romano 2001). Any semipermeable phase that separates two homogeneous phases but allows selective permeation of specific molecules through it can be termed as a membrane. However, the challenges related to the fouling of membranes, its scalability, durability, recyclability, automatization and manufacturing cost are extensively researched to enhance its utilization in large-scale practical scenarios without any compromise in its separation efficiency. Volatile fatty acids (sometimes referred as short-chain fatty acids) generated by the fermentation (i.e., anaerobic breakdown of wastes, specially lignocellulosic biomass) or as a byproduct of petrochemical industry are considered to be an important raw material for various industries including food, pharmaceuticals, cosmetics and even in developing biopolymers and biofuels/biogas (Morgan-Sagastume et al. 2011, Singhania et al. 2013). Separation and purification of these volatiles are often achieved by employing membrane-based processes such as micro/nano-filtration, reverse/forward osmosis, membrane distillation, electrodialysis and pervaporation.

The knowledge of material science and interdisciplinary approaches like exploiting nanotechnology has helped to design biomimetic and novel membranes with appreciable operational performance efficiency in oil-water separation. These membranes are mostly polymer-based, ceramic-based or nanomaterial-based functional hybrid membranes. Currently, the pressure-driven filtration process using polymeric membranes with specific pore geometry is studied meticulously to design efficient substitutes for existing water purification strategies. Some of the downsides of using filtration membranes (especially polymeric) are as follows (Zhang et al. 2014):

(a) The hydrophobic nature of membranes reduces the associated water permeation flux which is further lessened after getting clogged by oil droplets.
(b) Nonselective wetting of membranes by water and oil needs to be addressed. Hence, membrane surfaces with appropriate wettability should be developed by regulating the chemical and topographical properties.

This chapter introduces the mechanism of wetting by a liquid drop on a solid surface. Proper design and fabrication of membranes also require insight into the mechanism of demulsification and coalescence of oil droplets on the membrane surface. Subsequently, it reviews the complications encountered in membrane separation processes along with current breakthroughs and research prospects.

15.2 Theory of wettability

Surface free energy describes the phenomenon of wettability and its quantification can be done by measuring the contact angle formed by the liquid droplet on the solid surface (Mittal 2003). Considering a surface, the correlation between its roughness, energy and wettability contributes to its superhydrophobic nature (Feng and Jiang 2006, Gao et al. 2013). Several theories relating to the wetting mechanism of a solid surface by a liquid drop include the Young's model, the Wenzel's model and the Cassie–Baxter's model (Fig. 1).

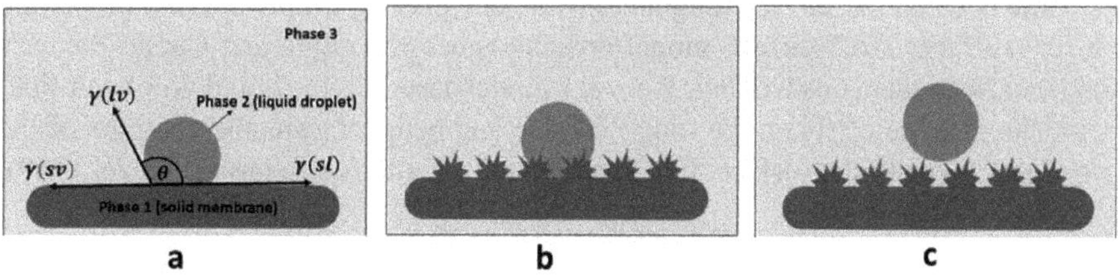

Fig. 1: Schematic representation of the three models relating the mechanism of wetting by a liquid drop on a solid surface – a) Young's model, b) Wenzel's model and c) Cassie-Baxter model.

15.2.1 *Young's model*

Wettability can be described by the Young's equation given below (Gennes 1985):

$$cos\theta = \frac{\gamma(sv) - \gamma(sl)}{\gamma(lv)}$$

where θ indicates the contact angle, γ(sv), γ(sl), γ(lv) indicate the interfacial surface tensions that exist between the boundaries of solid–vapor, solid–liquid and liquid–vapor interfaces, respectively.

However, Young's equation is limited to smooth solid surfaces only. It considers an immobile liquid droplet on an ideal, homogeneous (i.e., having uniform surface properties) and perfectly flat surface. The whole system holds three interfaces among three phases of liquid, vapor and solid. The measure of the contact angle is governed by the thermodynamic equilibrium of the free surface energies at these interfaces (de Leon and Advincula 2015). If the liquid droplet constitutes water molecules, the solid surface exhibiting a critical angle less than 90° is hydrophilic while hydrophobic surfaces will have a critical angle greater than 90°.

Young's equation can also be used to find the contact angle for a system consisting of three interfaces formed by a smooth solid surface and two immiscible liquids. Let us consider a case of an underwater smooth surface on which an oil droplet is placed. The above equation can be adjusted as follows to give the contact angle of oil associated with that underwater surface (de Leon and Advincula 2015):

$$cos\theta\,(ow) = \frac{\gamma(ao)\cos\theta\,(\mathrm{o}) - \gamma(aw)\cos\theta\,(w)}{\gamma(ow)}$$

where γ(ao), γ(aw) and γ(ow) are the interfacial tensions that exist between air-oil, air-water and oil-water, respectively. θ(o) is the contact angle of oil on the surface placed in air, θ(w) is the contact angle of water on the surface placed in air and θ(ow) is the contact angle of oil in the underwater surface.

15.2.2 *Wenzel's model*

Wettability is modified, i.e., either hydrophilicity or hydrophobicity of the surface increases if the surface exhibits some roughness on it. Young's model is applicable for surfaces with no roughness. However, any surface, seemingly smooth on a macroscopic level, is actually rough on a microscopic level. The two models that take into account the surface roughness while deciphering the mechanism of wetting on real surfaces are Wenzel's model or homogeneous wetting and Cassie-Baxter's model or heterogeneous wetting. Wenzel's model describes the liquid droplet to sink/permeate into the clefts on the rough surface and hence it trails the contours of the grooves present on the surface. It is given by the equation as follows (Wenzel 1949):

$$\cos\theta(w) = r\cos\theta$$

where $\theta(w)$ indicates the apparent contact angle formed on a rough surface under Wenzel's state, θ is the contact angle under Young's state, i.e., surface (of the same material) with no roughness, and r is the surface roughness factor.

For a surface with no roughness (i.e., ideally flat), r is taken to be 1 and its value increases as roughness increases. According to Wenzel's model, if the contact angle under Young's state is less than 90°, then increasing the surface roughness will decrease the apparent contact angle. However, in the cases where Young's contact angle is greater than 90°, the apparent contact angle increases with roughness factor (Chen and Qingxia 2020).

15.2.3 *Cassie–Baxter's model*

Considering Wenzel's theory of wettability, it is observed that for the values of r greater than 1.7, the model deviates (Jopp et al. 2004). For the more rough surfaces, permeation of the liquid droplet through the dents becomes difficult and air pockets get trapped between them. Cassie-Baxter's model describes this phenomenon of heterogeneous wetting where the surface is non-uniform and has varying surface composition/chemistry. The apparent contact angle or the Cassie-Baxter contact angle (θ_{CB}) is given by the following equation (Cassie and Baxter 1944):

$$\cos\theta_{CB} = f_{(SL)}(1 + \cos\theta) - 1$$

where $f_{(SL)}$ indicates the fraction of the surface in contact with the liquid and θ is the Young's contact angle.

With the increase in roughness, there is an increase in θ_{CB} when $\theta > 90°$ and θ_{CB} decreases when $\theta < 90°$. Hence, a surface shows super-hydrophobicity if $f_{(SL)}$ is very small or the surface material has a very high Young's contact angle.

The modified Cassie-Baxter's equation to determine the contact angle of an oil droplet placed on an underwater rough surface (i.e., θ_{ow}) is given by:

$$\cos\theta_{ow} = f_{(SL)}\cos\theta_{ow} + f_{(SL)} - 1$$

15.3 Membrane classification based on its wettability

The difference in interfacial tension between oil and water owing to their immiscibility is considered to develop promising materials for membranes with superwetting nature, i.e., having extreme affinity to either oil or water (Chu et al. 2015). These membranes are capable of selectively separating oil-in-water or water-in-oil emulsions with appreciable membrane separation performance. Different types of wettable filtration membranes exist that include superhydrophobic/superoleophilic membranes, superhydrophilic/superoleophobic membranes, Janus membranes and membranes with switchable wettability, each having diverse fabrication techniques.

15.3.1 Superhydrophobic/Superoleophilic membranes

We are all aware of the lotus effect in which water droplets retain their spherical structure and slide over the lotus leaves without wetting them. Using the technology of scanning electron microscopy, Barthlott and Neinhuis studied the surface structure of lotus leaves. They revealed that the self-cleansing and water-repellent properties of lotus leaves are the outcomes of the presence of micro and nanostructured hydrophobic units throughout the surface (Barthlott and Neinhuis 1997). Hence, the nature of super-hydrophobicity encountered in different plants' parts is due to the structural assembly of hydrophobic (wax-like) materials (Koch and Ensikat 2008). This finding provides a direction to research and fabricates superhydrophobic surfaces artificially by modifying the chemical composition of the surface and incorporating surface roughness. The membranes exhibiting super-hydrophobicity or super-oleophilicity allow selective permeation of oil through it and thus can be used to remove oil from oil-water mixture. The first artificial superhydrophobic/superoleophilic film was prepared by spraying polytetrafluoroethylene (i.e., PTFE having a low–surface energy) over a stainless steel mesh and drying it to form hierarchical micro/nanostructures of hydrophobic PTFE units. The contact angle measured using a water droplet was found to be more than 150° while the oil droplet spread easily over the polymeric mesh and permeated through it within a time of 240 ms (Feng et al. 2004).

Instead of assembling hydrophobic units on a surface to make the hierarchical structure, one can combine the fabrication and surface modification processes to develop membranes with super-hydrophobicity. As an example, the phase inversion technique is used to synthesize superhydrophobic/superoleophilic poly(vinylidene fluoride) (PVDF) membranes in which microparticles of PVDF are formed due to microphase separation. The synthesized membrane was exploited in the separation of oil/water mixtures and water-in-oil emulsions with droplets in the nano/sub-micron range (Zhang et al. 2013). The superhydrophobic/superoleophilic membranes are not suitable in the separation of oil-water mixtures where the process is influenced by gravity. This is because water, being heavier, will settle down and prevent the oil layer to permeate through the membrane.

Membrane filtration, with the use of any external forces or a process that is pressure-driven, is not favored since gravity-driven filtration is cost-effective and always easy to scale up. Recently, with the advancements of 3D printers, polydimethylsiloxane (PDMS) membrane is printed using PDMS ink comprising nano-sized hydrophobic silica. The fabricated membrane exhibited appropriate topography to possess superhydrophobic/superwettable properties with increased mechanical durability. It was reported to separate several oil-water mixtures with efficiency above 90% (Lv et al. 2017).

15.3.2 Superhydrophilic membranes

The surface tension of water is generally greater than organic compounds/liquids. Oil droplets and other organic compounds (e.g., surfactants used in oil-water mixtures) tend to adhere to the membranes and get absorbed causing fouling of membranes. To prevent this fouling, hydrophilic membranes can be exploited for underwater use or membranes with synchronized superhydrophilic and superoleophobic units are designed. A surface showing hydrophilicity in air can be oleophobic when placed underwater; hence, it will get wet by water but will not get contaminated by oil. This scenario of superoleophobicity is often seen in aquatic animals. In a research, copolymerization of dopamine and the hydrolysis of tetraethoxysilane was done to fabricate PVDF membranes incorporated with nanoparticles. It exhibited superoleophobicity when placed underwater, has ultrahigh water permeation rate, prevents antifouling and is operationally very stable (negligible effect of continuous rinsing and bending under cryogenic condition) (Wang et al. 2015). Dopamine acts as a binder to the hydrophilic functional groups present on the surface of membranes. Sodium periodate, when used with catechol coating solutions, can efficiently and inexpensively make superhydrophilic/underwater superoleophobic PVDF membranes with high water permeability and oil separation efficiency (Chen and Liu 2019). Sodium periodate initiates the oxidation of catechol and makes the polymerization process extremely fast. Graphene oxide is also considered in several studies to design membranes due to its favorable properties such as large surface area, appreciable flexibility and high Young's modulus. A free-standing graphene oxide membrane with intercalated palygorskite nanorods was developed by employing vacuum-assisted filtration self-assembly in which nanosheets of graphene oxide were arranged via π–π stacking in the vacuum filtration process. It reported a water flux of 267 $L.m^{-2}.h^{-1}$ and a separation efficiency of greater than 99.9% in hexadecane-in-water emulsions with antifouling activity (Zhao et al. 2016).

15.3.3 Janus membranes

Membranes can be fabricated such that they exhibit the opposite nature of wettability on their two surfaces. These Janus membranes allow the liquid to permeate unidirectionally. A composite of polymers hydrophobic poly(styrene) and hydrophilic poly(N,N-dimethylaminoethyl methacrylate) with carbon nanotubes is used to fabricate a hybrid Janus membrane that showed high separation efficiency in removing oil from oil-water mixtures (Gu et al. 2014). This coating and peeling strategy proved to be effective and economic (with no requirements to high-end instruments and expensive reagents) in synthesizing multi-functional Janus membranes for simultaneous water collection and decontamination, selective droplet permeation, and in microreaction (Wang et al. 2017).

15.3.4 Membranes with switchable wettability

Extensive research on designing filtration membranes has developed smart membranes that can change their nature of wettability (i.e., from superhydrophobic to superhydrophilic or vice-versa). This change is controlled by the application of an external stimulus (e.g., change in pH, temperature or certain irradiation) is applied (Kwon et al. 2012). The underlying surface chemistry is considered along with its topographical analysis to impart such special property to the membranes. These smart membranes have arrangements of hydrophilic and hydrophobic units along their surface that respond to a particular stimulus and get exposed selectively. Hence, they aid in the automatization of the process by changing their functional and structural characteristics in response to an external stimulus. An electrospun composite membrane with high free surface energy was fabricated having 2 layers on a non-woven surface: a layer of nano/micro-sized beads of polyvinylidene fluoride (PVDF)-silica composite and a PVDF nanofibrous intermediate layer. It was used in a cross-flow filtration process to treat oil-water mixtures including surfactant-free/surfactant-stabilized emulsions under ultralow pressure of 0.1 atm (i.e., absence of any external driving pressure across the membrane). The results indicate that the associated flux and separation efficiency is around 2000 L/m^2.h and 99.99%, respectively, with high durability (no significant effect of hot water, strong acids or bases and 50 hours of continuous operation) and antifouling property of the designed in-air superamphiphilic/underwater superoleophobic/under-oil superhydrophobic membrane (Liao et al. 2017). In another research work, modification of nylon membrane by Poly(N-isopropylacrylamide) is done to develop membranes that change the nature of wettability upon application of a thermal stimulus. The membrane is exploited in the separation process of various stabilized emulsions. It exhibited hydrophilicity/underwater-superoleophobicity at temperatures below 25°C and hence can be used in oil-in-water emulsions while its superhydrophobic/superoleophilic nature at temperatures above 25°C makes it suitable for separating water-in-oil emulsions. The inexpensive hydrothermal fabrication process employed in this study and the membrane recyclability makes it a promising alternative in treating wastewater or oil spills (Zhang et al. 2018).

15.4 Materials for membranes

15.4.1 Polymers

Oleophilic membranes are prone to fouling and low flux rate or poor permeability. Hence, blending and various surface modification techniques are employed to adjust the hydrophilicity and impart antifouling nature to the polymeric membranes (Hilal et al. 2015). While a porous polymeric film is prepared using the phase inversion technique, hydrophilic constituents (e.g., hydrophilic polymers, amphiphilic copolymers and inorganic nanoparticles) are blended with the polymer solution to achieve the desired modifications (Liu et al. 2011). But the membrane stability needs to be checked since the tendency of hydrophilic additives to favor interactions with water molecules will eventually release them from the polymer matrix (since only weak interactions exist between the additives and the polymer). Polyvinylidene fluoride and polysulfone are the frequently used polymers in membrane filtration to treat oily wastewater. Blends of PVDF prepared using poly(methyl methacrylate), sulfonated polycarbonate, or sulfonated poly(ether ether ketone) increased the hydrophilicity of filtration membranes, but in some cases the formation of microvoids was seen (Bowen et al. 2005, Masuelli et al. 2009). Owing to the large surface area and availability of functional groups, inorganic nanoparticles of Al_2O_3, TiO_2 and SiO_2 are often blended to form nano-hybrid composite membranes with desired properties (Jhaveri and Murthy 2016). In a study, an ultrafiltration membrane was designed by blending polymer polysulfone with TiO_2 nanoparticles to separate kerosene-emulsified wastewater. The membrane possessed good antifouling property but exhibited aggregation of nanoparticles in the polymer matrix due to their inadequate dispersion in the polymer solution. Also, the nanoparticles were released from the polymeric membranes due to the absence of strong interactions with the polymer, thus reducing the membrane stability (Yang et al. 2007). In order to bypass the disadvantages associated with this direct blending method, precursors of nanoparticles can be blended. A published work blended tetraethyl orthosilicate with the polymer cellulose acetate. SiO_2 nanoparticles are formed using tetraethyl orthosilicate as its precursor by *in situ* acid or alkali treatment during phase inversion. This method portrays evenly dispersed SiO2 nanoparticles in the cellulose acetate polymeric matrix with negligible aggregates of nanoparticles (Chen et al. 2010).

Surface modification is very useful in enhancing membrane filtration performance. The introduction of hydrophilic polymer chains on the membrane surface can be done by surface grafting. Reactive moieties are generated or immobilized on the surface by suitably treating the membrane (such as exposure to low-temperature plasma, ultraviolet (UV), gamma or electron beam radiation and chemical treatments) and then growing the surface-grafted monomers to form a layer of its own (Liu et al. 2007, Wavhal and Fisher 2002). Hydrophilic surface coats can also be easily prepared by dipping the membranes, using sprays or direct adsorption of water-soluble/amphiphilic polymers on the membrane surface. Due to the simplicity of these processes in developing such hydrophilic layers, it is industrially adopted. However, these layers seem to be unstable under operation conditions (Liu et al. 2011). Polyvinyl alcohol and chitosan are commonly used commercial polymers for physical modification of the membrane surface. Chemical means of surface modification strongly immobilize polymers with hydrophilic groups (e.g., poly(2-hydroxy-ethyl methacrylate), poly(acrylic acid), etc.) via covalent bonds. This ensures hydration of the top layer preventing the fouling of membranes by oil droplets. Zwitterionic polyelectrolyte-grafted membranes are fabricated nowadays for the separation of oil/water mixtures and various emulsions. Poly(3-(N-2-methacryloxyethyl-N,N-dimethyl) ammonatopropanesultone) was grafted onto a PVDF membrane employing a surface-initiated atom transfer radical polymerization process. Its nature of superhydrophilicity and underwater superoleophobicity was established with a reported water contact angle of 11° and an underwater oil contact angle of above 150° (Zhu et al. 2013).

15.4.2 Ceramics

Ceramic-based filtration membranes offer the advantage of high membrane stability: chemically, thermally and mechanically. These membranes generally have a rough surface indicating a larger surface area (Li et al. 2020). Hence, they are prone to fouling and low permeation rate due to the range of pore size these membranes possess (Faibish and Cohen 2001, Hua et al. 2007, Vasanth et al. 2011). Zirconia membranes are widely used to treat oil-water mixtures. Zirconia nanoparticle modified Al_2O_3 microfiltration membrane was prepared using $ZrCl_4$ as a precursor. It showed better membrane performance than the unmodified Al_2O_3 microfiltration membrane, due to the presence of hydrophilic ZrO_2. Targeting oily wastewater treatment, other ceramic membranes such as zeolite-based and their polymeric modifications are extensively reported in various studies (Cui et al. 2008, Faibish and Cohen 2001). A 19-channel zirconia-based ultrafiltration membrane (mean pore size of 0.05 μm) was employed in the clarification of the heated limed sugarcane juice. Analysis of the results from energy-dispersive X-ray spectrometry (EDX) and Fourier-transform infrared spectrometry (FTIR) reported the major foulants to be polysaccharides, proteins, aliphatics, sucrose, phenols, phosphorus, silicon, and some metals, namely calcium, magnesium, aluminum, potassium, and sodium. A cleanup process involving 1% NaOH, 0.5% NaClO and 0.5% HNO_3 solution was followed to achieve the flux recovery ratio greater than 96.6% with high repeatability of the protocol (Li et al. 2018).

15.4.3 *Nanomaterial-based membranes*

With the advancement in research, the multidisciplinary approach involving nanotechnology with material science proves to be a promising step in membrane technology. We have already discussed several examples where we witnessed the use of nanoparticles to enhance membrane performance. According to the permeation theory, the rate of filtration of a membrane is directly proportional to the square of its effective pore size and inversely proportional to its thickness; hence, optimization of these 2 parameters is crucial for designing membranes (Obotey Ezugbe and Rathilal 2020). Fabrication of ultrathin membranes with improved antifouling and high filtration efficiency has become possible by incorporating nanomaterials in their structure. 1-D nanostructures like carbon nanotubes and nanofibers are popular candidates while designing ultrathin but porous membranes (Du et al. 2020, Zhijiang et al. 2018). A superhydrophobic, highly stable and electrically conductive nanofiber composite was prepared by incorporating carbon nanotubes ultrasonically on the polymer nanofiber and then subsequently performing methyltrichlorosilane modification. The membrane efficiency in treating oil-water mixtures was found to be independent of pH (Huang et al. 2019).

15.5 Mechanism underlying fouling and antifouling techniques

Fouling of membranes refers to the continuous decrease in flux rate caused by the accumulation of feed constituents (i.e., foulants) on its surface/matrix. This increases the resistance to permeate through the membrane. International Union for Pure and Applied Chemistry (IUPAC) states membrane fouling as follows: "Fouling is the process resulting in loss of performance of a membrane due to the deposition of suspended or dissolved substances on its external surfaces, at its pore openings, or within its pores" (Koros et al. 1996). Different fouling mechanisms are identified:

clogging of membrane pores, adsorption inside the pores, concentration polarization and gel layer formation. Fouling due to concentration polarization and gel layer formation is reversible fouling where hydraulic cleaning of membranes can be done, while adsorption of foulants and pore-clogging contributes to irreversible fouling (Guo et al. 2012).

An approach to antifouling of membranes takes into account its surface hydrophilicity. Improving hydrophilicity will create a pure water layer over the membrane surface and arrest the adsorption or deposition of hydrophobic feed constituents. The surface charge also plays an important role in controlling fouling. The repulsive force that exists between two like-charged units can be exploited to adopt a suitable antifouling technique. Hence, membranes should be fabricated in accordance with the electrostatic nature of foulants. The grafting of hydrophilic polyethylene glycol on the membrane surface is reported to prevent the adsorption of macromolecules (e.g., proteins) by virtue of steric repulsion between them (Zhao et al. 2018). We have highlighted other antifouling strategies in the section of surface modification of polymeric membranes earlier.

Antifouling modifiers can be of two types based on its material: polymer or organic type and inorganic type of modifiers. Organic modifiers (like hydrophilic polymers, stimuli-responsive or amphiphilic copolymers, charged polymers, various surfactants, organic acids) are sometimes blended directly to the polymeric solution to achieve the antifouling property of the membranes (Singh and Purkait 2016). Some pH-responsive copolymers of acrylic acid or methacrylic acid are poly(acrylonitrile-co-acrylic acid), poly(acrylonitrile-acrylic acid-vinyl pyrrolidnone) and poly(N-isopropylacrylamide-co-methacrylic acid-co-methyl methacrylate) while examples of amphiphilic modifiers used in fabricating antifouling membranes are β-cyclodextrin polyurethane and polystyrene-β-poly (ethylene glycol) methacrylate (Fang et al. 2020, Liu et al. 2015, Zhao et al. 2011). Composite membranes of polymers and inorganic constituents (especially nanoparticles) are also studied to strategize the antifouling of membranes. We have discussed the use of carbon nanotubes and some nanoparticles of TiO_2, SiO_2, $Mg(OH)_2$ and ZnO in designing membranes with enhanced antifouling properties.

15.6 Conclusion

Traditional techniques employed in wastewater recycling for oil-water separation such as using gravity, centrifugation, coagulation and flocculation by chemical treatments, magnetic separations or bioremediation are being replaced by innovative alternatives involving multidisciplinary approaches. The use of membranes stands out to be a promising solution in separating oil-water mixture or surfactant-stabilized emulsions due to its simplicity in functioning, low energy consumption and ease of scaling up. Separation and purification of volatile fatty acids are often achieved by employing membrane-based processes such as micro/nano-filtration, reverse/forward osmosis, membrane distillation, electrodialysis and pervaporation. This chapter introduces the three models relating the mechanism of wetting by a liquid drop on a solid surface—Young's model, Wenzel's model and Cassie-Baxter model. Recently, superwettable membranes (namely superhydrophobic/superoleophilic membranes, superhydrophilic/superoleophobic membranes, Janus membranes and membranes with switchable wettability) are capable of selectively separating oil-in-water or water-in-oil emulsions with appreciable membrane separation performance. The growing knowledge of material science and interdisciplinary approaches like exploiting nanotechnology has helped to design biomimetic and novel membranes. Surface modification of different polymeric, ceramic or nanoparticle-based hybrid membranes is studied extensively to enhance the permeation flux and impart antifouling property. Subsequently, the chapter reviews the complications encountered in membrane separation processes along with current breakthroughs and research prospects.

Acknowledgment

The authors acknowledge TEQIP-III, Jadavpur University for providing the manpower, necessary resources and support.

References

AG, M., AM, E.S. and MS, S. 2013. Current situation of water pollution and its effect on aquatic life in Egypt. Egypt. J. Occup. Med. 37: 95–115.

Ahmad, A.L., Majid, M.A. and Ooi, B.S. 2011. Functionalized PSf/SiO2 nanocomposite membrane for oil-in-water emulsion separation. Desalination. 268: 266–269.

Al-Shamrani, A.A., James, A. and Xiao, H. 2002. Destabilisation of oil–water emulsions and separation by dissolved air flotation. Water Res. 36: 1503–1512.

Barthlott, W. and Neinhuis, C. 1997. Purity of the sacred lotus, or escape from contamination in biological surfaces. Planta. 202: 1–8.

Bowen, W.R., Cheng, S.Y., Doneva, T.A. and Oatley, D.L. 2005. Manufacture and characterisation of polyetherimide/sulfonated poly (ether ether ketone) blend membranes. J. Membr. Sci. 250: 1–10.

Cassie, A.B.D. and Baxter, S. 1944. Wettability of porous surfaces. Trans. Faraday Soc. 40: 546–551.

Chen, W., Su, Y., Zhang, L., Shi, Q., Peng, J. and Jiang, Z. 2010. *In situ* generated silica nanoparticles as pore-forming agent for enhanced permeability of cellulose acetate membranes. J. Membr. Sci. 348: 75–83.

Chen, Y. and Liu, Q. 2019. Oxidant-induced plant phenol surface chemistry for multifunctional coatings: Mechanism and potential applications. J. Membr. Sci. 570: 176–183.

Chen, Yulan and Qingxia Liu. 2020. Progress and prospects in membrane technology for oil/water separation. Multidisciplinary Advances in Efficient Separation Processes, 73–87.

Chu, Z., Feng, Y. and Seeger, S. 2015. Oil/water separation with selective superantiwetting/superwetting surface materials. Angew. Chem. Int. Ed. 54: 2328–2338.

Cui, J., Zhang, X., Liu, H., Liu, S. and Yeung, K.L. 2008. Preparation and application of zeolite/ceramic microfiltration membranes for treatment of oil contaminated water. J. Membr. Sci. 325: 420–426.

de Leon, A. and Advincula, R.C. 2015. Chapter 11—Conducting Polymers with Superhydrophobic Effects as Anticorrosion Coating. pp. 409–430. *In*: Tiwari, A., Rawlins, J. and Hihara, L.H. (Eds.). Intelligent Coatings for Corrosion Control. Butterworth-Heinemann, Boston, https://doi.org/10.1016/B978-0-12-411467-8.00011-8.

Drioli, E. and Romano, M. 2001. Progress and new perspectives on integrated membrane operations for sustainable industrial growth. Ind. Eng. Chem. Res. 40: 1277–1300.

Du, L., Quan, X., Fan, X., Wei, G. and Chen, S. 2020. Conductive CNT/nanofiber composite hollow fiber membranes with electrospun support layer for water purification. J. Membr. Sci. 596: 117613.

Faibish, R.S. and Cohen, Y. 2001. Fouling and rejection behavior of ceramic and polymer-modified ceramic membranes for ultrafiltration of oil-in-water emulsions and microemulsions. Colloids Surf. Physicochem. Eng. Asp. 191: 27–40.

Fang, C., Zhang, X., Gong, X., Feng, W., Zhu, L. and Matsuyama, H. 2020. Enhancing membrane surface antifouling by implanting amphiphilic polymer brushes using a swelling induced entrapment technique. Colloids Surf. B Biointerfaces. 195: 111212. https://doi.org/10.1016/j.colsurfb.2020.111212.

Feng, L., Zhang, Z., Mai, Z., Ma, Y., Liu, B., Jiang, L. et al. 2004. A super-hydrophobic and super-oleophilic coating mesh film for the separation of oil and water. Angew. Chem. 116: 2046–2048.

Feng, X.J. and Jiang, L. 2006. Design and creation of superwetting/antiwetting surfaces. Adv. Mater. 18: 3063–3078.

Gao, C., Sun, Z., Li, K., Chen, Y., Cao, Y., Zhang, S. et al. 2013. Integrated oil separation and water purification by a double-layer TiO 2-based mesh. Energy Environ. Sci. 6: 1147–1151.

Gennes, P. de. 1985. Wetting: statics and dynamics. Rev. Mod. Phys. 57: 827–863.

Gu, J., Xiao, P., Chen, J., Zhang, J., Huang, Y. and Chen, T. 2014. Janus polymer/carbon nanotube hybrid membranes for oil/water separation. ACS Appl. Mater. Interfaces 6: 16204–16209. https://doi.org/10.1021/am504326m.

Guo, W., Ngo, H.-H. and Li, J. 2012. A mini-review on membrane fouling. Bioresour. Technol. 122: 27–34.

Hilal, N., Ismail, A.F. and Wright, C. 2015. Membrane fabrication. CRC Press.

Hua, F.L., Tsang, Y.F., Wang, Y.J., Chan, S.Y., Chua, H. and Sin, S.N. 2007. Performance study of ceramic microfiltration membrane for oily wastewater treatment. Chem. Eng. J. 128: 169–175.

Huang, X., Li, B., Song, X., Wang, L., Shi, Y., Hu, M. et al. 2019. Stretchable, electrically conductive and superhydrophobic/superoleophilic nanofibrous membrane with a hierarchical structure for efficient oil/water separation. J. Ind. Eng. Chem. 70: 243–252. https://doi.org/10.1016/j.jiec.2018.10.021.

Iglesias, A., Garrote, L., Flores, F. and Moneo, M. 2007. Challenges to manage the risk of water scarcity and climate change in the Mediterranean. Water Resour. Manag. 21: 775–788.

Jhaveri, J.H. and Murthy, Z.V.P. 2016. Nanocomposite membranes. Desalination Water Treat. 57: 26803–26819.

Jopp, J., Grüll, H. and Yerushalmi-Rozen, R. 2004. Wetting behavior of water droplets on hydrophobic microtextures of comparable size. Langmuir. 20: 10015–10019.

Kant, R. 2012. Textile dyeing industry an environmental hazard. J. Nat. Sci. 4(1): 22–26.

Koch, K. and Ensikat, H.-J. 2008. The hydrophobic coatings of plant surfaces: epicuticular wax crystals and their morphologies, crystallinity and molecular self-assembly. Micron. 39: 759–772.

Koros, W.J., Ma, Y.H. and Shimidzu, T. 1996. Terminology for membranes and membrane processes (IUPAC Recommendations 1996). Pure Appl. Chem. 68: 1479–1489.

Kwon, G., Kota, A.K., Li, Y., Sohani, A., Mabry, J.M. and Tuteja, A. 2012. On-demand separation of oil-water mixtures. Adv. Mater. 24: 3666–3671.

Li, C., Sun, W., Lu, Z., Ao, X. and Li, S. 2020. Ceramic nanocomposite membranes and membrane fouling: A review. Water Res. 175: 115674.

Li, W., Ling, G., Lei, F., Li, N., Peng, W., Li, K. et al. 2018. Ceramic membrane fouling and cleaning during ultrafiltration of limed sugarcane juice. Sep. Purif. Technol. 190: 9–24. https://doi.org/10.1016/j.seppur.2017.08.046.

Liao, Y., Tian, M. and Wang, R. 2017. A high-performance and robust membrane with switchable super-wettability for oil/water separation under ultralow pressure. J. Membr. Sci. 543: 123–132. https://doi.org/10.1016/j.memsci.2017.08.056.

Liu, F., Du, C.-H., Zhu, B.-K. and Xu, Y.-Y. 2007. Surface immobilization of polymer brushes onto porous poly (vinylidene fluoride) membrane by electron beam to improve the hydrophilicity and fouling resistance. Polymer. 48: 2910–2918.

Liu, F., Hashim, N.A., Liu, Y., Abed, M.M. and Li, K. 2011. Progress in the production and modification of PVDF membranes. J. Membr. Sci. 375: 1–27.

Liu, Y., Su, Y., Zhao, X., Li, Y., Zhang, R. and Jiang, Z. 2015. Improved antifouling properties of polyethersulfone membrane by blending the amphiphilic surface modifier with crosslinked hydrophobic segments. J. Membr. Sci. 486: 195–206.

Lv, J., Gong, Z., He, Z., Yang, J., Chen, Y., Tang, C. et al. 2017. 3D printing of a mechanically durable superhydrophobic porous membrane for oil–water separation. J. Mater. Chem. A 5: 12435–12444.

Mairal, A.P., Ng, A., Wijmans, J.G., Pinnau, I. and Ly, J.H. 2006. Treatment of shipboard-generated oily wastewaters.

Masuelli, M., Marchese, J. and Ochoa, N.A. 2009. SPC/PVDF membranes for emulsified oily wastewater treatment. J. Membr. Sci. 326: 688–693.
Mittal, K.L. 2003. Contact Angle, Wettability and Adhesion, Volume 3. CRC Press.
Morgan-Sagastume, F., Pratt, S., Karlsson, A., Cirne, D., Lant, P. and Werker, A. 2011. Production of volatile fatty acids by fermentation of waste activated sludge pre-treated in full-scale thermal hydrolysis plants. Bioresour. Technol. 102: 3089–3097.
Obotey Ezugbe, E. and Rathilal, S. 2020. Membrane technologies in wastewater treatment: a review. Membranes 10: 89.
Singh, R., Purkait, M.K. 2016. Evaluation of mPEG effect on the hydrophilicity and antifouling nature of the PVDF-co-HFP flat sheet polymeric membranes for humic acid removal. J. Water Process Eng. 14: 9–18.
Singhania, R.R., Patel, A.K., Christophe, G., Fontanille, P. and Larroche, C. 2013. Biological upgrading of volatile fatty acids, key intermediates for the valorization of biowaste through dark anaerobic fermentation. Bioresour. Technol. 145: 166–174.
Vasanth, D., Pugazhenthi, G. and Uppaluri, R. 2011. Fabrication and properties of low cost ceramic microfiltration membranes for separation of oil and bacteria from its solution. J. Membr. Sci. 379: 154–163.
Wang, Z., Jiang, X., Cheng, X., Lau, C.H. and Shao, L. 2015. Mussel-inspired hybrid coatings that transform membrane hydrophobicity into high hydrophilicity and underwater superoleophobicity for oil-in-water emulsion separation. ACS Appl. Mater. Interfaces. 7: 9534–9545.
Wang, Z., Yang, X., Cheng, Z., Liu, Y., Shao, L. and Jiang, L. 2017. Simply realizing "water diode" Janus membranes for multifunctional smart applications. Materials Horizons. 4(4): 701–708.
Wavhal, D.S. and Fisher, E.R. 2002. Hydrophilic modification of polyethersulfone membranes by low temperature plasma-induced graft polymerization. J. Membr. Sci. 209: 255–269.
Wenzel, R.N. 1949. Surface roughness and contact angle. J. Phys. Chem. 53: 1466–1467.
Yang, Y., Zhang, H., Wang, P., Zheng, Q. and Li, J. 2007. The influence of nano-sized TiO2 fillers on the morphologies and properties of PSF UF membrane. J. Membr. Sci. 288: 231–238.
Yu, L., Han, M. and He, F. 2017. A review of treating oily wastewater. Arab. J. Chem. 10: S1913–S1922.
Zhang, W., Liu, N., Zhang, Q., Qu, R., Liu, Y., Li, X. et al. 2018. Thermo-driven controllable emulsion separation by a polymer-decorated membrane with switchable wettability. Angew. Chem. Int. Ed. 57: 5740–5745. https://doi.org/10.1002/anie.201801736.
Zhang, W., Shi, Z., Zhang, F., Liu, X., Jin, J. and Jiang, L. 2013. Superhydrophobic and superoleophilic PVDF membranes for effective separation of water-in-oil emulsions with high flux. Adv. Mater. 25: 2071–2076.
Zhang, W., Zhu, Y., Liu, X., Wang, D., Li, J., Jiang, L. et al. 2014. Salt-induced fabrication of superhydrophilic and underwater superoleophobic PAA-g-PVDF membranes for effective separation of oil-in-water emulsions. Angew. Chem. Int. Ed. 53: 856–860.
Zhao, X., Su, Y., Chen, W., Peng, J. and Jiang, Z. 2011. pH-responsive and fouling-release properties of PES ultrafiltration membranes modified by multi-functional block-like copolymers. J. Membr. Sci. 382: 222–230. https://doi.org/10.1016/j.memsci.2011.08.014.
Zhao, X., Su, Y., Liu, Y., Li, Y. and Jiang, Z. 2016. Free-standing graphene oxide-palygorskite nanohybrid membrane for oil/water separation. ACS Appl. Mater. Interfaces 8: 8247–8256.
Zhao, X., Zhang, R., Liu, Y., He, M., Su, Y., Gao, C. et al. 2018. Antifouling membrane surface construction: Chemistry plays a critical role. J. Membr. Sci. 551: 145–171.
Zhijiang, C., Cong, Z., Ping, X., Jie, G. and Kongyin, Z. 2018. Calcium alginate-coated electrospun polyhydroxybutyrate/carbon nanotubes composite nanofibers as nanofiltration membrane for dye removal. J. Mater. Sci. 53: 14801–14820.
Zhu, Y., Zhang, F., Wang, D., Pei, X.F., Zhang, W. and Jin, J. 2013. A novel zwitterionic polyelectrolyte grafted PVDF membrane for thoroughly separating oil from water with ultrahigh efficiency. J. Mater. Chem. A. 1: 5758–5765.

Index

P

R

S

T

W

About the Editors

Dr. Jitendra Kumar Pandey

Dr. Pandey is working as a senior professor in UPES, Dehradun, India. He earned his Ph.D. from National Chemical Laboratory Pune, India. Dr. Pandey serves as postdoctoral scientist on various countries like South Korea, Japan, and Germany. His current research interest includes nanomaterials, bionanocomposites, and sustainable development. He has published more than 100 research and reviews in peer-reviewed journals and edited more than 10 books. His articles are highly cited.

Dr. Syed M. Tauseef

Dr. Tauseef is working in UPES, Dehradun, India as a full professor and associate Dean (R&D). He earned his Ph.D. from Pondicherry University. Dr. Tauseef also holds B.S. and M.S. degrees from Shiraz University, Iran in Chemical engineering. He is currently working on biogas energy production systems, large scale vermireactor design & operation, simulation of accidents in chemical process industries using CFD. He has published 50+ articles in peer-reviewed journals and 10+ book chapters, and edited 7+ books.

Dr. Suvendu Manna

Dr. Manna is an assistant professor in UPES Dehradun, India since 2020. He has 3 years of post-doctoral experience. Currently Dr. Manna is working on the application of nanomaterials for waste remediation. His team has been trying to extract valuable resources including energy from wastes. His team has also been working on exploring new microbial species from the Himalayan regions. Dr. Manna has published 44 journal articles, 18 book chapters, edited 5 books, and filed for 5 patents.

Dr. Ravi Kumar Patel

Dr. Patel has been working towards building a vibrant startup ecosystem and helping startups in technology transformation. He is involved in organizing innovation programs and activities in accordance with the mission and goal of the incubator. He has supported the expansion of programs and outreach, developed relationships with other incubators and government agencies for tech and funding support for incubated startups, and mentored incubated startups for tech development. Dr. Patel has completed his doctorate in 3D printing, materials engineering, and water treatment from the University of Petroleum and Energy Studies, Dehradun (UPES), India. He also has expertise in the domain of robotics and automation.

Mr. Vishal Kumar Singh

Mr. Singh is an Assistant Manager-EHS (OI) at SGS, India. He is currently involved in development of safety culture in the multi laboratories PAN India and performing safety audit across multi laboratories. He is the lead auditor for ISO 45001 and ISO 14001. Currently, Mr. Singh is pursuing Ph.D. from UPES, Dehradun, India. His current research interest is on the extraction of valuable metals from industrial solid waste using biological intervention.

Mr. Ankit Dasgotra

Mr. Dasgotra is a Ph.D. student in UPES, Dehradun. He is a mechanical engineer, having specifications in thermodynamic at master's level. Currently, working as an interdisciplinary researcher in the field of fire and process safety. He is working on fire safety simulation using CFD modelling. Mr. Dasgotra published 10+ peer reviewed journals articles and book chapters.

For Product Safety Concerns and Information please contact our EU
representative GPSR@taylorandfrancis.com
Taylor & Francis Verlag GmbH, Kaufingerstraße 24, 80331 München, Germany

www.ingramcontent.com/pod-product-compliance
Lightning Source LLC
Chambersburg PA
CBHW060810220326
41598CB00022B/2583
* 9 7 8 1 0 3 2 0 0 9 1 4 8 *